Summary of Floods in the United States, January 1992 Through September 1993

Edited by C.A. PERRY and L.J. COMBS

U.S. GEOLOGICAL SURVEY WATER-SUPPLY PAPER 2499

U.S. DEPARTMENT OF THE INTERIOR
BRUCE BABBITT, Secretary

U.S. GEOLOGICAL SURVEY
CHARLES G. GROAT, Director

For sale by the
U.S. Geological Survey
Information Services
Box 25286
Federal Center
Denver, CO 80225

Library of Congress Cataloging in Publication Data

Summary of floods in the United States, January 1992 through September 1993 / edited by C.A. Perry and L.J. Combs.
286 p. cm.—(U.S. Geological Survey water-supply paper ; 2499)
Includes bibliographical references and index.
Supt. of Docs. no.: I 19.13:2499
1. Floods—United States. I. Perry, Charles A. II. Combs, L.J.
III. Series.
GB1399.3.S862 1998
551.48'9'0973—dc21 97–50582
CIP

CONTENTS

FIGURES

FIGURES—Continued

FIGURES—Continued

TABLES

TABLES—Continued

TABLES—Continued

CONVERSION FACTORS AND VERTICAL DATUM

Multiply	By	To obtain
acre	4,047	square meter
acre-foot	1,233	cubic meter
cubic foot per second	0.02832	cubic meter per second
cubic foot per second per square mile	0.01093	cubic meter per second per kilometer
foot	0.3048	meter
foot per hour	0.3048	meter per hour
inch	25.4	millimeter
mile	1.609	kilometer
mile per hour	1.609	kilometer per hour
square foot	0.09290	square meter
square mile	2.590	square kilometer

Temperature can be converted to degrees Celsius (°C) or degrees Fahrenheit (°F) by the equations:

$$°C = 5/9\ (°F - 32)$$
$$°F = 9/5\ (°C) + 32.$$

Sea level: In this report, "sea level" refers to the National Geodetic Vertical Datum of 1929—a geodetic datum derived from a general adjustment of the first-order level nets of the United States and Canada, formerly called Sea Level Datum of 1929.

GLOSSARY

Although much of the terminology used in this report is widely understood, some terms have specialized meanings in hydrology or are unfamiliar outside of hydrologic usage. The definitions given here are from Langbein and Iseri (1960), with slight modifications, and explain the terms as they are generally used by hydrologists in the U.S. Geological Survey.

Absorption The entrance of water into the soil or rocks by all natural processes. It includes the infiltration of precipitation or snowmelt.

Bank The margins of a channel. Banks are called right or left as viewed facing the direction of the flow.

Cubic feet per second A unit expressing rates of **discharge**. One cubic foot per second is equal to the **discharge** of a stream of rectangular cross section, 1 foot wide and 1 foot deep, flowing water an average velocity of 1 foot per second.

Current meter An instrument for measuring the velocity of flowing water. The U.S. Geological Survey uses a rotating cup meter.

Discharge In its simplest concept, **discharge** means outflow; therefore, the use of this term is not restricted as to course or location, and it can be applied to describe the flow of water from a pipe or from a **drainage basin**. If the **discharge** occurs in some course or channel, it is correct to speak of the **discharge** of a canal or of a river.

Drainage area The **drainage area** of a stream at a specified location is that area, measured in a horizontal plane, that is enclosed by a drainage divide.

Drainage basin A part of the surface of the Earth that is occupied by a drainage system, which consists of a surface stream or a body of impounded **surface water** together with all tributary surface streams and bodies of impounded **surface water**.

Flood An overflow or inundation that comes from a river or other body of water (Barrows, 1948, p. 4) and causes or threatens damage. Any relatively high **streamflow** overtopping the natural or artificial banks in any reach of a stream (Leopold and Maddock, 1954, p. 249–251).

Flood plain The lowland that borders a river, usually dry but subject to flooding (Hoyt and Langbein, 1955, p. 12).

Flood stage The **stage** at which overflow of the natural banks of a stream begins to cause damage in the reach in which the elevation is measured.

Isohyetal map A map or chart showing lines that join points that received the same amount of precipitation.

Overland flow The flow of rainwater or snowmelt over the land surface toward stream channels.

Regulation The artificial manipulation of the flow of a stream.

Reservoir A pond, lake, or basin, either natural or artificial, for the storage, regulation, and control of water.

Runoff That part of the precipitation that appears in surface streams.

Stage The height of a water surface above an established datum plane (also gage height).

Stage-discharge curve A graph showing the relation between the gage height, usually plotted as ordinate, and the amount of water flowing (**discharge**) in a channel, expressed as volume per unit of time, usually plotted as abscissa.

Stage-discharge relation The relation expressed by the **stage-discharge curve**.

Streamflow The **discharge** that occurs in a natural channel. Although the term **discharge** can be applied to the flow of a canal, the word "**streamflow**" uniquely describes the **discharge** in a surface stream course. The term "**streamflow**" is more general than **runoff**, as **streamflow** may be applied to **discharge** whether or not it is affected by diversion or **regulation**.

Streamflow-gaging station A gaging station where a record of **discharge** of a stream is obtained.

Surface runoff That part of the **runoff** that travels over the soil surface to the nearest stream channel. It also is defined as that part of the **runoff** of a **drainage basin** that has not passed beneath the surface following precipitation.

Surface water Water on the surface of the Earth.

Water equivalent of snow The amount of water that would be obtained if the snow should be completely melted. Water content may be merely the amount of liquid water in the snow at the time of observation (Wilson, 1942, p. 153–154).

Water year In U.S. Geological Survey reports, **water year** is the 12-month period, October 1 through September 30. The **water year** is designated by the year in which it ends. Thus, the year ending September 30, 1993, is called the "1993 **water year**."

Summary of Floods in the United States, January 1992 Through September 1993

Edited by C.A. Perry *and* L.J. Combs

Abstract

This volume contains a summary of the flooding in the upper Mississippi River Basin during the spring and summer of 1993 and 36 articles describing severe, widespread, or unusual flooding in the United States from January 1, 1992, to the end of the 1993 water year, September 30, 1993. Each flood is described to an extent commensurate with its significance and the availability of data on the hydrology and the damages. Each article includes one or more maps showing the general area of flooding and the sites for which data are presented. Most articles include tables of data that allow the reader to compare the described flood with past floods at selected flood-determination sites. The articles generally do not attempt to analyze the floods or draw definitive conclusions, except for a few cases in which the author had sufficient information for an analysis to be made.

INTRODUCTION

This report summarizes information on floods in the United States from January 1, 1992, through September 30, 1993. The floods reported were unusual hydrologic events during which large areas were affected, great damage resulted, or record-high stages or discharges occurred and for which sufficient data were available for the preparation of an informative article. The States in which the floods described in this volume occurred are shown in figure 1. Also shown is the year(s) of occurrence.

A flood may be defined as any abnormally high streamflow that overtops natural or artificial banks of a stream. Every year, a large number of floods occur that are not reported in national or regional media.

Innumerable combinations of variable meteorologic and physiographic factors produce floods of all degrees and severity. Some meteorologic factors that affect floods are the form, amount, duration, and intensity of precipitation; the amount of previous precipitation, which would affect the moisture absorption of the soil; the air temperature, which may result in frozen soil or may determine the rate of snowmelt; and the direction of storm movement. The principal physiographic features of a drainage basin that determine floodflows are drainage area, elevation, character of soil, shape, slope, direction of slope, and vegetative or other land cover. With the exception of vegetative cover and soil preconditions, the physiographic features are fixed for any given drainage basin. The combination of the magnitude and intensity of meteorologic phenomena, the antecedent moisture conditions, and the effect of inherent physiographic features on runoff determines what the magnitude of a flood will be.

Flood damages frequently are difficult to assess. Dollar amounts given in this report should be used as a general indication of flood losses rather than as definite values. Even if detailed surveys and estimates have been made, there is little consistency among methods used and types of losses included. Some estimates may exclude certain locations (such as mountainous areas) or types of loss (either insured or uninsured) or type of property (either private or public). Some estimates include traffic interruptions and flood-mitigation costs; others include strictly physical damage. Estimates may be based on replacement costs or on depreciated values. For floods not described in detailed published reports, the only damage estimates available usually are the preliminary figures contained in newspapers, National Oceanic and Atmospheric Administration

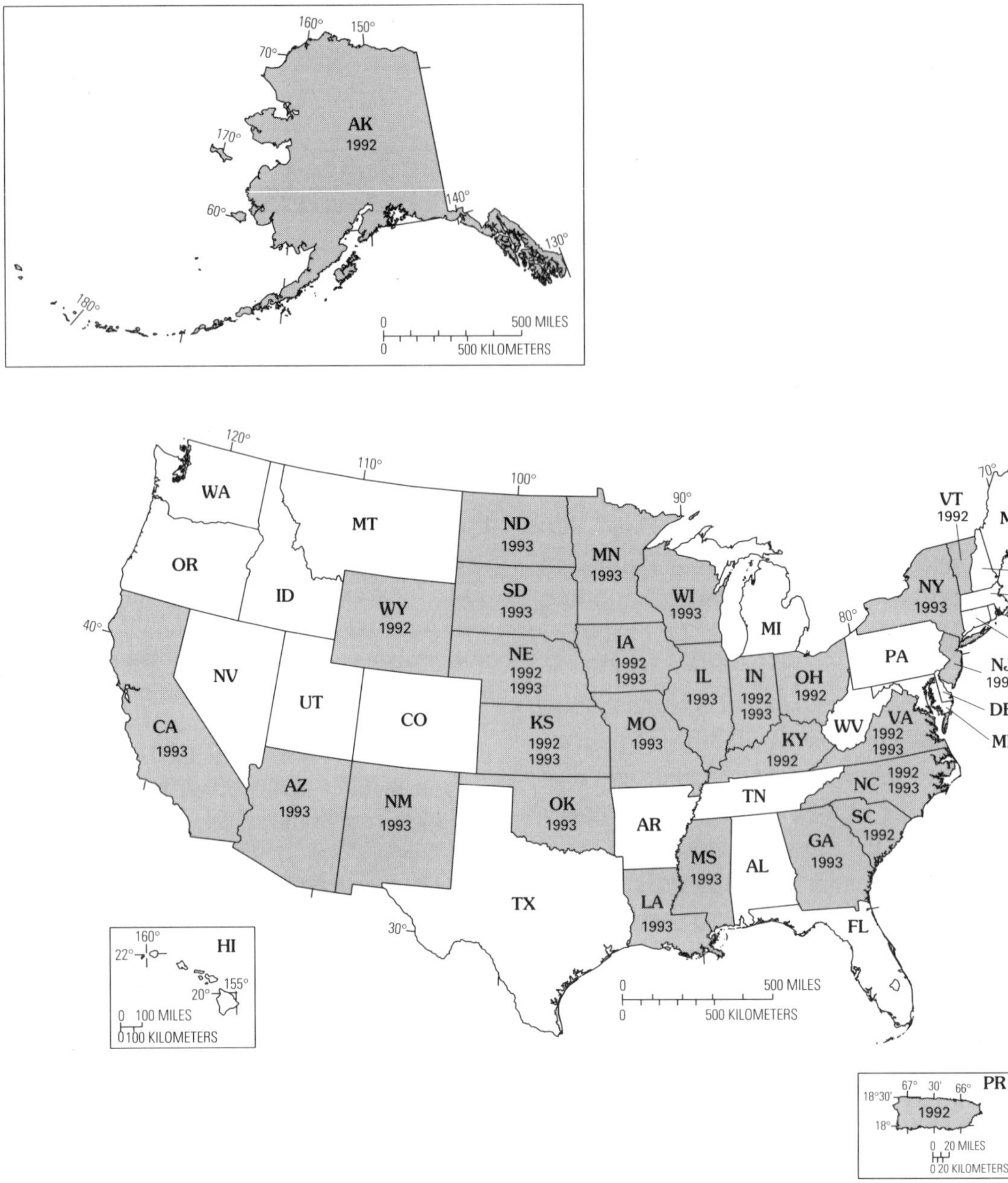

Figure 1. States and years in which floods reported in this volume occurred in the conterminous United States, Alaska, Hawaii, and Puerto Rico during 1992 and 1993.

(NOAA) climatological data, or other sources published shortly after the flood. A statement that a disaster declaration was issued indicates that the damage was severe and that financial aid to victims was authorized by the governmental entity making the declaration.

Many of the articles in this volume give the amount of rainfall and duration of the storm associated with the flooding. Recurrence intervals for these storms may be determined from a rainfall-frequency atlas of the United States (U.S. Weather Bureau, 1961) or from a simplified set of equal-rainfall maps and charts contained in a report by Rostvedt (1965).

Continuing investigation of surface-water resources in the flooded areas reported by this volume is performed by the U.S. Geological Survey in cooperation with State agencies, the U.S. Army Corps of Engineers, the Bureau of Reclamation, and other Federal or local agencies. NOAA, in addition to collecting and compiling data on meteorological phenomena, also collects data on stream stages in some areas. The data presented herein were collected, computations were made, and most of the text was written by U.S. Geological Survey personnel located in offices in or near the flooded areas.

Previous Reports

During the 1950's and 1960's, the U.S. Geological Survey summarized floods of each year in an annual series of Water-Supply Papers entitled, "Summary of Floods in the United States." A summary was published for each calendar year from 1950 through 1969. Water-Supply Paper 1137–I, the first in the series (U.S. Geological Survey, 1954), states the purpose of the series as being:

> "To assemble in a single volume information relating to all known severe floods in the United States whether local or of wide areal extent. For floods that are described in ... other publications of the Geological Survey or in reports by other Federal and State agencies, only very brief mention including references to the reports containing detailed descriptions, will be given here. Local floods for which no individual reports have been prepared are described briefly."

In the first volume of that Water-Supply Paper series, each flood was described in a maximum of three or four paragraphs. Later volumes contained longer articles including maps.

The series was discontinued after the 1969 volume; however, in 1987 a program was begun to prepare and publish summaries for 1970 and succeeding years. Much of the following explanation is paraphrased from Byron N. Aldridge (U.S. Geological Survey, written commun., 1993) and from the published report for 1968 (Rostvedt, 1972).

Determination of Flood Stages and Discharges

The usual method of determining stream discharges at a streamflow-gaging station is the application of a stage-discharge relation to a known stage. This relation usually is defined by current-meter measurements made through as wide a range of stage as possible (fig. 2). If the maximum discharge exceeds the range of the current-meter measurements, short extensions may be made to a graph of the stage-discharge relation by logarithmic extrapolation, by

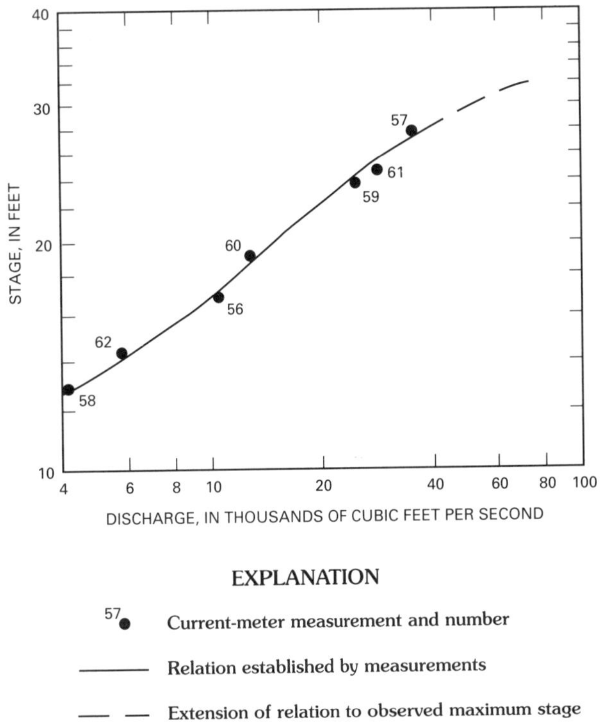

Figure 2. Example of stage-discharge relation and upward extension (from Jordan and Combs, 1997).

velocity-area studies, or by the use of other measurable hydraulic factors (Kennedy, 1983).

Maximum discharges that are greatly above the range of the defined stage-discharge relation at streamflow-gaging stations and maximum discharges at miscellaneous sites that have no developed stage-discharge relation generally are determined by various types of indirect measurements. In addition, adverse conditions often make it impossible to obtain current-meter measurements at some sites during major floods. Maximum discharges at these sites are determined, after the floods have subsided, by indirect methods, which involve determination of water-surface elevations from high-water marks, surveying cross sections, and computing discharge from hydraulic equations rather than from direct measurement of stream velocity by use of a current meter. Indirect methods are described by Dalrymple and Benson (1967), Hulsing (1967), Matthai (1967), Bodhaine (1968), and Benson and Dalrymple (1987).

The accuracy of indirect measurements depends on onsite conditions and the experience of data-collection personnel who select sites and make the surveys, and generally is poorer than for current-meter measurements. The indirect measurements used in determining maximum discharges for floods are not identified as such in this volume. Information as to the source and quality of discharge data in this volume can be obtained from the U.S. Geological Survey office in the State in which the reported flood-determination site is located.

Explanation of Data

Floods are described in this volume in chronological order. Because the type and the amount of information differ for the floods, no consistent form can be used to report the events.

The data for each flood include: (1) a description of the storm, the flood, and the flood damage; (2) a map of the flood area showing flood-determination sites and, for some storms, precipitation data sites or lines of equal precipitation; (3) rainfall amounts and intensities; (4) and maximum stages and discharges for the streams affected.

When considerable rainfall data are available, they are presented in tabular form and show daily or storm totals. When sufficient data are available to determine the pattern and distribution of rainfall, an isohyetal map may be shown.

A summary table of maximum stages and discharge is given for each flood, except where the number of flood-determination sites in the article is small and for which the information is included in the text description. In the summary table (table 1), the first three columns identify the site, which may be a continuous-record streamflow-gaging station, a partial-record station, or another site at which data have been obtained. The number in the first column identifies the site on a map that accompanies each article. The second column gives the U.S. Geological Survey station number (downstream-order number) if such a number has been assigned. The third column gives the name of the streamflow-gaging station or flood-determination site.

Drainage area in the summary table is the total area, as measured on a flat projection map, that would contribute surface runoff to the indicated site. The contributing drainage area may be smaller than the total drainage area if the total area includes areas of extremely rapid infiltration rates that do not produce surface runoff, or closed subbasins that retain all their inflow.

The column headed "Period" shows the calendar years prior to the described flood for which the stage or discharge shown in the seventh and eighth columns are known to be a maximum. For most sites, this period corresponds to the period of systematic collection of streamflow data. For other sites, written or oral history may indicate that a flood stage was the highest since people have observed the stream or was the highest since some known date.

The sixth column shows the calendar year in which the maximum stage and discharge for the indicated period occurred. The seventh and eighth columns show the stage and discharge of that maximum. Separate listings are made when maximum stage and maximum discharge did not occur concurrently. An effort was made to use stages that were measured relative to the datum in use at the time of the flood being described or to indicate by a footnote that a different datum was used.

The last four columns present data for the maximums during the described flood or floods. The data include the date on which the maximum occurred, maximum stage, and maximum discharge and, where available, the recurrence interval of the discharge.

The probability of a given discharge being equaled or exceeded in any given year frequently is used as an indication of a flood's relative magnitude and for comparison with floods at other sites. The relative magnitude also can be expressed in terms of recurrence interval, which is the reciprocal of the flood probability. A third way of expressing the relative flood magnitude is the percent chance of occurrence, which is 100 times the flood probability. A discharge that will be equaled or exceeded on an average (over a long period of time) of once in 10 years has a recurrence interval of 10 years, is termed a "10-year flood," has a probability of 0.10, and has a 10-percent chance of occurring in any given year. A 100-year flood has a recurrence interval of 100 years, a probability of 0.01, and a 1-percent chance of occurring in any given year. Because recurrence interval is used most commonly by Federal agencies (for example, in the context of flood insurance), it is used in this volume even though percent chance avoids the unintended connotations of regularity of occurrence that accompany the term "recurrence interval."

Table 1. Example of summary table presented in flood articles

[mi^2, square miles; ft, feet above an arbitrary datum; ft^3/s, cubic feet per second; >, greater than; <, less than; --, not determined or not applicable. Source: Recurrence intervals calculated from U.S. Geological Survey data. Other data from U.S. Geological Survey reports or data bases]

				Maximum prior to [month] 1992				Maximum during [month] 1992			
Site no. (fig. #)	Station no.	Stream and place of determination	Drainage area (mi^2)	Period	Year	Stage (ft)	Discharge (ft^3/s)	Day	Stage (ft)	Discharge (ft^3/s)	Discharge recurrence interval (years)
1	05551212	Hypothetical Creek near Town	21.0	1971–92	1987	11.1	--	10	12.22	4,200	25
2	05555000	Hypothetical River at City	1,212	1939, 1955–92	1939	12.12	28,200	12	21.21	82,800	>100
3	06930030	Hypothetical River near Metropolis	3,333	1919–92	1943	33.33 --	-- 99,900	13	25.55	33,000	<2
4	--	Hypothetical Ditch at Village	--	1992	--	--	--	10	--	3,800	--

Equivalence of flood probability and percent-chance values to selected recurrence-interval values is as follows:

Probability	Percent chance	Recurrence interval
0.50	50	2
.20	20	5
.10	10	10
.04	4	25
.02	2	50
.01	1	100

In addition to probability or percent chance of a given magnitude of discharge occurring in any one year, the probability or percent chance of occurrence during a given period of consecutive years also can be calculated. Results of such calculations for selected combinations of recurrence interval and length of period are as follows (* means greater than 99.9 but less than a 100-percent chance):

Recurrence interval	Percent chance for indicated time period, in years				
	5	10	50	100	500
2	97	99.9	*	*	*
10	41	65	99.5	*	*
50	10	18	64	87	*
100	5	10	39	63	99.3

Recurrence intervals during any given flood may differ from site to site because of nonuniform distribution of runoff and uncertainty in the computed recurrence values. Operational patterns for reservoirs generally are not defined adequately to permit recurrence intervals to be computed for maximum discharges on regulated streams.

Another method of indicating a flood's relative magnitude is by comparison of its maximum discharge and the stream's drainage area with values on a regional "envelope curve." A flood-envelope curve is one drawn on a graph in which maximum known discharges are plotted against the drainage area of each stream site (fig. 3). The envelope curve is a smooth curve drawn to equal or exceed all the plotted discharges in relation to the drainage areas. Envelope curves are given for 17 regions of the conterminous United States in Crippen and Bue (1977). This method is better than the formerly used calculation of "unit dis-

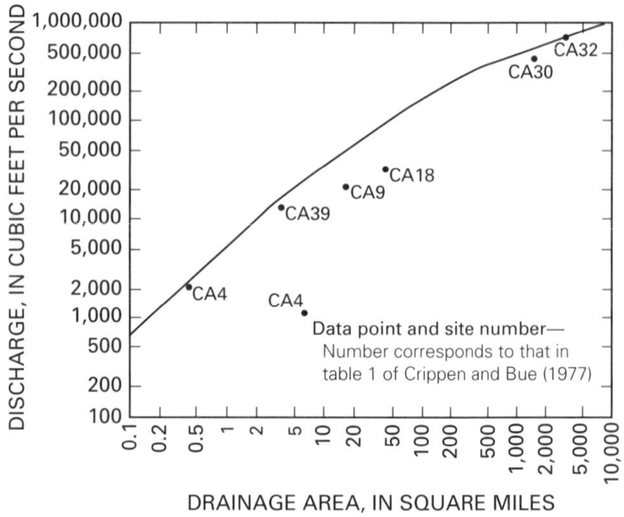

Figure 3. Maximum discharge versus drainage area and envelope curve for a region (modified from Crippen and Bue, 1977, p. 15).

charge" (division of the discharge by the drainage area) because unit discharges for greatly different sizes of drainage area are not comparable. If the unit discharges for a very small and a very large drainage area are the same, the unit discharge is much more unusual for the large drainage area.

REFERENCES

Barrows, H.K., 1948, Floods, their hydrology and control: New York, McGraw-Hill Book Co., 432 p.

Benson, M.A., and Dalrymple, Tate, 1987, General field and office procedures for indirect measurements: U.S. Geological Survey Techniques of Water-Resources Investigations, book 3, chap. A1, 29 p.

Bodhaine, G.L., 1968, Measurement of peak discharge at culverts by indirect methods: U.S. Geological Survey Techniques of Water-Resources Investigations, book 3, chap. A3, 60 p.

Crippen, J.R., and Bue, C.D., 1977, Maximum floodflows in the conterminous United States: U.S. Geological Survey Water-Supply Paper 1887, 52 p.

Dalrymple, Tate, and Benson, M.A., 1967, Measurement of peak discharge by slope-area method: U.S. Geological Survey Techniques of Water-Resources Investigations, book 3, chap. A2, 12 p.

Hoyt, W.G., and Langbein, W.B., 1955, Floods: Princeton, N.J., Princeton University Press, 469 p.

Hulsing, Harry, 1967, Measurement of peak discharge at dams by indirect methods: U.S. Geological Survey Techniques of Water-Resources Investigations, book 3, chap. A5, 29 p.

Jordan, P.R., and Combs, L.J., eds., 1997, Summary of floods in the United States during 1990 and 1991: U.S. Geological Survey Water-Supply Paper 2474, 257 p.

Kennedy, E.J., 1983, Computation of continuous records of streamflow: U.S. Geological Survey Techniques of Water-Resources Investigations, book 3, chap. A13, 53 p.

Langbein, W.B., and Iseri, K.T., 1960, General introduction and hydrologic definitions, *in* Manual of hydrology, part 1, General surface-water techniques: U.S. Geological Survey Water-Supply Paper 1541–A, 29 p.

Leopold, L.B., and Maddock, Thomas, Jr., 1954, The flood control controversy: New York, Ronald Press Co., 278 p.

Matthai, H.F., 1967, Measurement of peak discharge at width contractions by indirect methods: U.S. Geological Survey Techniques of Water-Resources Investigations, book 3, chap. A4, 44 p.

National Weather Service, 1992a, Climatological data (by States): Asheville, N.C., National Oceanic and Atmospheric Administration, Weather Service (various months).

———1992b, Hourly precipitation data (by States): Asheville, N.C., National Oceanic and Atmospheric Administration, Weather Service (various months).

———1992c, Storm data and unusual weather phenomena: Asheville, N.C., National Oceanic and Atmospheric Administration, National Climate Data Center, v. 34, nos. 1–12 (various pages).

———1993a, Climatological data (by States): Asheville, N.C., National Oceanic and Atmospheric Administration, Weather Service (various months).

———1993b, Hourly precipitation data (by States): Asheville, N.C., National Oceanic and Atmospheric Administration, Weather Service (various months).

———1993c, Storm data and unusual weather phenomena: Asheville, N.C., National Oceanic and Atmospheric Administration, National Climate Data Center, v. 35, nos. 1–12 (various pages).

Rostvedt, J.O., 1965, Summary of floods in the United States during 1960: U.S. Geological Survey Water-Supply Paper 1790–B, 147 p.

———1972, Summary of floods in the United States during 1968: U.S. Geological Survey Water-Supply Paper 1970–B, 73 p.

U.S. Army Corps of Engineers, 1994, Annual flood damage report to Congress for fiscal year 1993: Washington, D.C., U.S. Army Corps of Engineers, Engineering Division, 17 p.

U.S. Geological Survey, 1954, Summary of floods in the United States during 1950: U.S. Geological Survey Water-Supply Paper 1137–I, p. 957–991.

U.S. Weather Bureau, 1961, Rainfall frequency atlas of the United States: U.S. Weather Bureau Technical Paper 40, 115 p.

Wilson, W.T., 1942, Melting characteristics of snow and its contribution to runoff, *in* June 30–July 2, 1941, Hydrology Conference Proceedings: University Park, Pennsylvania State College, School of Engineering Technical Bulletin 27, p. 153–165.

Floods of the Upper Mississippi River Basin, Spring and Summer 1993

During spring and summer 1993, the hydrologic effects of extended rainfall throughout the upper midwestern United States were severe and widespread. Record flooding inundated much of the upper Mississippi Basin. The magnitude of the damages—in terms of property, disrupted business, and personal trauma—was unmatched by any other flood disaster in United States history. Property damage alone was estimated to exceed $10 billion (Parrett and others, 1993). Damaged highways and submerged roads disrupted overland transportation throughout the flooded region. The Mississippi and the Missouri Rivers were closed to navigation before, during, and after the flooding. Millions of acres of productive farmland remained under water for weeks during the growing season. Rills and gullies in many tilled fields resulted from the severe erosion that occurred throughout the midwestern United States farm belt. The banks and channels of many rivers were severely eroded, and sediment was deposited over large areas of the basin's flood plain. Record flows submerged many areas that had not been affected by previous floods. Industrial and agricultural areas were inundated, which caused concern about the transport and fate of industrial chemicals, sewage effluent, and agricultural chemicals in the floodwaters. The extent and duration of the flooding caused numerous levees to fail.

Precipitation in the Upper Mississippi River Basin During the Floods of 1993

By Kenneth L. Wahl, Kevin C. Vining, *and* Gregg J. Wiche[1]

INTRODUCTION

Excessive precipitation produced severe flooding in a nine-State area in the upper Mississippi River Basin during spring and summer 1993. Following a spring that was wetter than average, weather patterns that persisted from early June through July caused the upper Midwest to be deluged with an unusually large amount of rainfall. Monthly precipitation data were examined at 10 weather-station locations in the flood-affected region to illustrate precipitation patterns and amounts in the flood-affected area. During 1993, all 10 of the selected locations received greater-than-normal rainfall for January through June, 8 of the 10 stations received more than 200 percent of the normal rainfall for July, and 3 received more than 400 percent of the normal rainfall for July. (The average rainfall for the 30-year period 1961–90 is termed "normal" rainfall.) May through August 1993 was the wettest or nearly the wettest such period on record at many locations in the flooded area. All 10 selected locations received more rainfall during the first 9 months of 1993 than generally is received during a year.

MONTHLY PRECIPITATION

In early June, a weather pattern (fig. 4) developed that was characterized by a strong low-pressure trough over the western United States and a corresponding large high-pressure system positioned over the southeastern United States. The jetstream dipped south over the western United States and flowed northeasterly across the upper Midwest. Southeastern high pressure blocked the eastward movement of storms, thus creating a convergence zone between the warm, moist air from the Gulf of Mexico and the much cooler and drier air from Canada, which resulted in thunderstorms. This pattern persisted through most of June and July (National Weather Service, 1993b). As a result, the upper Midwest within this convergence zone was deluged with rain, while the southeastern and the eastern United States from Florida and Alabama to Delaware, under the influence of the high-pressure system, was hot, humid and had little rainfall. Slight movements in the atmospheric pattern determined the timing and location of the excessive rainfall throughout the upper Midwest.

The persistence of this weather pattern caused unusually large amounts of rain to fall over the upper Midwest. These large amounts and the wetter-than-normal spring produced flooding throughout the upper Mississippi River Basin. The rains were extraordinary in the areal extent and in the amounts accumulated. Precipitation for January 1 through July 31 totaled more than 20 inches over most of the flood-affected area and was more than 40 inches in areas of northeast Kansas and east-central Iowa (fig. 5*A*; Parrett and others, 1993). Most of the area received from 150 to 200 percent of the 1961–90 normal total amounts for January through July (fig. 5*B*; Wahl and others, 1993).

Annual precipitation over the nine-State area affected by flooding (fig. 5*A*, 5*B*) generally averages slightly more than 30 inches but ranges from about 16 inches in south-central North Dakota to about 40 inches in southern Missouri. Although precipitation is about evenly divided between the first and last halves of the year, the accumulation rates are not uniform throughout the year. Normally, 45 percent of the annual precipitation falls between April 1 and July 31; June precipitation represents about 15 percent (2 to 5 inches) of the average annual precipitation total.

Precipitation amounts recorded throughout the upper Mississippi River Basin during the first 9 months of 1993 generally were substantially greater than normal (January–September 1961–90). Although May-through-August precipitation was much greater than normal, little evidence early in the year indicated that precipitation amounts in 1993 would be above normal. January-through-March precipitation in the States of the upper Mississippi River Basin was near normal to slightly above normal. Because precipitation for those 3 months is often in the form of snow and gener-

[1]This article is based on previously published reports (Wahl and others, 1993, 1994).

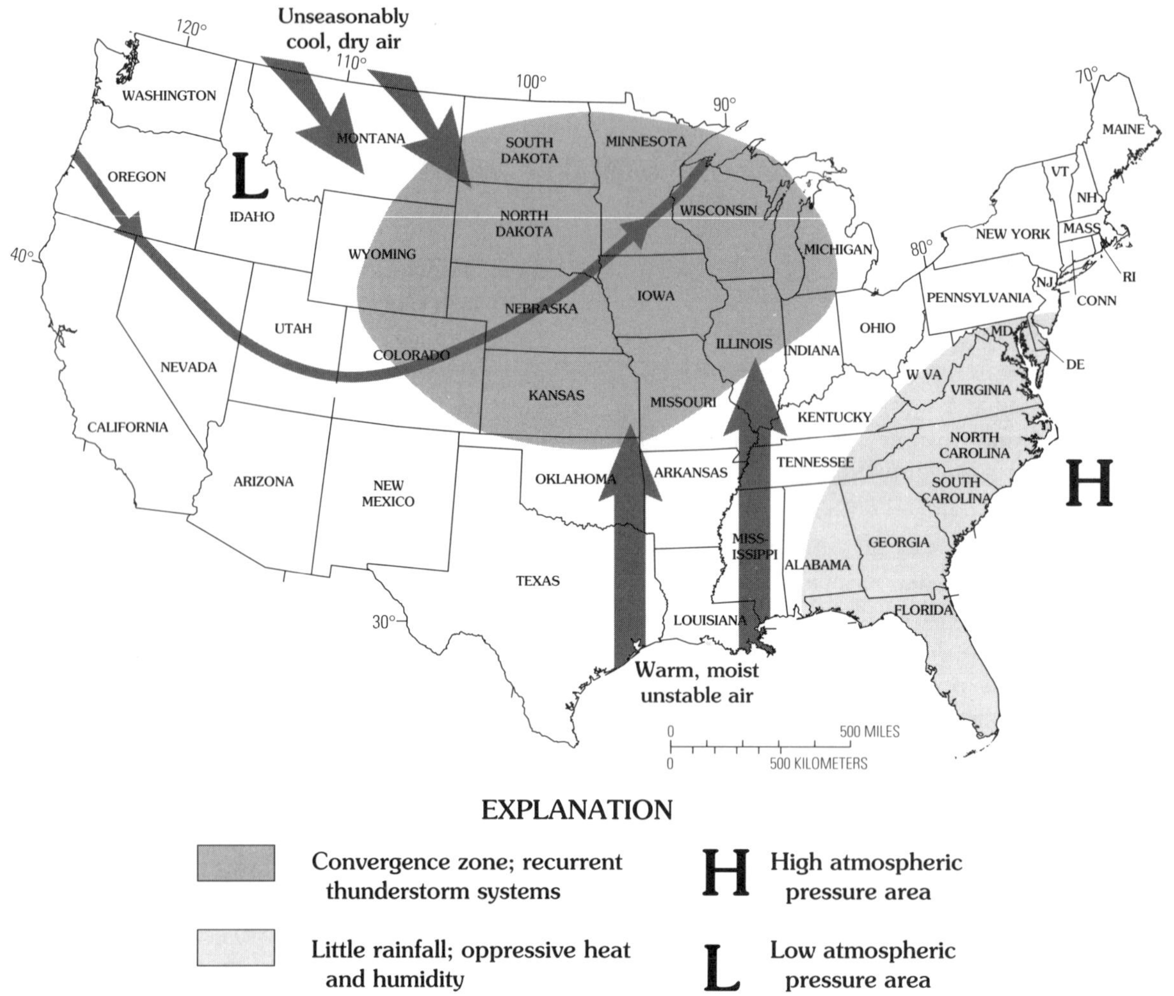

Figure 4. Dominant weather patterns over the United States for June through July 1993 (from National Weather Service, 1993b).

ally totals less than 6 inches of moisture, it caused little concern for potential flooding. However, that situation began to change in April.

Precipitation in April and May over the area ranged from near normal to much greater than normal, but the areas of greatest precipitation differed for the 2 months. April rainfall was nearly twice the normal amounts in parts of Wisconsin and in Missouri but was only moderately above normal in much of the remainder of the flood-affected area. By contrast, May rainfall was more than twice the normal amounts for the month over an area that extended from southeastern South Dakota across Iowa to eastern Kansas. The largest storms, though, were still to come.

Monthly precipitation data for January through September 1993 and the normal (average) amounts for the period 1961–90 (National Oceanic and Atmospheric Administration, 1992) are compared in figure 6 for 10 weather-station locations in the upper Mississippi River Basin (table 2). These locations—Bismarck, North Dakota; Sioux Falls, South Dakota; Minneapolis, Minnesota; Des Moines and Cedar Rapids, Iowa; Madison, Wisconsin; Peoria, Illinois; Manhattan, Kansas; and Kansas City and St. Louis, Missouri—are illustrative of precipitation patterns in the flood-affected area.

Precipitation during 1993 generally was much greater than normal for May through August over the entire area. The July rainfall totals, however, were particularly impressive. The significance of the July 1993 rainfall amounts can best be understood by comparing the 1993 values for January through June, July, and January through September with the 30-year normal amounts for those periods (fig. 7). All 10 of the selected locations received more than 100 percent of the normal precipitation for January through June 1961–90, and 8 locations received more than 130 per-

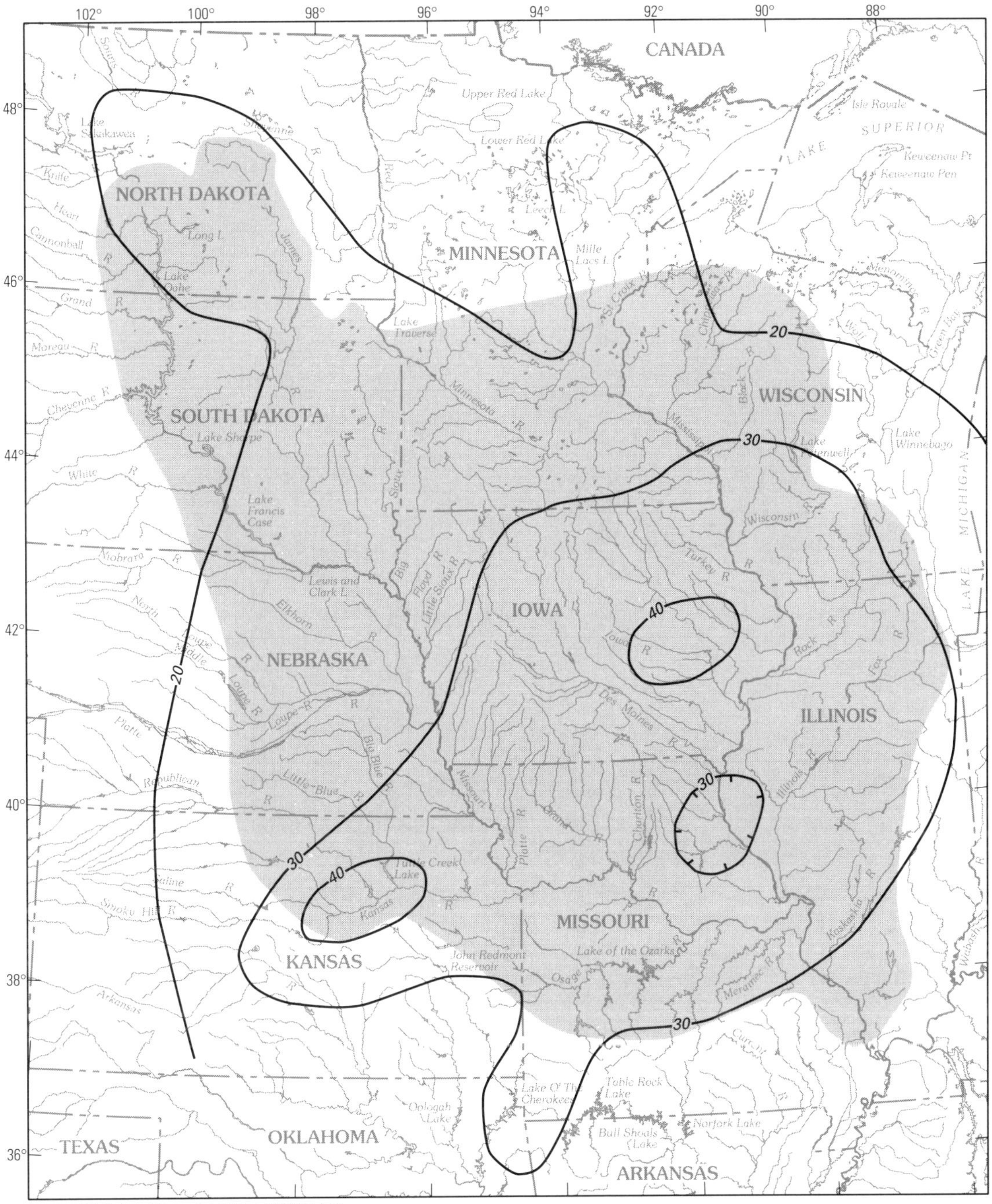

Base from U.S. Geological Survey digital data, 1:2,000,000, 1994
Albers Equal-Area Conic projection
Standard parallels 29°30' and 45°30', central meridian -96°00'

0 100 200 MILES
0 100 200 KILOMETERS

A.

EXPLANATION

Area of flooding streams

—30— **Line of equal total precipitation for January through July 1993**—Hachures indicate closed areas of lesser precipitation. Interval 10 inches

Figure 5. Areal distribution of total precipitation in the area of flooding in the upper Mississippi River Basin, January through July 1993. *A*, In inches (from Parrett and others, 1993); *B*, In percentage of 30-year precipitation normal for January through July 1961–90 (from Wahl and others, 1993). Data from National Weather Service (David Miscus, National Weather Service, written commun., 1993).

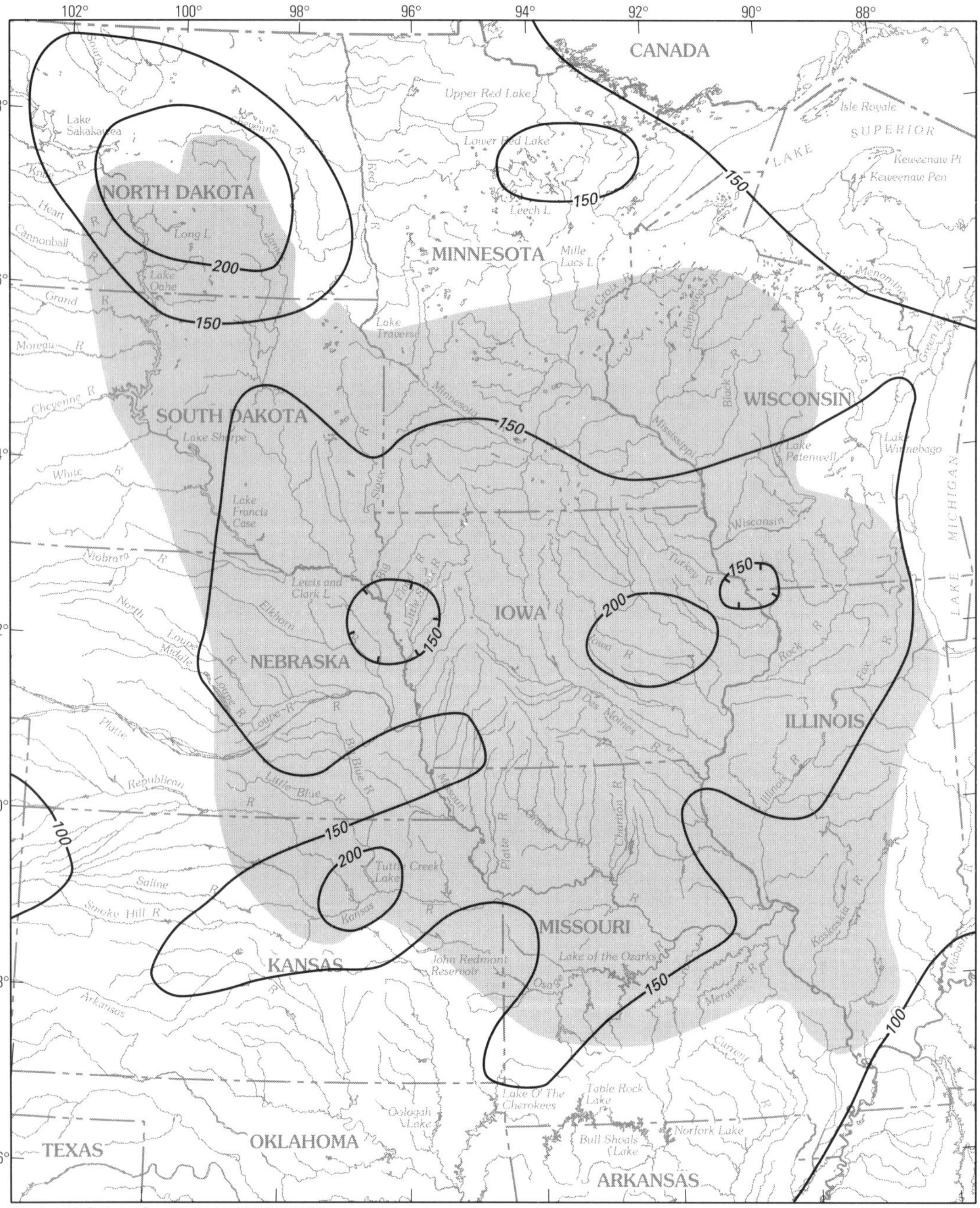

Base from U.S. Geological Survey digital data, 1:2,000,000, 1994
Albers Equal-Area Conic projection
Standard parallels 29°30' and 45°30', central meridian -96°00'

0 100 200 MILES
0 100 200 KILOMETERS

B.

EXPLANATION

Area of flooding streams

—200— **Line of equal total precipitation for January through July 1993 as percentage of the 30-year precipitation normal for January through July 1961–90**—Hachures indicate closed areas of lesser precipitation. Interval 10 inches

Figure 5. Areal distribution of total precipitation in the area of flooding in the upper Mississippi River Basin, January through July 1993. *A*, In inches (from Parrett and others, 1993); *B*, In percentage of 30-year precipitation normal for January through July 1961–90 (from Wahl and others, 1993). Data from National Weather Service (David Miscus, National Weather Service, written commun., 1993)—Continued.

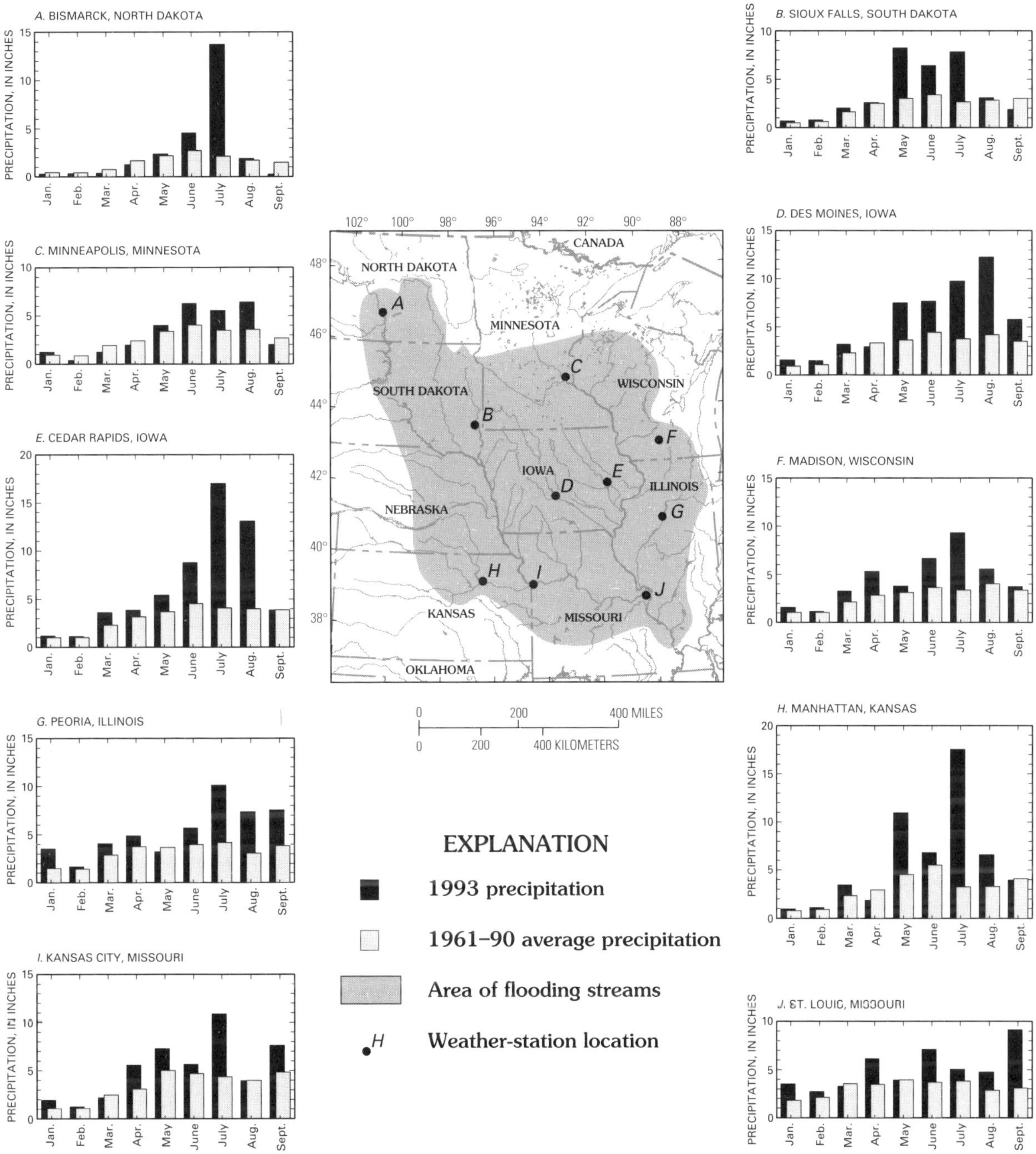

Figure 6. Monthly precipitation for January through September 1993 and 30-year monthly normals (January through September 1961–90) at 10 weather-station locations in the upper Mississippi River Basin. Data from National Weather Service.

cent of the normal precipitation for that period. By the end of June, the soil was saturated; then came the July 1993 deluge. Of the 10 locations, 8 received more than 200 percent of the normal rainfall for July (1961–90). Only Minneapolis (158 percent) and St. Louis (131 percent) did not receive 200 percent of the normal rainfall in July. Three locations, Bismarck, Cedar Rapids, and Manhattan, received from about 400 to about 650 percent of the normal rainfall amounts for July.

Total January-through-September rainfall for 1993 for each of the 10 locations was compared to the 1961–90 normals for the 9-month period and the annual total (fig. 8). All 10 locations received more rain in the first 9 months of 1993 than generally is

Table 2. Summary of 1993 precipitation and 30-year averages for January–September for 10 weather-station locations in the upper Mississippi River Basin

[Averages are for the 30-year period 1961–90. Data from National Oceanic and Atmospheric Administration, 1992 and 1993]

	Bismarck, ND	Sioux Falls, SD	Minneapolis, MN	Des Moines, IA	Cedar Rapids, IA	Madison, WI	Peoria, IL	Manhattan, KS	Kansas City, MO	St. Louis, MO
January										
1993, inches	0.29	0.70	1.25	1.59	1.18	1.60	3.55	0.98	1.96	3.54
Average, inches	.45	.51	.95	.96	1.01	1.07	1.51	.81	1.09	1.81
1993, percent	64	137	132	166	117	150	235	121	180	196
February										
1993, inches	.33	.81	.39	1.52	1.11	1.18	1.68	1.12	1.28	2.75
Average, inches	.43	.64	.88	1.11	1.02	1.08	1.42	.93	1.10	2.12
1993, percent	77	127	44	137	109	109	118	120	116	130
March										
1993, inches	.38	2.04	1.25	3.22	3.63	3.29	4.08	3.46	2.15	3.31
Average, inches	.77	1.64	1.94	2.33	2.32	2.17	2.91	2.36	2.51	3.58
1993, percent	49	124	64	138	156	152	140	147	88	92
April										
1993, inches	1.26	2.61	1.99	2.96	3.86	5.33	4.89	1.88	5.59	6.16
Average, inches	1.67	2.52	2.42	3.36	3.19	2.86	3.77	2.95	3.12	3.50
1993, percent	75	104	82	88	121	186	130	64	179	176
May										
1993, inches	2.36	8.26	4.02	7.51	5.43	3.81	3.25	10.99	7.30	3.94
Average, inches	2.18	3.03	3.39	3.66	3.71	3.14	3.70	4.56	5.04	3.97
1993, percent	108	273	119	205	146	121	88	241	145	99

Table 2. Summary of 1993 precipitation and 30-year averages for January–September for 10 weather-station locations in the upper Mississippi River Basin—Continued

	Bismarck, ND	Sioux Falls, SD	Minneapolis, MN	Des Moines, IA	Cedar Rapids, IA	Madison, WI	Peoria, IL	Manhattan, KS	Kansas City, MO	St. Louis, MO
June										
1993, inches	4.57	6.43	6.28	7.68	8.79	6.67	5.70	6.83	5.67	7.12
Average, inches	2.72	3.40	4.05	4.46	4.55	3.66	3.99	5.54	4.72	3.72
1993, percent	168	189	155	172	193	182	143	123	120	191
July										
1993, inches	13.75	7.86	5.58	9.75	17.03	9.34	10.15	17.56	10.90	5.06
Average, inches	2.14	2.68	3.53	3.78	4.11	3.39	4.20	3.28	4.38	3.85
1993, percent	643	293	158	258	414	276	242	535	249	131
7 months (Jan.–July)										
1993, inches	22.94	28.71	20.76	34.23	41.03	31.22	33.30	42.82	34.91	31.88
Average, inches	10.36	14.42	17.16	19.66	19.91	17.37	21.50	20.43	21.96	22.55
1993, percent	221	199	121	174	206	180	155	210	159	141
August										
1993, inches	1.89	3.10	6.45	12.24	13.09	5.57	7.38	6.61	3.98	4.78
Average, inches	1.72	2.85	3.62	4.20	4.01	4.04	3.10	3.32	4.01	2.85
1993, percent	110	109	178	291	326	138	238	199	99	168
September										
1993, inches	.26	1.88	2.04	5.79	3.88	3.74	7.56	3.99	7.63	9.16
Average, inches	1.49	3.02	2.72	3.53	3.90	3.37	3.87	4.13	4.86	3.12
1993, percent	17	62	75	164	99	111	195	97	157	294
9 months (Jan.–Sept.)										
1993, inches	25.09	33.69	29.25	52.26	58.00	40.53	48.24	53.42	46.52	45.82
Average, inches	13.57	20.29	23.50	27.39	27.82	24.78	28.47	27.88	30.83	28.52
1993, percent	185	166	124	191	208	164	169	192	151	161
Annual average, inches	**15.47**	**23.86**	**28.32**	**33.12**	**33.72**	**30.88**	**36.25**	**33.82**	**37.62**	**37.51**

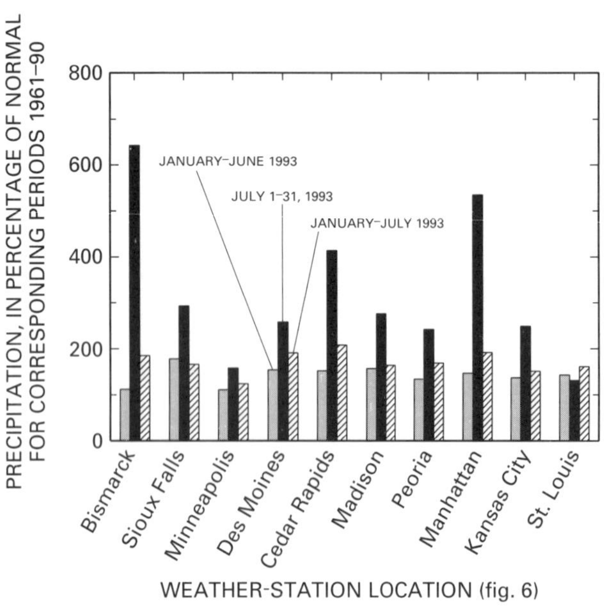

Figure 7. Seasonal precipitation for 1993 at 10 weather-station locations in the upper Mississippi River Basin. Data from National Weather Service.

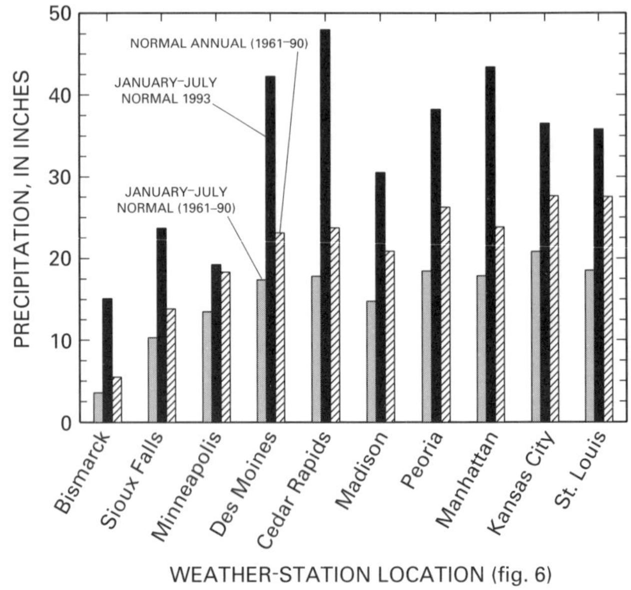

Figure 8. Precipitation for January through September 1993, normal precipitation for January through September 1961–90, and normal annual precipitation for 1961–90 at 10 weather-station locations in the upper Mississippi River Basin. Data from National Weather Service.

received during a year. Except for Minneapolis, the 9-month totals exceeded 150 percent of normal for the period and ranged from 122 to 172 percent of the normal annual totals; Minneapolis received 124 percent of normal for January through September.

The National Weather Service computed statewide-average precipitation by month for 1895 to the present. They reported that the statewide averages for July 1993 were among the three wettest years since 1895 for eight of the nine States in the flood-affected area (National Weather Service, 1993b). Statewide average precipitation totals for the combined months of April through August 1993 were the greatest since 1895 in four States (North Dakota, Minnesota, Wisconsin, and Iowa) and were among the greatest five years in the other five States of the flood-affected area. Rainfall totals for the 4-month period May through August computed for this report for Bismarck, Cedar Rapids, and Manhattan were compared to period-of-record maximums for that 4-month period (table 3). At Bismarck and Cedar Rapids, May through August 1993 was the wettest such period in more than 100 years of recordkeeping. In fact, the precipitation for May through August 1993 at Cedar Rapids was 59 percent greater than for the second wettest year, 1969. At Manhattan, May through August 1993 was the second wettest such period in 104 years of record; in 1951, the 4-month rainfall total was only 2 percent greater than that of 1993.

DAILY ACCUMULATIONS

The rates of accumulation for 1993 daily precipitation are compared with normal values in figure 9. The normal values are the accumulated daily averages for 1961–90 smoothed to pass through the month-end accumulated totals. The precipitation totals for January through September 1993 at Bismarck, Manhattan, and Cedar Rapids were about 200 percent of normal for January through September 1961–90, but the rates at which the precipitation accumulated were different (fig. 9). Although the rate of accumulation at Bismarck was about normal through June, a dryer-than-normal January through March caused the amount of precipitation received to be slightly less than normal until the end of June. Three large storms, June 29 through July 1, July 15 and 16, and July 21 and 22, combined to produce the large seasonal totals. There was little precipitation after the middle of August. Precipitation at Manhattan followed a pattern similar to that of Bismarck, except that several large storms came earlier in the year. Manhattan's large seasonal total resulted primarily from precipitation during four distinct periods—March 29 through 31, May 7 through 11, July 1 and 2, and July 18 through 22. However, precipitation continued to accumulate at greater-than-normal rates through August and September.

Table 3. Five greatest May–August precipitation totals for Bismarck, North Dakota, Cedar Rapids, Iowa, and Manhattan, Kansas

[Totals, in inches; length of records, in parentheses. Data from National Weather Service]

Bismarck (119 years)		Cedar Rapids (112 years)		Manhattan (104 years)	
Year	Total	Year	Total	Year	Total
1993	**22.57**	**1993**	**44.34**	1951	43.05
1915	17.59	1969	27.88	**1993**	**42.21**
1914	17.57	1902	27.63	1902	34.71
1879	15.60	1990	27.55	1908	34.69
1927	15.14	1924	25.27	1915	31.22

The rate of accumulation of precipitation at Cedar Rapids was different from that at either Bismarck or Manhattan. Although large storms, such as that for July 4 and 5, contributed to the excessive moisture, the rate of accumulation was greater than normal after mid-March. Unlike Bismarck and Manhattan, the rate reflected the accumulation of many small-to-moderate precipitation amounts and shows the effects of widespread storms over the entire area. The above-normal accumulation rates continued into late August, but September precipitation was near normal.

DISTRIBUTION OF JULY PRECIPITATION

Much of the severe flooding in the upper Mississippi River Basin during 1993 was the culmination of the wet spring and a series of storms during July. Daily rainfall totaled more than 4.00 inches at many locations during July. Thus, flooding was affected not only by wet antecedent conditions and large rainfall totals, but also by the way July daily rainfall was distributed. Maximum 1- and 3-day rainfall totals for July at Bismarck, Manhattan, and Cedar Rapids are similar (table 4). Maximum 5-day rainfall totals are similar at Bismarck and Cedar Rapids, but the maximum 5-day rainfall total at Manhattan was about 1.6 inches greater than at Bismarck and Cedar Rapids (table 4).

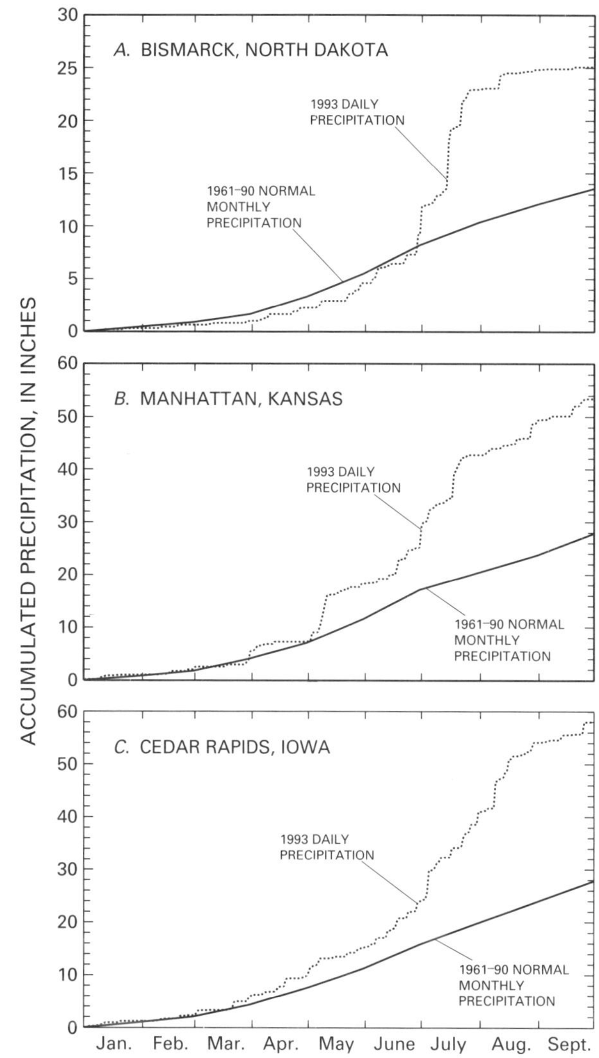

Figure 9. Accumulated daily precipitation for January through September 1993 and accumulated normal daily precipitation for January through September 1961–90 for (*A*) Bismarck, North Dakota, (*B*) Manhattan, Kansas, and (*C*) Cedar Rapids, Iowa.

Table 4. Maximum 1-, 3-, and 5-day rainfall totals for July 1993 at Bismarck, North Dakota, Manhattan, Kansas, and Cedar Rapids, Iowa

[Data from National Weather Service]

	Rainfall totals (inches)		
Location	1-day	3-day	5-day
Bismarck......	4.32	5.74	6.18
Manhattan.....	4.81	5.56	7.70
Cedar Rapids...........	4.18	5.55	6.01

Effects of Reservoirs in the Kansas, Missouri, and Mississippi River Basins on the 1993 Floods

By Charles A. Perry[1]

INTRODUCTION

The floods of 1993 were of historic magnitude as water in the Mississippi and Missouri Rivers reached levels that exceeded many of the previous observed maximums. Although large parts of the flood plains of both rivers upstream from St. Louis, Missouri, were inundated, water levels would have been even higher had it not been for the large volume of runoff retained in flood-control reservoirs. Most of the total flood-control storage available upstream from St. Louis is located along the main stem and tributaries of the Missouri River; the largest concentration of reservoirs is located within the Kansas River Basin (fig. 10). The Kansas River Basin accounts for about 10 percent (60,000 square miles) of the drainage area of the Missouri River Basin, and reservoirs control streamflow from 85 percent (50,840 square miles) of the drainage area of the Kansas River Basin. Analyses of flood discharges in the Kansas River indicate that reservoirs reduced flooding along the Kansas and the lower Missouri Rivers.

Flood discharges from the Mississippi and the Missouri Rivers combined for a historic peak of 1,080,000 cubic feet per second on the Mississippi River at St. Louis, Missouri, on August 1, 1993. Historic streamflow records show that this discharge was the largest since 1861 and has been exceeded only by an estimated discharge of 1,300,000 cubic feet per second for the flood of 1844. Discharge for the flood of 1903, which had been estimated to be 1,019,000 cubic feet per second was slightly less than that of 1993. However, changes in the upper Mississippi River Basin that have been made in the last 50 years, such as the construction of many flood-control reservoirs, reduced the magnitude of the maximum discharge of the 1993 flood at St. Louis.

[1]This article is based on a previously published report (Perry, 1994).

FLOOD-CONTROL RESERVOIRS

The function of flood-control reservoirs is to temporarily store a part of the flood discharge for later release so that the flood peak downstream will be reduced. In an uncontrolled stream, the flood discharges of the tributary streams are added to the discharge in the main stem. As a result, the total flood volume increases in the downstream direction, as does the maximum discharge (fig. 11*A*). In the case of the controlled stream, all or part of the flood discharge is stored in a reservoir for later release at a reduced flow rate (fig. 11*B*). Downstream from the reservoir, additional flood discharges in the tributaries enter the main stem, which add uncontrolled flood discharges to the controlled discharge. In the actual operation of a flood-control reservoir, the uncontrolled flood discharges from the drainage area downstream from a reservoir need to be considered before reservoir releases are made. If uncontrolled flood discharge from areas downstream from the reservoir produces a flood on the main stem, then reservoir releases can be reduced to near zero to minimize additional flooding downstream, provided storage capacity is available in the reservoir.

Most flood-control reservoirs in the upper Mississippi River Basin are of the multipurpose type, which are used to store water for irrigation, power generation, navigation, public-water supply, and recreation. The flood-control, or flood-storage capacity, pool of a reservoir always is above the multipurpose pool level (fig. 12). All reservoirs with provision for flood control are operated so that a minimum amount of water in the flood-control pool is maintained prior to flooding to maximize flood protection. The flood-reduction potential of a reservoir is compromised if additional floodwater must be stored before the previously stored water can be released.

Flood-control reservoirs are constructed with an emergency spillway to protect the dam from being overtopped, which can cause severe damage to or failure of the dam. Flow through the spillway can be uncontrolled or can be controlled by gates that regulate the releases up to a certain elevation in the reservoir.

Base from U.S. Geological Survey digital data, 1:2,000,000, 1994
Albers Equal-Area Conic projection
Standard parallels 29°30' and 45°30', central meridian -96°00'

0 100 200 MILES
0 100 200 KILOMETERS

EXPLANATION

- Boundary of Mississippi River Basin upstream from Thebes, Illinois
- Boundary of drainage basin
- Boundary of drainage subbasin
- 32 Reservoir reference number used in table 5
- ▲A Selected streamflow-gaging station and reference letter used in table 6

Figure 10. Location of flood-control reservoirs and selected streamflow-gaging stations in the Kansas, Missouri, and Mississippi River Basins.

Once the water level in the reservoir rises to the top of the closed spillway gates or the sill of an uncontrolled spillway, water stored above this elevation in the reservoir is in the surcharge pool. Outflow of surcharge in the reservoir is determined by the depth of water and the geometry of the spillway or the spillway gate opening.

FLOOD STORAGE IN AND EFFECTS OF RESERVOIRS ON FLOOD DISCHARGES, MISSOURI RIVER BASIN

There are 34 major flood-control reservoirs within the Missouri River Basin that drain areas greater than 100 square miles (table 5). Table 5 also includes 11 res-

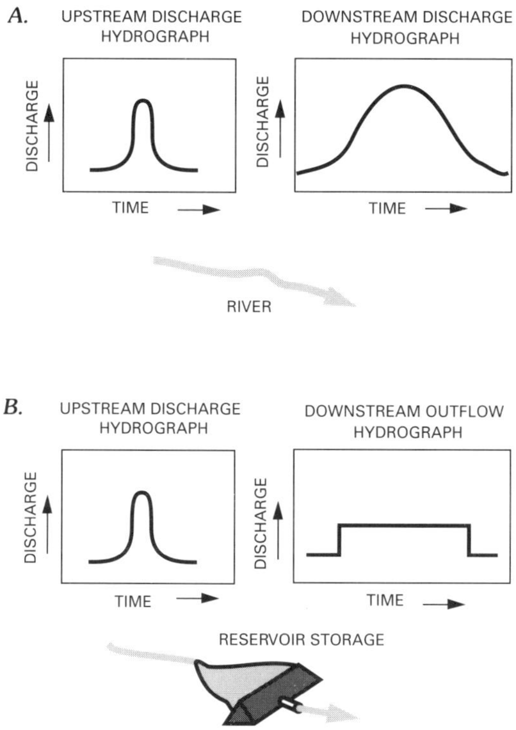

Figure 11. Hypothetical hydrographs for *(A)* an uncontrolled stream and *(B)* reservoir-regulated outflow.

ervoirs in the Mississippi River Basin upstream from its confluence with the Missouri River. Of the reservoirs in the Missouri River Basin, water levels in 13 reached historic elevations, 3 came within 1 foot of their records, and 6 exceeded their spillway elevations. Water levels in reservoirs on tributaries of the Mississippi River upstream from its confluence with the Missouri River, including Saylorville Lake, Coralville Reservoir, and Lake Red Rock, all in Iowa, also reached record elevations. Several reservoirs in the Arkansas River Basin, just south of the Kansas River Basin, had record and near-record water-level elevations. The number of reservoirs with record water-level elevations is an indication of the magnitude and wide extent of the floods of 1993.

Missouri River Main Stem

The six-reservoir system on the main stem of the Missouri River from Montana through North Dakota and South Dakota is used for power generation, storage for navigation and public-water supply, and flood control. When the 1993 water year began (October 1, 1992), the total storage content in the reservoir system

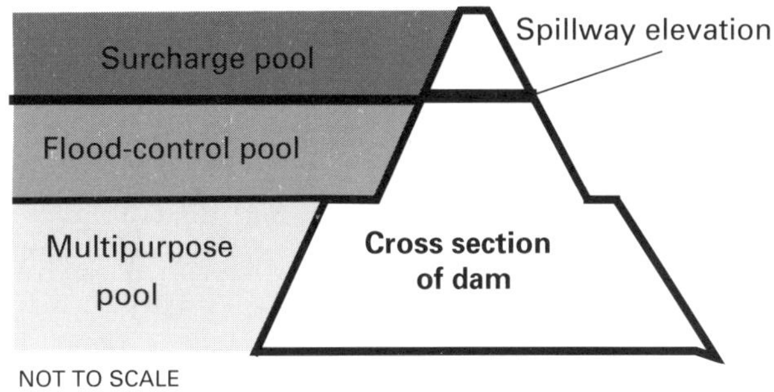

Figure 12. Typical reservoir showing location of surcharge, flood-control, and multipurpose-pool levels.

was about 43,900,000 acre-feet. By April 1, 1993, the total system content had increased to 45,468,000 acre-feet (U.S. Army Corps of Engineers, written commun., 1993). The additional 1,568,000 acre-feet resulted from runoff produced by melting snowpack in the mountains during winter and early spring. The total increase in storage contents of the six-reservoir system from April 1 to midnight August 1, 1993, was 10,293,000 acre-feet and the result of excessive snowmelt and rainfall during this period. During July 1993 alone, reservoirs on the main stem Missouri River stored nearly 5,369,000 acre-feet of floodwater. If this water had been released at a constant rate, the daily average discharge of the Missouri River downstream during July would have been about 87,000 cubic feet per second larger than the observed average discharge of 291,000 cubic feet per second on the Missouri River at Kansas City, Missouri (map reference P, fig. 10).

Kansas River Basin

The Kansas River Basin is about 60,000 square miles in area, of which streamflow from 85 percent, or 50,840 square miles, of the basin is controlled by reservoirs. Except for the main-stem Missouri River reservoir system, the Kansas River Basin is the largest basin under flood control in the Mississippi River Basin. Eighteen reservoirs, which have a total flood-control capacity of 7,390,000 acre-feet, provide flood protection within the basin and along the Missouri River downstream. From April 1 to August 1, 1993, the reservoir system in the Kansas River Basin stored 4,500,000 acre-feet of water. Of this amount, 4,027,000 acre-feet were stored during July alone. If this water had been released at a constant rate, the average discharge of the Kansas River downstream during July would have been about 65,500 cubic feet per sec-

Table 5. Upper Mississippi River Basin flood-control reservoir information, 1993

[mi^2, square miles; acre-ft, acre-feet; ft above SL, feet above sea level or the National Geodetic Vertical Datum of 1929; –, data not available]

Map reference number (fig. 10)	Station number	Reservoir name	Drainage area (mi^2)	Year storage began	Multipurpose pool (acre-ft)	Flood storage (acre-ft)	Surcharge storage (acre-ft)	Spillway elevation (ft above SL)	Previous record elevation (ft above SL)	Previous record date (month/day/year)	1993 Maximum elevation (ft above SL)	1993 Maximum elevation date (month/day)	Change April 1 to August 1 (acre-ft)
						Missouri River Basin							
	MONTANA												
1	06131500	Fort Peck	57,500	1937	15,773,000	2,657,000	980,000	2,250.0	2,251.6	7/15/75	2,228.4	9/30	2,085,000
	NORTH DAKOTA												
2	06338000	Sakakawea	181,400	1953	18,750,000	4,250,000	--	1,854	1,854.6	7/5/75	1,837.4	9/25	4,453,000
	SOUTH DAKOTA												
3	06439980	Oahe	243,500	1958	20,140,000	2,390,000	1,100,000	1,620.0	1617.9	8/22/75	1,611.6	9/12	3,426,000
4	06442700	Sharpe	249,300	1963	1,465,000	260,000	175,000	1,423.0	1,421.9	4/22/71	1,421.2	6/13	9,000
5	06452500	Francis Case	263,500	1952	1,336,000	3,498,000	982,000	1,375.0	1,364.2	6/2/62	1,361.0	7/28	333,000
6	06467000	Lewis and Clark	279,500	1955	156,000	321,000	64,000	1,210.0	1,210.7	4/1/60	1,208.9	7/15	-13,000
	COLORADO												
7	06826000	Bonny	1,820	1950	41,320	128,800	--	3,710.0	3,678.10	5/17/57	3,671.92	6/6	-940
	NEBRASKA												
8	06829000	Swanson	8,620	1953	116,100	137,900	107,600	2,773.0	2,757.42	8/2/62	2,752.29	6/14	2,160
9	06832000	Enders	950	1950	36,010	38,510	6,210	3,127.0	3,118.20	3/25/60	3,101.62	6/24	-1,230
10	06837390	Hugh Butler	730	1961	31,470	54,890	76,240	2,604.9	2,584.14	9/8/78	2,580.63	7/30	4,210
11	06842000	Harry Strunk	880	1949	32,230	57,080	106,690	2,386.2	2,374.10	3/23/60	2,371.40	7/28	10,220
12	06849000	Harlan Co.	20,750	1952	319,800	509,000	46,800	1,973.5	1,955.67	4/6/60	1,953.62	9/8	126,300
	KANSAS												
13	06847950	Keith Sebelius	683	1964	35,930	98,800	58,280	2,331.4	2,304.59	6/27/67	2,297.10	9/29	2,620
14	06853900	Lovewell	345	1957	41,690	50,460	94,140	1,595.3	1,595.01	10/13/73	[1]1,595.34	7/22	24,400
15	06857050	Milford	24,880	1967	415,400	673,600	780,000	1,176.2	1,170.08	10/17/73	[1]1,181.94	7/25	805,000
16	06861500	Cedar Bluff	5,530	1950	185,060	191,900	493,400	2,166.0	2,154.90	7/2/51	2,119.79	9/29	46,000
17	06865000	Kanopolis	7,857	1948	55,200	356,700	705,000	1,507.0	1,506.98	7/14/51	1,505.85	7/25	246,000

Table 5. Upper Mississippi River Basin flood-control reservoir information, 1993—Continued

Map reference number (fig. 10)	Station number	Reservoir name	Drainage area (mi^2)	Year storage began	Multipurpose pool (acre-ft)	Flood storage (acre-ft)	Surcharge storage (acre-ft)	Spillway elevation (ft above SL)	Previous record elevation (ft above SL)	Previous record date (month/day/year)	1993 Maximum elevation (ft above SL)	1993 Maximum elevation date (month/day)	Change April 1 to August 1 (acre-ft)
						Missouri River Basin—Continued							
	KANSAS—Continued												
18	06868100	Wilson	1,917	1964	242,500	1,245,000	179,500	1,582.0	1,528.06	4/26/87	[1]1,548.27	8/6	388,000
19	06871700	Kirwin	1,367	1955	99,700	214,900	198,400	1,757.3	1,732.15	6/10/61	[1]1,733.47	9/29	54,000
20	06873100	Webster	1,150	1956	76,470	184,300	140,900	1,923.7	1,899.66	6/10/61	[1]1,903.90	9/29	63,700
21	06874200	Waconda	5,076	1969	241,400	722,300	165,000	1,488.3	1,471.32	4/27/87	[1]1,487.02	7/29	566,000
22	06886900	Tuttle Creek	9,628	1962	388,600	1,937,000	860,100	1,136.0	1,127.90	10/18/73	[1]1,137.76	7/22	1,615,000
23	06890898	Perry	1,117	1969	225,000	517,500	36,160	920.6	917.07	10/19/73	[1]920.94	7/25	443,000
24	06891478	Clinton	367	1977	129,100	268,400	285,800	903.4	886.72	6/4/82	[1]887.57	7/31	84,400
25	06910997	Melvern	349	1972	154,400	258,600	507,600	1,057.0	1,049.07	6/2/82	1,048.31	7/29	107,000
26	06912490	Pomona	322	1963	66,640	176,500	255,400	1,003.0	990.24	6/2/82	[1]992.67	7/31	91,000
27	06914995	Hillsdale	144	1981	76,270	83,570	155,800	931.0	928.49	10/20/86	926.70	7/28	55,000
	IOWA												
28	06903880	Rathbun	549	1969	199,800	345,700	--	926.0	924.46	7/22/82	[1]927.20	7/28	295,000
	MISSOURI												
29	06821140	Smithville	213	1981	144,600	101,800	182,200	876.2	873.17	11/16/85	[1]874.3	7/28	72,600
30	06906190	Longbranch	109	1978	34,640	30,600	98,590	801.0	799.56	7/28/81	799.0	7/26	9,000
31	06918990	Stockton	1,160	1969	912,100	779,600	--	892.0	885.94	4/28/73	884.5	9/29	74,000
32	06921325	Pomme de Terre	611	1960	241,500	407,200	--	874.0	862.35	4/30/73	[1]864.6	9/27	21,000
33	06922440	Harry S Truman	11,500	1977	1,303,000	4,006,000	2,911,000	739.6	738.69	10/11/86	735.2	8/3	3,078,000
34	06925500	Lake of the Ozarks	14,000	1931	1,927,000	400,000	--	660.0	665.45	5/22/43	659.92	6/17	122,000
						Upper Mississippi River Basin							
	IOWA												
35	05453510	Coralville	3,115	1958	40,300	428,700	--	712.0	711.85	7/21/69	[1]716.7	7/19	119,000
36	05481630	Saylorville	5,823	1977	74,000	496,000	960,000	884.0	889.25	6/22/84	[1]892.0	7/13	74,400

Table 5. Upper Mississippi River Basin flood-control reservoir information, 1993—Continued

Map reference number (fig. 10)	Station number	Reservoir name	Drainage area (mi^2)	Year storage began	Multipurpose pool (acre-ft)	Flood storage (acre-ft)	Surcharge storage (acre-ft)	Spillway elevation (ft above SL)	Previous record elevation (ft above SL)	Previous record date (month/day/year)	1993 Maximum elevation (ft above SL)	1993 Maximum elevation date (month/day)	Change April 1 to August 1 (acre-ft)
						Upper Mississippi River Basin—Continued							
	IOWA—Continued												
37	05483470	Panorama	433	1979	19,700	--	--	1,048.0	1050.10	10/15/89	--	--	--
38	05488100	Red Rock	12,323	1969	89,000	1,701,000	--	780.0	779.61	6/25/84	[1]782.7	7/13	221,600
	MISSOURI												
39	05507700	Mark Twain	2,318	1984	--	--	--	--	630.56	11/22/85	[1]636.77	9/27	584,000
	MINNESOTA												
40	05201000	Winnibigoshish	1,442	1884	229,100	439,600	--	1,300.9	1,303.39	7/30/03	1,298.83	7/28	103,100
41	05206000	Leech	1,163	1902	336,400	352,600	--	1,295.7	1,297.88	6/30/16	1,295.52	7/26	179,600
42	05210500	Pokegama	3,265	1889	27,630	52,480	--	1,274.4	1,275.28	5/23/86	1,275.08	7/17	23,320
43	05218500	Sandy	421	1911	35,500	37,540	--	1,221.3	1,224.82	5/19/50	1,218.08	7/14	14,600
44	05230500	Pine River	562	1886	65,430	53,270	--	1,234.8	1,234.56	7/10/16	1,229.85	7/11	14,800
45	05291500	Big Stone	--	1937	--	--	--	[2]7.00	[2]12.73	4/17/52	[2]9.88	7/28	--

[1]New record elevation.
[2]Gage height above arbitrary datum.

ond larger than the observed average discharge of 76,800 cubic feet per second on the Kansas River at DeSoto, Kansas (map reference N, fig. 10). About one-half of the 4,027,000 acre-feet were stored in Milford and Tuttle Creek Lakes (reservoir reference numbers 15 and 22 in figure 10). Both lakes filled their flood-control pools and were required to store floodwater in their surcharge pools. Tuttle Creek Lake stored 97,000 acre-feet in its surcharge pool, and Milford Lake stored 207,000 acre-feet in its surcharge pool.

Chariton River Basin

The Chariton River is a tributary of the Missouri River and flows from Iowa through northern Missouri. Lake Rathbun in Iowa and Lake Longbranch in Missouri (reservoir reference numbers 28 and 30 in figure 10) are flood-control reservoirs in the Chariton River Basin and stored 269,000 and 9,000 acre-feet, respectively, during July 1993. The water level in Lake Rathbun reached a record elevation of 927.20 feet above sea level on July 28, thus requiring the storage of 27,000 acre-feet of water in its surcharge pool.

Osage River Basin

Streamflow from the nearly 15,000-square-mile Osage River Basin is almost completely controlled by Melvern, Pomona, and Hillsdale Lakes in Kansas (reservoir reference numbers 25, 26, and 27 in figure 10) and Stockton, Pomme de Terre, and Harry S Truman Lakes and Lake of the Ozarks in Missouri (reservoir reference numbers 31, 32, 33, and 34 in figure 10). The reservoir system in this basin stored 3,547,000 acre-feet of water from April 1 to August 1, 1993; of this total, 3,289,000 acre-feet were stored during July. The effect of Harry S Truman Lake on discharge in the Osage River was significant because the lake stored more than 3,000,000 acre-feet of water during July. The storage in Harry S Truman Lake and that of the other reservoirs in the Osage River Basin system reduced the average discharge of the Osage River at its confluence with the Missouri River for July by 53,500 cubic feet per second.

COMBINED EFFECT OF FLOOD-CONTROL RESERVOIRS ON MISSOURI RIVER DISCHARGE

As severe as the flooding was during 1993, stream and river levels could have been even higher had a system of flood-control reservoirs not been in place throughout the Missouri River Basin. About 10,300,000 acre-feet of potential floodwater were stored in the upper Missouri River main-stem reservoirs in Montana, North Dakota, and South Dakota from April 1 to August 1, 1993. In the downstream sections of the Missouri River Basin, the quantity of water stored from April 1 to August 1 in reservoirs on the Kansas River was 4,500,000 acre-feet, while reservoirs in the Platte, the Chariton, and the Osage River Basins stored 3,900,000 acre-feet. If the total 18,700,000 acre-feet stored in the system had been allowed to flow to St. Louis, the average discharge of the Missouri River would have been 77,300 cubic feet per second greater for this 4-month period. During July alone, the combined storage of about 13,000,000 acre-feet in the Missouri River Basin—5,400,000 acre-feet in the Missouri River main-stem reservoirs, about 4,000,000 acre-feet in the Kansas River Basin reservoirs, and about 3,600,000 acre-feet in the reservoirs of the Platte, the Chariton, and the Osage River Basins—reduced the average discharge of the Missouri River at Hermann, Missouri (map reference Q, fig. 10), from about 587,000 to 376,000 cubic feet per second, which is a difference of 211,000 cubic feet per second. An analysis of the storage of flood volumes in the Missouri River Basin from April 1 to September 1, and specifically during July, enables a comparison of discharges at various points along the river and tributaries with and without the protection of the reservoirs.

The discharges of streams and the changes of storage in reservoirs in the Kansas River Basin during July 1993 were analyzed to estimate the discharges that would have occurred in the absence of the reservoir system. Floodwater that was stored in a particular reservoir was routed down the river valley under high-discharge conditions and added to the observed discharge downstream. This simulation process was iterative because several streams had more than one reservoir. Routing times were determined from observed high discharges before reservoir construction. The Muskingum routing method (Viessman and others, 1972) was used to allow for flood-discharge storage along the river valley as the flood discharges moved downstream. Using this method, daily mean discharges were

Table 6. Flood discharge in selected streams in the Kansas and the Missouri River Basins, July 1993

[mi^2, square miles; ft^3/s, cubic feet per second; --, data unavailable or not applicable; <, less than]

Site letter (fig. 10)	Station no.	Stream and place of determination	Drainage area (mi^2)	Percentage of total Kansas River Basin	Area where discharge is uncontrolled (mi^2)	Percentage of basin where discharge is uncontrolled	Observed instantaneous maximum discharge, July 1993		Uncontrolled instantaneous maximum discharge[1]		Simulated uncontrolled maximum daily average discharge[2]	
							ft^3/s	Date	ft^3/s	Date	ft^3/s	Date
					Kansas							
A	06856000	Republican River at Concordia	23,560	39	2,517	11	38,500	7/23	42,400	7/23	33,800	7/23
B	06857100	Republican River below Milford Dam	24,890	42	10	0	33,700	7/26	--	--	67,300	7/9
C	06866500	Smoky Hill River at Mentor	8,358	14	501	6	10,700	7/22	30,300	7/25	29,500	7/24
D	03269500	Saline River at Tescott	2,820	5	903	32	10,700	7/25	52,900	7/25	45,600	7/22
E	06876900	Solomon River at Niles	6,770	11	1,694	25	17,900	7/22	74,000	7/24	62,700	7/23
F	06877600	Smoky Hill River at Enterprise	19,260	32	4,410	23	45,600	7/22	155,000	7/24	122,000	7/23
G	06879100	Kansas River at Ft. Riley	44,870	75	5,130	11	87,600	7/25	200,000	7/24	189,000	7/24
H	06887000	Big Blue River near Manhattan	9,640	16	12	0	60,000	7/26	--	--	107,000	7/5
I	06887500	Kansas River at Wamego	55,280	92	5,912	11	171,000	7/26	258,000	7/25	240,000	7/25
J	06889000	Kansas River at Topeka	56,720	95	7,352	13	166,000	7/26	261,000	7/26	245,000	7/25
K	06890900	Delaware River below Perry Dam	1,117	2	0	0	5,000	7/26	--	--	28,500	7/6
L	06891000	Kansas River at Lecompton	58,460	98	7,975	14	175,000	7/27	265,000	7/26	240,000	7/25
M	06891500	Wakarusa River near Lawrence	425	<1	58	14	260	7/27	--	--	8,100	7/23
N	06892350	Kansas River at Desoto	59,756	100	8,904	15	172,000	7/27	266,000	7/27	252,000	7/10
					Missouri							
O	06818000	Missouri River at St. Joseph	420,300	--	140,500	34	335,000	7/26	461,000	7/26	--	--
P	06893000	Missouri River at Kansas City	485,200	--	154,900	32	541,000	7/28	713,000	7/27	--	--
Q	06934500	Missouri River at Hermann	524,200	--	179,400	34	750,000	7/31	852,000	7/31	--	--

[1]Data supplied by the U.S. Army Corps of Engineers, Kansas City District. Values computed by use of the BENEFITS computer program (U.S. Army Corps of Engineers, written commun., 1993).
[2]Values computed by the Muskingum routing method described in Viessman and others (1972).

estimated for selected gaging stations in the Kansas River Basin by using daily reservoir storage and daily observed stream discharges. The computer program BENEFITS (U.S. Army Corps of Engineers, written commun., 1993) was used to estimate the uncontrolled instantaneous maximum discharge at selected gaging stations on the Kansas and the Missouri Rivers. The uncontrolled instantaneous maximum discharges are compared with the observed instantaneous maximum discharges and the simulated uncontrolled maximum daily mean discharges (table 6).

The total effect of the Kansas River Basin reservoirs can be seen in the analysis of the flood discharges on the Kansas River at DeSoto, Kansas (fig. 13). The simulation of uncontrolled discharges resulted in the highest daily mean discharge of 252,000 cubic feet per second on July 10. A secondary simulated uncontrolled discharge of 233,000 cubic feet per second would have occurred on July 26. An observed instantaneous maximum discharge of 172,000 cubic feet per second occurred on July 27, while the instantaneous uncontrolled discharge was 266,000 cubic feet per second on July 27. Many other cities and hundreds of thousands of acres of farmland along the tributaries and main stem of the Kansas River benefited from the flood-control reservoirs as flood discharges were reduced by 30 to 70 percent.

All simulated uncontrolled discharges on the Kansas River would have been contained by the Federal levee system, except in Kansas City where backwater from the flooding Missouri River on July 27 might have caused the river stage to overtop the levee system there. However, without the control of reservoirs on the main-stem Missouri River, the combined uncontrolled discharges of the Kansas and the Missouri Rivers would have overtopped the Kansas City levees (Flood Insurance Administration, 1981).

RESERVOIR-LEVEL MAINTENANCE

To maintain storage capacity in flood-control reservoirs, stored floodwater is released as soon as the river downstream can accept it without additional flooding, as indicated by figure 14, which shows water-level fluctuations during the 1993 water year at selected reservoirs in the Kansas River Basin. Water levels in many of the reservoirs in the Kansas River Basin at the beginning of the 1993 water year were above multipurpose-pool elevation, but all were lowered during the 1992–93 winter. However, the snowmelt and precipitation of February through May 1993 resulted in fluctuations and steadily increasing discharge in streams in the Kansas River Basin as summer approached. An example is Tuttle Creek Lake, where, beginning in February, monthly increases in storage were followed by controlled releases to lower the lake level back to multipurpose-pool elevation. At the same time, other reservoirs in the Kansas River Basin were releasing stored water, and many uncontrolled streams were flooding. This combination resulted in many streams being at bankfull capacities for extended periods of time.

This cycle of precipitation, flooding, and resulting releases of water from reservoirs was interrupted during July when intense rains fell somewhere in the basin nearly every day of the month. With most uncontrolled streams at or above flood stage and the lower Missouri and Mississippi Rivers flooding, the flood-storage capacity of the Kansas River Basin reservoir system was nearly completely filled. Some floodwater was released as water levels in Tuttle Creek and Milford Lakes reached surcharge storage elevations, but the reservoir system performed effectively to reduce the flooding.

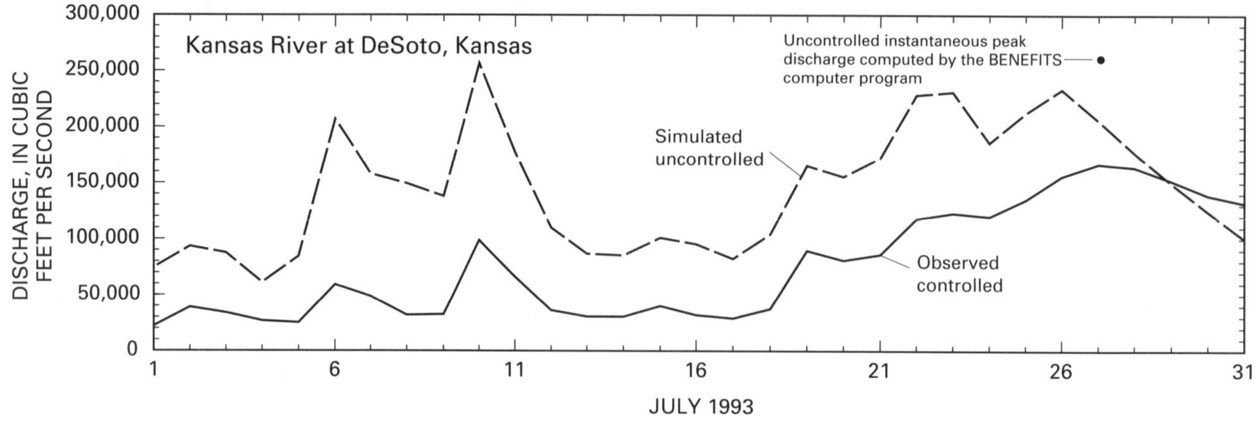

Figure 13. Observed controlled and simulated uncontrolled discharges in the Kansas River at DeSoto, Kansas, July 1993.

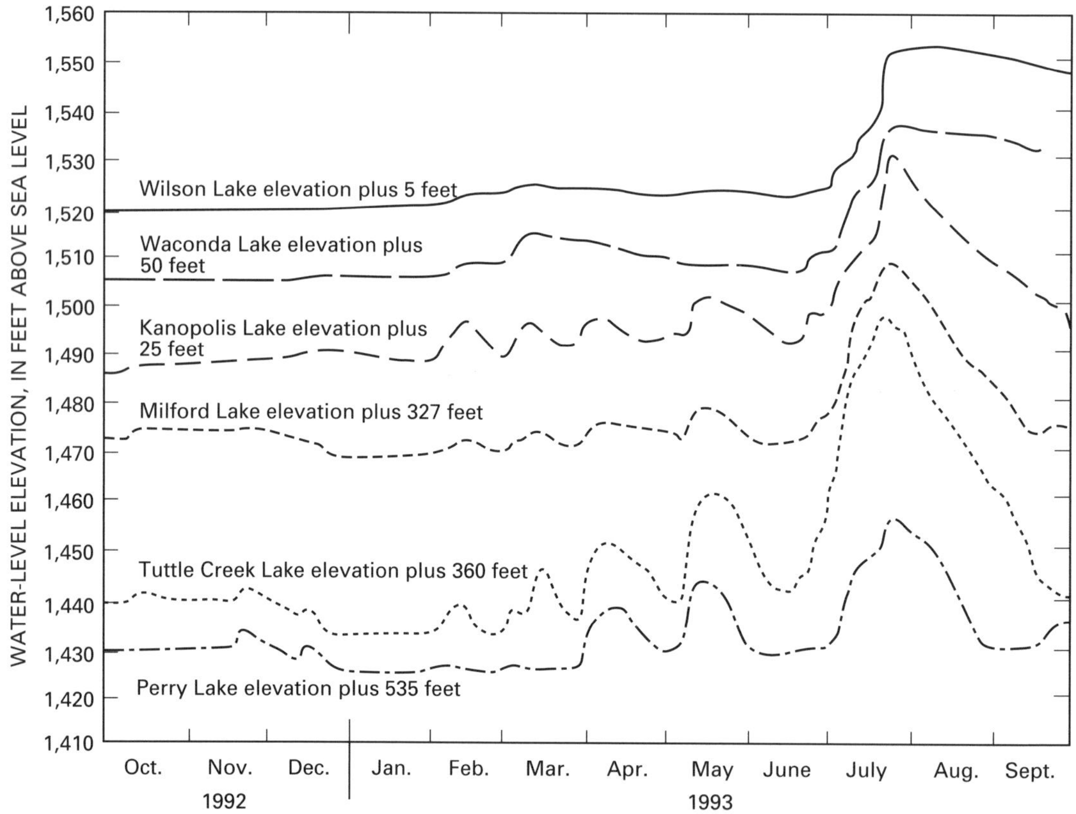

Figure 14. Water-level elevations of selected reservoirs in the Kansas River Basin, 1993 water year.

Floods in the Upper Mississippi River Basin, 1993

By Charles Parrett, Nick B. Melcher, *and* Robert W. James, Jr.[1]

INTRODUCTION

From spring through summer of 1993, severe flooding in the upper Mississippi River Basin resulted from intense, persistent, widespread rainfall from January through September. The flooding was unusual because it came so late in the spring-summer runoff season and because of the large number of streamflow-gaging stations that had record or near-record maximum discharges. Record maximum discharges were recorded from mid-June through early August at many U.S. Geological Survey (USGS) streamflow-gaging stations in the Minnesota River Basin in Minnesota; in the Skunk, the Des Moines, the Little Sioux, and the Nishnabotna River Basins in Iowa; on the Mississippi River at Keokuk, Iowa; in the James River Basin in North and South Dakota; in the Platte River Basin in Nebraska; in the Kansas River Basin in Kansas; in the Grand River Basin in Missouri; and along the Missouri River from St. Joseph to Booneville, Missouri. Unusually high flood discharges were recorded at other locations throughout the area of flooding. The flooding also was unusual for its long duration and widespread and severe damage. At St. Louis, Missouri, the Mississippi River reached flood stage on June 26 and remained above flood stage until late August. Millions of acres of agricultural and urban lands in the upper Mississippi Basin were inundated for weeks, and unofficial damage estimates exceeded $10 billion (Parrett and others, 1993).

FLOOD RECURRENCE INTERVAL

For comparative purposes, flood-maximum discharges are referenced to a specific recurrence interval or probability of occurrence. The recurrence interval is the average number of years between occurrences of annual maximum discharges that equal or exceed a specified discharge. For example, a discharge that has a 100-year recurrence interval is so large that an equal or greater annual maximum discharge is expected, on average, only once in any 100-year period. Because of the random nature of flood events, the times between annual maximum discharges of a certain magnitude are far from uniform; a large flood in 1 year does not preclude the occurrence of an even larger flood the next year. In any given year, the annual maximum discharge has 1 chance in 100 of equaling or exceeding the 100-year flood (U.S. Interagency Advisory Committee on Water Data, 1982).

Recurrence intervals for the 1993 flood peaks presented in this report are generally determined by using the most current published USGS flood-frequency reports for States in the area of flooding. Recurrence intervals for the 1993 maximum discharges on the Kansas River, the Missouri River, and the Mississippi River are based on unpublished flood-frequency analyses completed by the U.S. Army Corps of Engineers (Gary Dyhouse, St. Louis District, U.S. Army Corps of Engineers, written commun., 1993; Jerry Buehre, Kansas City District, U.S. Army Corps of Engineers, written commun., 1993).

CHRONOLOGY OF THE SPRING AND SUMMER FLOODING

The magnitude and timing of several rainstorms during late June and July, combined with wet antecedent climatic conditions, were the principal causes of the severe flooding in the upper Mississippi River Basin. To illustrate the effect of the timing of runoff from these storms on the maximum discharge in the Mississippi River, the maximum discharges and their dates of occurrence for selected streamflow-gaging stations in the general area of flooding are shown in figure 15.

During June 17–18, 2 to 7 inches of rain fell throughout southern Minnesota, northern Iowa, and southwestern Wisconsin. Runoff from this storm caused flooding on the Minnesota and the Mississippi Rivers in Minnesota and the Chippewa and the Black Rivers in Wisconsin. As a result of these floodwaters, the discharge of the Mississippi River at Clinton, Iowa, peaked on July 8, 1993.

[1]This article is based on a previously published report (Parrett and others, 1993).

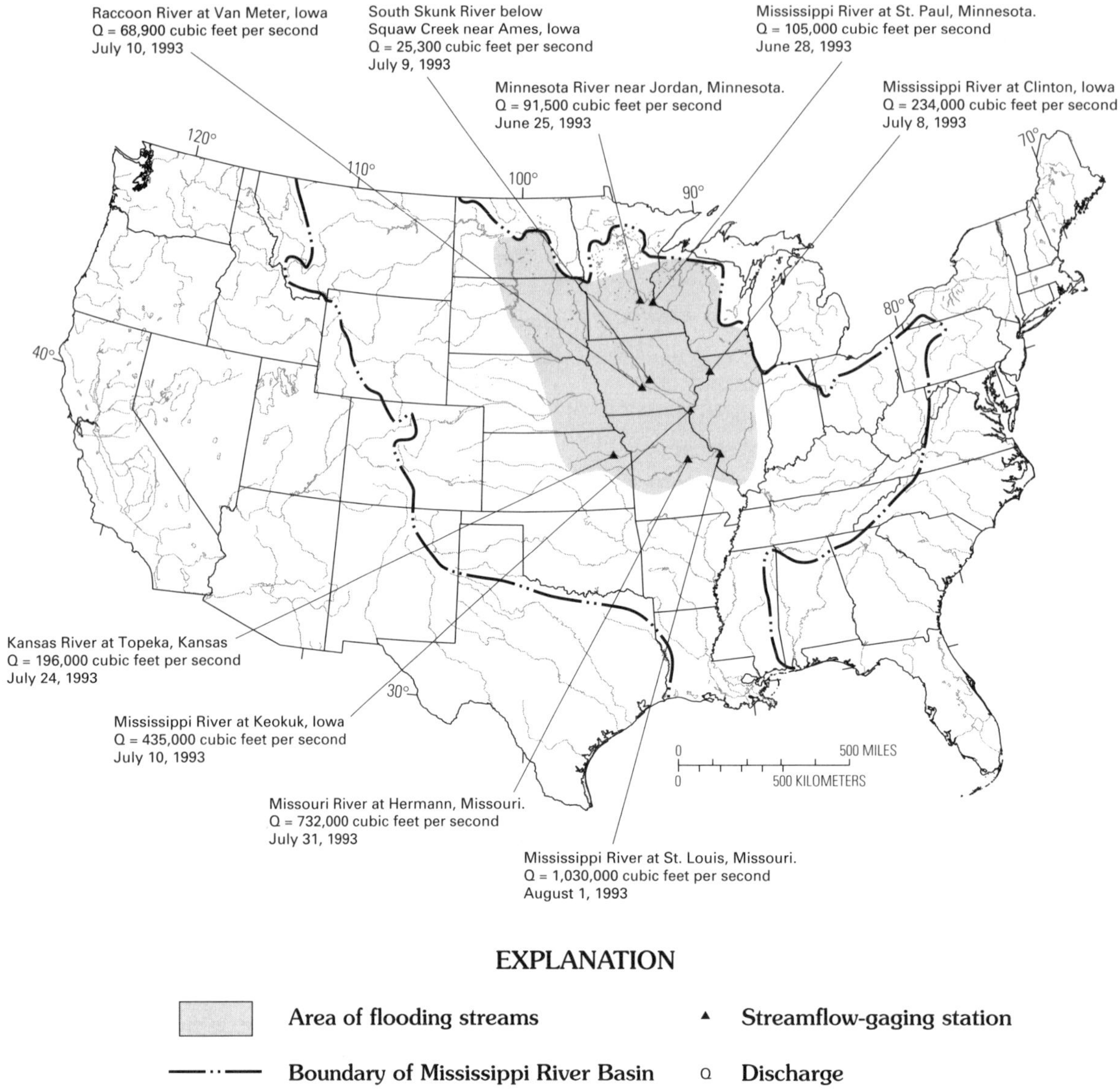

Figure 15. Maximum discharges (Q) and dates of occurrence for the 1993 flood at selected streamflow-gaging stations in the upper Mississippi River Basin (Parrett and others, 1993).

Two separate storms during early July caused large-scale flooding in Iowa. During the first storm on July 5, 2 to 5 inches of rain fell in central Iowa and caused lowland flooding on the Iowa, the Skunk, and the Des Moines Rivers. During the second storm on July 8–9, 2 to 8 inches of rain fell in central Iowa. Rivers throughout central Iowa had not receded from the July 5 storm, and the three major reservoirs in this part of the State were at capacity. The runoff from this storm, combined with the runoff from the July 5 storm, caused record or near-record maximum discharges at streamflow-gaging stations throughout the Iowa, the Skunk, the Raccoon, and the Des Moines River Basins. The floodwaters from these rivers entered the Mississippi River at about the same time as the flood peak from the late June storm in northern basins reached Keokuk, Iowa. The coincident timing of the flood peaks from these tributary rivers increased the maximum discharge on the Mississippi River and aggravated flooding on the Mississippi River from Davenport, Iowa, to St. Louis, Missouri. The discharge on the Mississippi River at St. Louis that resulted from these combined floodwaters peaked on July 20.

On July 15–16, 2 to 7 inches of rain fell in eastern North Dakota and western Minnesota and caused flooding in the upstream reaches of the Minnesota River Basin in Minnesota and the James River Basin in North Dakota. Although maximum discharges from this storm were not as large in the downstream reaches of these basins as the maximum discharges of late June, the floodwaters from the James River added to the flooding of late July on the Missouri River.

From July 22 to 24, 2 to 13 inches of rain fell in parts of Nebraska, Kansas, Missouri, Iowa, and Illinois. The runoff from this storm caused record maximum discharges on the Platte River in Nebraska and contributed large flows to previously filled reservoirs in the Kansas River Basin in Kansas. Maximum discharges on the Kansas River were the largest since 1951, which is before significant river regulation began. Discharges also were near-record on the Nishnabotna River in Iowa and the Illinois River in Illinois.

Before the July 22 to 24 storm, the Missouri River was at or near flood stage as a result of large tributary inflows earlier in the month from the James River in North and South Dakota, the Big Sioux River in South Dakota, and the Little Sioux River in Iowa. As a result, floodwaters from the Platte and the Kansas Rivers caused record or near-record maximum discharges on the Missouri River at streamflow-gaging stations downstream from the confluence of the Platte River. The flood peak on the Missouri River reached Hermann, Missouri, on July 31. The maximum discharge from the Missouri River caused a second and greater maximum discharge at the streamflow-gaging station on the Mississippi River at St. Louis on August 1.

Flood conditions on the Mississippi River differed upstream and downstream from the confluence of the Ohio River. At Thebes, Illinois, just upstream from the confluence, severe flooding on the Mississippi River peaked on August 7. Downstream from the confluence, flooding on the Mississippi River was not severe because of less-than-average discharge contributed by the Ohio River and a substantially larger channel capacity in this reach of the Mississippi River. The discharge of the Ohio River was less than average during July and August as a result of generally dry conditions and low reservoir outflows throughout the Ohio River Basin.

Flood Discharges, Gage Heights, and Recurrence Intervals in the Upper Mississippi and Missouri River Basins by State

By Charles A. Perry

Flooding during the spring and summer of 1993 in the upper Mississippi and Missouri River Basins was widespread, encompassing nine States. Many streams in this nine-State area had historic floods, while some streams had only moderate flooding. Tables 7 through 15 include a compilation of flood information for selected streams within each of the states of Illinois, Iowa, Kansas, Minnesota, Missouri, Nebraska, North Dakota, South Dakota, and Wisconsin (U.S. Geological Survey, 1994). Figures 16–24 provide the location of the streamflow-gaging stations within each State. Only streams within the upper Mississippi and Missouri Basins are listed. Flooding outside of these basins or other than the spring and summer of 1993 are listed in the sections "Summary of Floods of 1992" and "Summary of Floods of 1993."

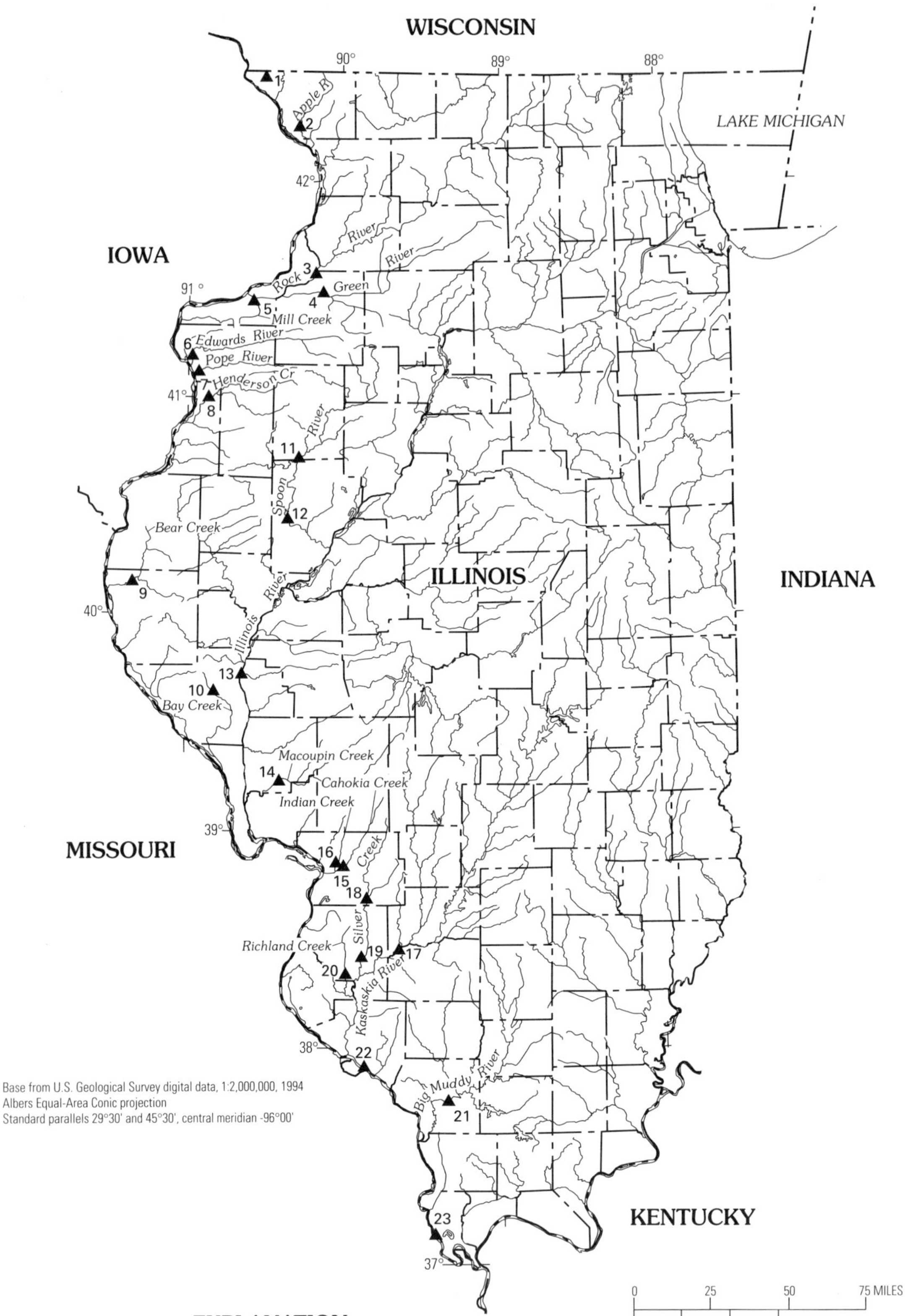

EXPLANATION

▲21 **Streamflow-gaging station and site number—**
Number corresponds to that in table 7

Figure 16. Location of selected streamflow-gaging stations in Illinois.

Table 7. Maximum stages and discharges prior to and during March 1–September 30, 1993, in Illinois

[mi², square miles; ft, feet above an arbitrary datum; ft³/s, cubic feet per second; >, greater than; <, less than; --, not determined or not applicable. Source: Recurrence intervals calculated from U.S. Geological Survey data. Other data from U.S. Geological Survey reports or data bases]

				Maximum prior to 1993				Maximum during March–September 1993			
Site no. (fig.16)	Station no.	Stream and place of determination	Drainage area (mi²)	Period	Year	Stage (ft)	Discharge (ft³/s)	Date (month/day)	Stage (ft)	Discharge (ft³/s)	Discharge, recurrence interval (years)
1	05414820	Sinsinawa River near Menominee, IL	39.6	1967–93	1969	13.34	11,600	07/05	13.20	10,900	>10
2	05419000	Apple River near Hanover, IL	247	1934–93	1946	[1]26.12	12,000	07/06	20.75	7,130	< 5
3	05446500	Rock River near Joslin, IL	9,549	1939–93	1993	18.35	[2]46,500	06/11	18.35	46,500	<25
4	05447500	Green River near Geneseo, IL	1,003	1936–93	1974	[1]18.59	12,100	06/09	15.15	9,600	10
5	05448000	Mill Creek at Milan, IL	62.4	1939–93	1973	12.65	9,300	06/25	10.29	7,680	< 20
6	05466500	Edwards River near New Boston, IL	445	1934–93	1973	23.33	18,000	08/16	22.37	8,590	< 15
7	05467000	Pope Creek near Keithsburg, IL	174	1934–93	1973	[3]29.08	8,900	07/24	29.08	7,270	> 50
8	05469000	Henderson Creek near Oquawka, IL	432	1934–93	1982	31.05	34,600	07/25	32.65	30,800	> 100
9	05495500	Bear Creek near Marcelline, IL	349	1944–93	1985	28.38	29,500	07/01	19.08	12,700	< 5
10	05512500	Bay Creek at Pittsfield, IL	39.4	1939–93	1965	14.77	12,600	09/14	14.92	13,700	> 25
11	05569500	Spoon River at London Mills, IL	1,072	1942–93	1974	28.03	41,000	07/25	25.92	22,600	<25
12	05570000	Spoon River at Seville, IL	1,636	1914–93	1924	30.77	37,300	07/26	33.10	34,700	50
13	05586100	Illinois River at Valley City, IL	26,743	1938–93	1943	28.61	123,000	08/01	[3]25.95	[2]92,400	--
14	05587000	Macoupin Creek near Kane, IL	868	1921–33, 1940–93	1943	28.50	40,000	04/16	[3]22.91	11,100	> 2
15	05587900	Cahokia Creek at Edwardsville, IL	212	1969–93	1979	24.74	8,200	04/25	19.23	5,670	< 5

Table 7. Maximum stages and discharges prior to and during March 1–September 30, 1993, in Illinois—Continued

Site no. (fig.16)	Station no.	Stream and place of determination	Drainage area (mi²)	Maximum prior to 1993				Maximum during March–September 1993			
				Period	Year	Stage (ft)	Discharge (ft³/s)	Date (month/day)	Stage (ft)	Discharge (ft³/s)	Discharge, recurrence interval (years)
16	05588000	Indian Creek at Wanda, IL	36.7	1940–93	1946	18.41	9,340	04/25	15.32	2,810	< 5
17	05594100	Kaskaskia River near Venedy Station, IL	4,393	1969–93	1990	25.11	48,600	04/17	21.51	15,200	--
18	05594450	Silver Creek near Troy, IL	154	1966–93	1979	[3]17.94	10,600	04/15	15.21	3,330	< 2
19	05594800	Silver Creek near Freeburg, IL	464	1970–93	1990	22.10	11,000	09/24	17.07	4,800	< 2
20	05595200	Richland Creek near Hecker, IL	129	1969–93	1972	42.88	14,900	09/23	41.76	9,020	< 10
21	05599500	Big Muddy River at Murphysboro, IL	2,169	1916–93	1983 1961	36.88 [3]37.97	[4]32,100 [5]33,300	09/29	[3,4]33.19	11,500	< 2
22	07020500	Mississippi River at Chester, IL	708,600	1927–93	[6]1844	[6]39.8	[6]1,350,000	08/06	49.59	950,000	10–50
23	07022000	Mississippi River at Thebes, IL	713,200	1932–93	[6]1844	[6]45.14	[6]1,375,000	08/07	45.50	996,000	10–50

[1]Ice jam.
[2]Affected by breaks in levee.
[3]Occurred on different date.
[4]Backwater from the Mississippi River.
[5]Prior to regulation of flow.
[6]Data from U.S. Army Corps of Engineers.

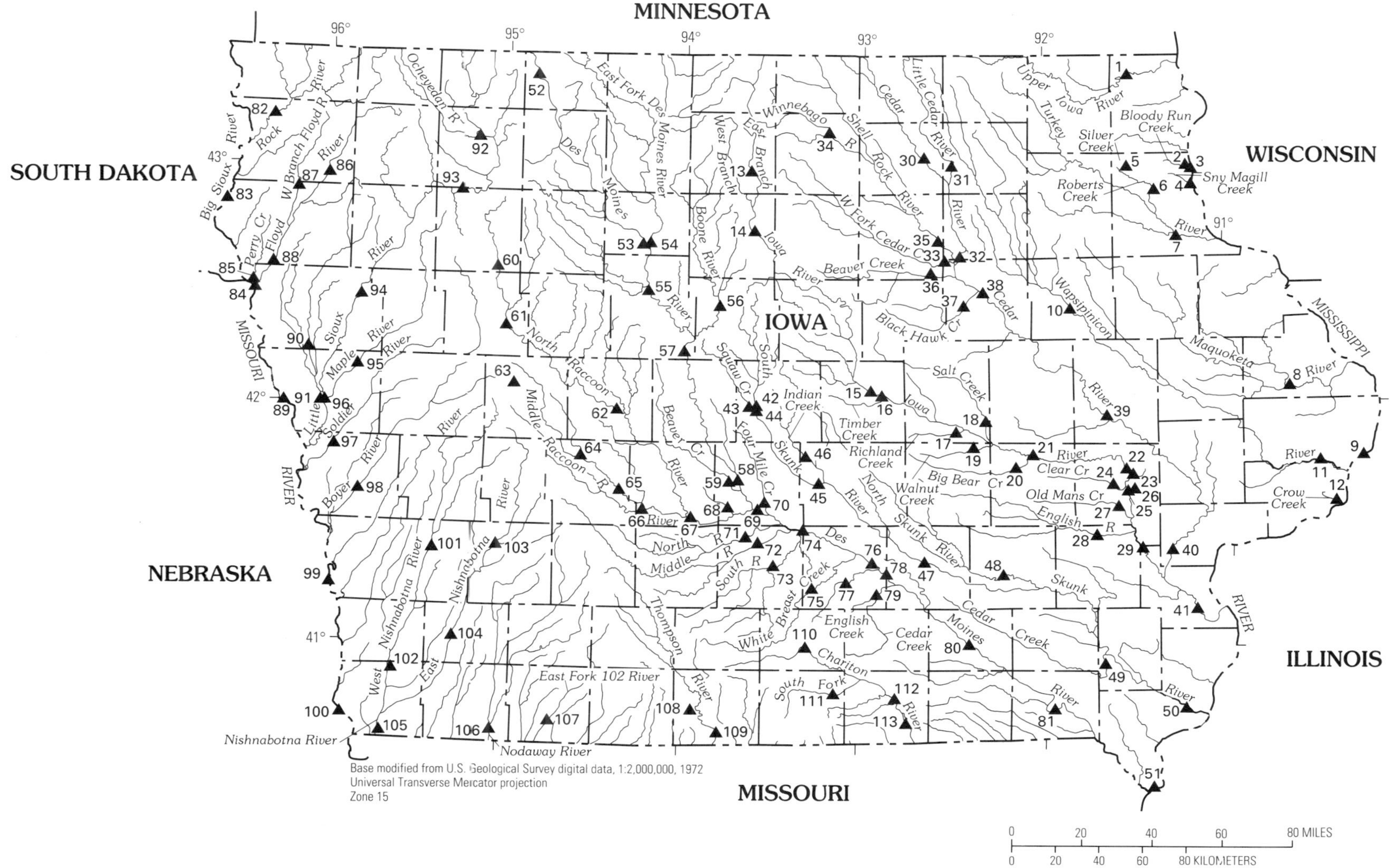

EXPLANATION

▲21 **Streamflow-gaging station and site number—**
Number corresponds to that in table 8

Figure 17. Location of selected streamflow-gaging stations in Iowa.

Table 8. Maximum stages and discharges prior to and during March 1–September 30, 1993, in Iowa

[mi^2, square miles; ft, feet above an arbitrary datum; ft^3/s, cubic feet per second; --, not determined or not applicable; <, less than; >, greater than. Source: Recurrence intervals calculated from U.S. Geological Survey data. Other data from U.S. Geological Survey reports or data bases]

				Maximum prior to 1993				Maximum during March–September 1993			
Site no. (fig. 17)	Station no.	Stream and place of determination	Drainage area (mi^2)	Period	[1]Year	Stage (ft)	Discharge (ft^3/s)	[1]Date (month/day)	Stage (ft)	Discharge (ft^3/s)	Discharge recurrence interval (years)
1	05388250	Upper Iowa River near Dorchester, IA	770	1975–93	1941	21.8	30,400	8/17	20.00	22,000	70
2	05389400	Bloody Run Creek near Marquette, IA	34.1	1991–93	1992	6.75	339	3/31	7.57	1,540	--
3	05389500	Mississippi River at McGregor, IA	67,500	1936–93	1965	25.38	[2]276,000	6/29	21.97	189,000	10–50
4	05411400	Sny Magill Creek near Clayton, IA	27.6	1991–93	1992	5.92	227	8/23	8.60	1,300	--
5	05412060	Silver Creek near Luana, IA	4.39	1986–93	1991	14.97	3,300	6/29	11.58	960	--
6	05412100	Roberts Creek above Saint Olaf, IA	70.7	1986–93	1991	27.88	19,600	3/31	16.72	2,120	--
7	05412500	Turkey River near Garber, IA	1,545	1913–16, 1919–27, 1929–30, 1932–93	1991	30.10	49,900	3/31	22.94	19,400	4
8	05418500	Maquoketa River near Maquoketa, IA	1,553	1913–93	1944	[3]24.70	48,000	7/6	32.86	35,300	19
9	05420500	Mississippi River at Clinton, IA	85,600	1873–1993	1965	24.65	[2]307,000	7/7	[4]22.98	239,000	--
10	05421000	Wapsipinicon River at Independence, IA	1,048	1933–93	1968	21.11	26,800	4/3	14.90	13,400	7
11	05422000	Wapsipinicon River near DeWitt, IA	2,330	1934–93	1990	14.19	31,100	7/8	12.88	22,300	15
	05422470	Crow Creek at Bettendorf, IA	17.8	1977–93	1990	11.03	7,700	3/22	7.74	1,140	2
13	05449000	East Branch Iowa River near Klemme, IA	133	1948-93	1954	[3,4]10.67	5,960	3/31	10.82	4,380	35
14	05449500	Iowa River near Rowan, IA	429	1940–93	1954	14.88	8,460	4/1	14.69	6,140	17
15	05451500	Iowa River at Marshalltown, IA	1,564	1902–03, 1914–27, 1932–93	1918	[4]20.47	42,000	8/17	20.77	20,400	15

Table 8. Maximum stages and discharges prior to and during March 1–September 30, 1993, in Iowa—Continued

Site no. (fig. 17)	Station no.	Stream and place of determination	Drainage area (mi^2)	Maximum prior to 1993: Period	[1]Year	Stage (ft)	Discharge (ft^3/s)	Maximum during March–September 1993: [1]Date (month/day)	Stage (ft)	Discharge (ft^3/s)	Discharge recurrence interval (years)
16	05451700	Timber Creek near Marshalltown, IA	118	1949–93	1977	17.69	12,000	7/9	17.03	8,870	25
17	05451900	Richland Creek near Haven, IA	56.1	1949–93	1991	26.71	12,200	7/17	22.84	4,900	16
18	05452000	Salt Creek near Elberon, IA	201	1945–93	1947	[4]20.00	35,000	7/9	20.85	36,600	90
19	05452200	Walnut Creek near Hartwick, IA	70.9	1949–93	1991	16.93	7,900	8/29	16.76	5,770	9
20	05453000	Big Bear Creek at Ladora, IA	189	1945–93	1960	[3,4]15.32	10,500	8/30	24.69	6,610	6
21	05453100	Iowa River at Marengo, IA	2,794	1956–93	1960	[4]19.79	30,800	7/19	20.31	38,000	90
22	05453520	Iowa River below Coralville Dam near Coralville, IA	3,115	1992–93	--	--	--	7/19	63.95	25,800	--
23	05454000	Rapid Creek near Iowa City, IA	25.3	1937–93	1965	[4]14.93	6,100	8/10	15.61	6,700	35
24	05454300	Clear Creek near Coralville, IA	98.1	1952–93	1990	16.36	10,200	7/6	14.74	6,760	20
25	05454500	Iowa River at Iowa City, IA	3,271	1903–93	1969	[3]13.93	15,000	8/10	28.52	28,200	>100
26	05455010	South Branch Ralston Creek at Iowa City, IA	2.94	1963–93	1972	9.47	1,070	8/10	9.43	994	--
27	05455100	Old Mans Creek near Iowa City, IA	201	1950–93	1962	[4]17.20	12,000	7/6	17.61	13,000	45
28	05455500	English River at Kalona, IA	573	1939–93	1965	21.45	20,000	7/6	22.55	36,100	[5]1.2
29	05455700	Iowa River near Lone Tree, IA	4,293	1956–93	1974	[4]20.27	35,700	7/7	22.94	57,100	>100
30	05457700	Cedar River at Charles City, IA	1,054	1964–93	1965	[4]21.64	21,000	8/16	21.44	26,400	25
31	05458000	Little Cedar River near Ionia, IA	306	1954–93	1961	15.58	10,800	8/16	18.99	14,000	45
32	05458500	Cedar River at Janesville, IA	1,661	1904–06, 1914–27, 1932–42, 1945–93	1961	16.33	37,000	8/18	15.74	35,000	40
33	05458900	West Fork Cedar River at Finchford, IA	846	1945–93	1951	[4]18.45	31,900	4/1	16.73	17,600	10
34	05459500	Winnebago River at Mason City, IA	526	1932–93	1933	15.70	10,800	4/1	12.59	7,190	10
35	05462000	Shell Rock River at Shell Rock, IA	1,746	1953–93	1961	16.26	33,500	4/1	15.21	20,800	9

Table 8. Maximum stages and discharges prior to and during March 1–September 30, 1993, in Iowa—Continued

Site no. (fig. 17)	Station no.	Stream and place of determination	Drainage area (mi^2)	Maximum prior to 1993: Period	[1]Year	Stage (ft)	Discharge (ft^3/s)	Maximum during March–September 1993: [1]Date (month/day)	Stage (ft)	Discharge (ft^3/s)	Discharge recurrence interval (years)
36	05463000	Beaver Creek at New Hartford, IA	347	1945–93	1947	13.50	18,000	3/31	13.45	14,700	15
37	05463500	Black Hawk Creek at Hudson, IA	303	1952–93	1969	18.23	19,300	7/9	16.94	9,670	11
38	05464000	Cedar River at Waterloo, IA	5,146	1940–93	1961	21.86	76,700	4/2	20.60	68,100	19
39	05464500	Cedar River at Cedar Rapids, IA	6,510	1902–93	1961	[4]20.00	73,000	4/4	19.27	71,000	35
40	05465000	Cedar River near Conesville, IA	7,785	1939–93	1961	[4]16.85	70,800	4/6	17.11	74,000	35
41	05465500	Iowa River at Wapello, IA	12,499	1914–93	1973	[4]28.91	92,000	7/8	[4]29.53	111,000	100
42	05470000	South Skunk River near Ames, IA	315	1920–27, 1932–93	1954	[4]13.90	8,630	8/16	14.23	11,200	[5]1.2
43	05470500	Squaw Creek at Ames, IA	204	1919–27, 1965–93	1990	15.97	12,500	7/9	18.54	24,300	[5]1.8
44	05471000	South Skunk River below Squaw Creek near Ames, IA	556	1952–79, 1991–93	1975	25.57	14,700	7/9	25.53	26,500	[5]1.5
45	05471050	South Skunk River at Colfax, IA	803	1985–93	1990	19.07	8,770	7/12	21.53	14,200	--
46	05471200	Indian Creek near Mingo, IA	276	1958–75, 1985–93	1991	19.16	23,500	7/9	18.64	18,600	[5]1.1
47	05471500	South Skunk River near Oskaloosa, IA	1,635	1945–93	1947	[4]23.05	20,000	7/15	24.78	20,700	30
48	05472500	North Skunk River near Sigourney, IA	730	1945–93	1960	25.33	27,500	7/6	24.68	17,500	18
49	05473400	Cedar Creek near Oakland Mills, IA	530	1977–93	1983	19.68	8,560	7/9	21.27	8,920	10–15
50	05474000	Skunk River at Augusta, IA	4,303	1913–93	1973	27.05	66,800	7/10	23.70	46,600	35
51	05474500	Mississippi River at Keokuk, IA	119,000	1878–1993	1851	21.0	360,000	7/10	27.58	446,000	>100
52	05476500	Des Moines River at Estherville, IA	1,372	1951–93	1969	17.68	16,000	6/30	15.38	9,330	20
53	05476750	Des Moines River at Humboldt, IA	2,256	1964–93	1969	15.40	18,000	7/13	15.22	19,000	80
54	05479000	East Fork Des Moines River at Dakota City, IA	1,308	1940–93	1954	24.02	17,400	4/1	23.35	16,200	25
55	05430500	Des Moines River at Fort Dodge, IA	4,190	1905–06, 1913–27, 1946–93	1965	[4]19.62	35,600	4/1	15.81	31,200	30

Table 8. Maximum stages and discharges prior to and during March 1–September 30, 1993, in Iowa—Continued

Site no. (fig. 17)	Station no.	Stream and place of determination	Drainage area (mi²)	Maximum prior to 1993: Period	[1]Year	Stage (ft)	Discharge (ft³/s)	Maximum during March–September 1993: [1]Date (month/day)	Stage (ft)	Discharge (ft³/s)	Discharge recurrence interval (years)
56	05481000	Boone River near Webster City, IA	844	1940–93	1954	18.55	20,300	4/1	15.32	13,100	13
57	05481300	Des Moines River near Stratford, IA	5,452	1920–93	1954	29.70	57,400	4/2	25.68	42,300	25
58	05481650	Des Moines River near Saylorville, IA	5,841	1961–93	1984	20.72	30,100	7/21	[4]24.22	45,700	>100
59	05481950	Beaver Creek near Grimes, IA	358	1960–93	1986	14.73	7,980	7/10	16.58	14,300	[5]1.2
60	05482135	North Raccoon River near Newell, IA	233	1982–93	1984	16.73	2,850	7/11	16.20	2,420	5
61	05482300	North Raccoon River near Sac City, IA	700	1958–93	1979	[4]20.14	13,100	7/11	17.55	6,550	4
62	05482500	North Raccoon River near Jefferson, IA	1,619	1940–93	1947	22.30	29,100	7/10	19.20	16,900	10
63	05483343	Hazelbrush Creek near Maple River, IA	9.22	1990–93	1991	13.59	957	7/9	14.77	1,120	--
64	05483450	Middle Raccoon River near Bayard, IA	375	1979–93	1986	24.70	12,300	7/9	29.02	27,500	[5]1.2
65	05483600	Middle Raccoon River near Panora, IA	440	1958–93	1986	15.50	15,300	7/9	20.04	22,400	100
66	05484000	South Raccoon River at Redfield, IA	994	1940–93	1958	29.04	35,000	7/10	26.98	44,000	[5]1.2
67	05484500	Raccoon River at Van Meter, IA	3,441	1915–93	1947	[4]22.69	41,200	7/10	26.34	70,100	[5]1.3
68	05484800	Walnut Creek at Des Moines, IA	78.4	1971–93	1986	18.32	12,500	8/29	17.56	6,460	8
69	05485500	Des Moines River below Raccoon River at Des Moines, IA	9,879	1940–93	1984	28.46	58,400	7/11	34.29	116,000	--
70	05485640	Fourmile Creek at Des Moines, IA	92.7	1971–93	1977	[4]14.84	5,380	7/9	14.02	4,210	6
71	05486000	North River near Norwalk, IA	349	1940–93	1947	25.30	32,000	8/31	23.40	8,690	8
72	05486490	Middle River near Indianola, IA	503	1940–93	1947	28.27	34,000	7/5	21.62	10,000	4
73	05487470	South River near Ackworth, IA	460	1940–93	1990	[4]32.85	38,100	7/6	30.02	32,200	35
74	05487500	Des Moines River near Runnells, IA	11,655	1985–93	1990	[4]79.20	88,300	7/11	82.88	134,000	--
75	05487980	White Breast Creek near Dallas, IA	342	1962–93	1982	33.45	37,300	7/6	30.20	25,500	50

Table 8. Maximum stages and discharges prior to and during March 1–September 30, 1993, in Iowa—Continued

Site no. (fig. 17)	Station no.	Stream and place of determination	Drainage area (mi^2)	Maximum prior to 1993: Period	[1]Year	Stage (ft)	Discharge (ft^3/s)	Maximum during March–September 1993: [1]Date (month/ day)	Stage (ft)	Discharge (ft^3/s)	Discharge recurrence interval (years)
76	05438110	Des Moines River near Pella, IA	12,330	1992–93	--	--	--	7/12	109.71	105,000	--
77	05438200	English Creek near Knoxville, IA	90.1	1985–93	1982	30.28	28,000	7/5	27.88	18,900	--
78	05438500	Des Moines River near Tracy, IA	12,479	1920–93	1984	18.11	42,600	7/12	24.16	109,000	--
79	05439000	Cedar Creek near Bussey, IA	374	1947–93	1982	34.61	96,000	7/5	28.53	36,100	35
80	05439500	Des Moines River at Ottumwa, IA	13,374	1917–93	1984	14.64	47,800	7/12	22.15	112,000	>100
81	05490500	Des Moines River at Keosauqua, IA	14,038	1903–06, 1910–93	1973	28.02	72,200	7/13	32.66	111,000	--
82	06433500	Rock River near Rock Valley, IA	1,592	1948–93	1969	[3]17.32	40,400	5/9	19.97	29,300	25
83	06435500	Big Sioux River at Akron, IA	8,424	1928–93	1969	22.99	80,800	5/10	23.05	66,700	50
84	06436000	Missouri River at Sioux City, IA	314,600	1897–1993	1960	[4]30.65	101,000	7/15	27.33	72,200	--
85	06600000	Perry Creek at 38th Street, Sioux City, IA	65.1	1945–69 1981–93	1990	28.54	8,670	3/8	12.14	1,120	<2
86	06600100	Floyd River at Alton, IA	268	1955–93	1983	[4]18.54	16,300	3/28	17.46	6,580	7
87	06600300	West Branch Floyd River near Struble, IA	180	1955–93	1962	[4]15.86	8,060	3/26	14.23	3,470	3
88	06600500	Floyd River at James, IA	886	1934–93	1953	25.30	71,500	3/29	22.34	9,680	6
89	06601200	Missouri River at Decatur, NE	316,200	1987–93	1990	[4]25.59	40,900	7/16	32.19	76,400	--
90	06602020	West Fork Ditch at Hornick, IA	403	1939–69, 1974–93	1962	[4]25.20	12,400	3/28	19.47	4,740	4
91	06602400	Monona-Harrison Ditch near Turin, IA	900	1958–93	1971	28.03	19,900	3/27	20.48	6,980	--
92	06605000	Ocheyedan River near Spencer, IA	426	1977–93	1983	10.49	6,450	7/1	11.28	6,170	8
93	06605850	Little Sioux River at Linn Grove, IA	1,548	1972–93	1984	19.58	13,100	7/2	20.63	16,100	14
94	06606600	Little Sioux River at Correctionville, IA	2,500	1918–25, 1928–32, 1936–93	1965	25.86	29,800	7/18	23.82	22,600	25
95	06607200	Maple River at Mapleton, IA	669	1941–93	1978	[4]22.10	20,800	7/9	9.67	7,130	2

Table 8. Maximum stages and discharges prior to and during March 1–September 30, 1993, in Iowa—Continued

Site no. (fig. 17)	Station no.	Stream and place of determination	Drainage area (mi^2)	Maximum prior to 1993: Period	[1]Year	Stage (ft)	Discharge (ft^3/s)	Maximum during March–September 1993: [1]Date (month/day)	Stage (ft)	Discharge (ft^3/s)	Discharge recurrence interval (years)
96	06607500	Little Sioux River near Turin, IA	3,526	1958–93	1983	[4]27.44	31,200	7/19	23.96	26,300	--
97	06608500	Soldier River at Pisgah, IA	407	1940–93	1950	28.17	22,500	7/9	27.56	23,400	20
98	06609500	Boyer River at Logan, IA	871	1918–25, 1937–93	1990	[4]25.22	30,800	7/9	22.19	26,200	20
99	06610000	Missouri River at Omaha, NE	322,800	1928–93	1952	[4]40.20	396,000	7/10	30.26	115,000	--
100	06807000	Missouri River at Nebraska City, NE	410,000	1929–93	1952	27.66	414,000	7/23	27.19	196,000	50–100
101	06807410	West Nishnabotna River at Hancock, IA	609	1959–93	1972	22.12	26,400	7/10	23.52	30,100	35
102	06808500	West Nishnabotna River at Randolph, IA	1,326	1948–93	1987	[4]24.80	40,800	7/23	23.60	22,100	4
103	06809210	East Nishnabotna River near Atlantic, IA	436	1960–93	1972	22.81	26,700	8/30	17.34	13,100	4
104	06809500	East Nishnabotna River at Red Oak, IA	894	1918–25, 1936–93	1972	[4]28.23	38,000	8/31	21.98	21,600	10
105	06810000	Nishnabotna River above Hamburg, IA	2,806	1922–23, 1928–93	1947	[4]28.27	55,500	7/25	30.56	37,700	45
106	06817000	Nodaway River at Clarinda, IA	762	1918–25, 1936–93	1947	25.30	31,100	7/22	22.95	28,000	13
107	06819185	East Fork 102 River at Bedford, IA	85.4	1983–93	1986	23.47	9,570	7/5	23.85	9,170	10–20
108	06897950	Elk Creek near Decatur City, IA	52.5	1967–93	1990	[4]28.22	18,000	7/5	29.93	32,800	35
109	06898000	Thompson River at Davis City, IA	701	1918–25, 1941–93	1992	24.29	57,000	7/5	20.53	30,300	80
110	06903400	Chariton River near Chariton, IA	182	1965–93	1992	29.32	37,700	7/5	22.60	14,900	25
111	06903700	South Fork Chariton River near Promise City, IA	168	1967–93	1992	34.84	70,600	7/5	25.09	16,900	13
112	06903900	Chariton River near Rathbun, IA	549	1956–93	1973	14.92	1,880	3/24	[4]13.12	2,040	--
113	06904010	Chariton River near Moulton, IA	740	1979–93	1982	36.83	11,200	7/8	34.18	6,500	--

[1]Unless noted otherwise with the stage, the maximum stage and maximum discharge occurred on the same date.
[2]Daily mean discharge.
[3]Gage at different datum.
[4]Maximum stage occurred on date different from date of maximum discharge.
[5]Ratio of 1993 peak discharge to 100-year recurrence-interval discharge.

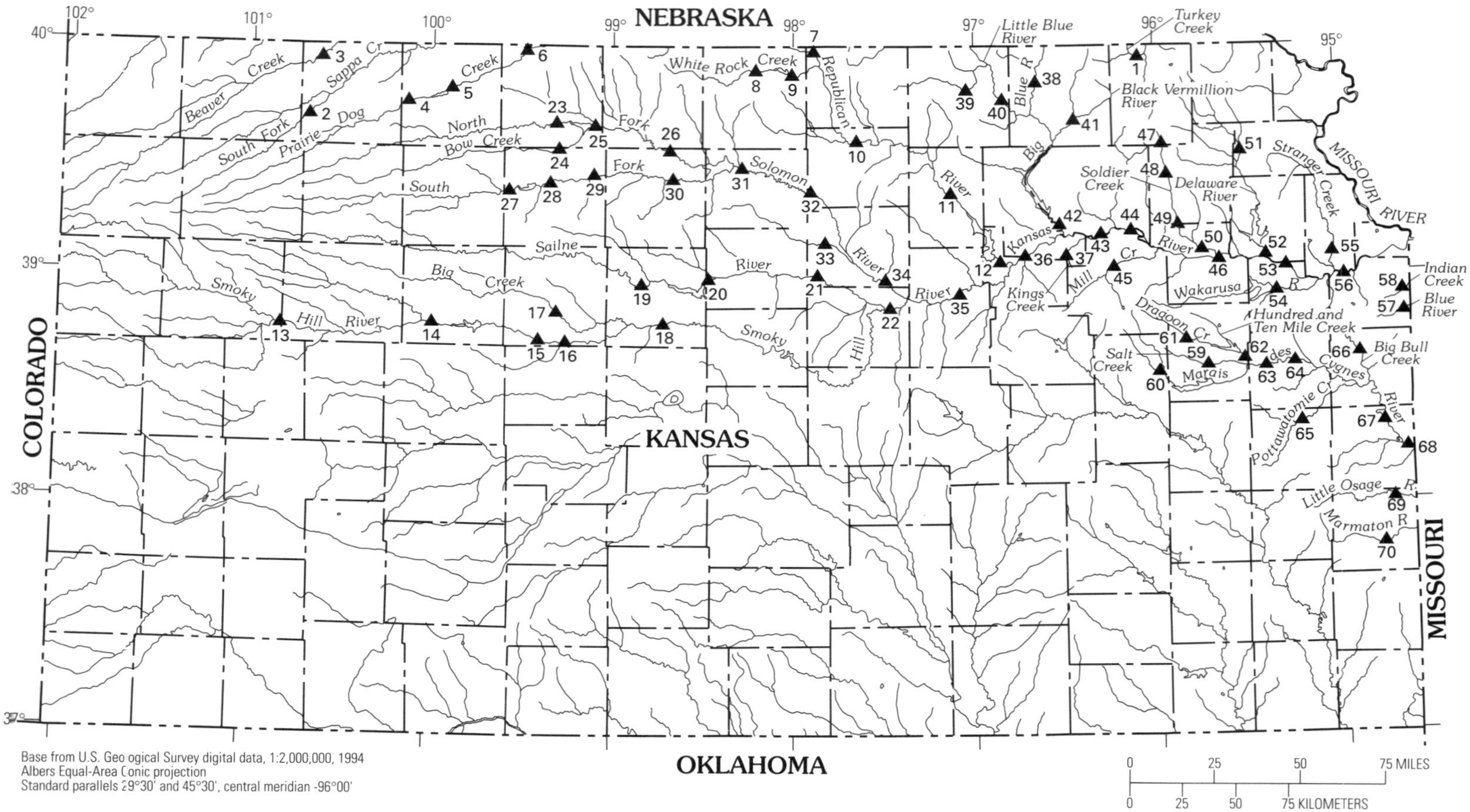

EXPLANATION

▲21 **Streamflow-gaging station and site number—**
Number corresponds to that in table 9

Figure 18. Location of selected streamflow-gaging stations in Kansas.

Table 9. Maximum stages and discharges prior to and during March 1–September 30, 1993, in Kansas[1]

[mi^2, square miles; ft, feet above an arbitrary datum; ft^3/s, cubic feet per second; <, less than; --, not determined or not applicable; >, greater than. Source: Recurrence intervals calculated from U.S. Geological Survey data. Other data from U.S. Geological Survey reports or data bases]

				Maximum prior to 1993				Maximum during March–September 1993			
Site no. (fig. 18)	Station no.	Stream and place of determination	Drainage area (mi^2)	Period	Year	Stage (ft)	Discharge (ft^3/s)	Date (month/day)	Stage (ft)	Discharge (ft^3/s)	Discharge recurrence interval (years)
1	06814000	Turkey Creek near Seneca, KS	276	1949–93	1973	24.77	21,400	7/7	24.06	15,800	5–10
2	06844900	South Fork Sappa Creek near Achilles, KS	446	1959–93	1975	10.47	5,310	3/7	8.86	314	<2
3	06846500	Beaver Creek at Cedar Bluffs, KS	1,618	1945–93	1960	18.71	7,940	6/27	9.59	588	2–5
4	06847900	Prairie Dog Creek at Keith Sebelius Lake, KS	590	1962–93	1972	12.81	8,880	3/8	12.34	1,130	2–5
5	06848000	Prairie Dog Creek at Norton, KS	684	1943–93	1953	25.60	37,500	7/8	5.59	62	<2
6	06848500	Prairie Dog Creek near Woodruff, KS	1,007	1928–32, 1944–93	1947	--	15,000	3/8	15.34	1,060	<2
7	06853500	Republican River near Hardy, NE	22,401	1931–93	1935	19.40	225,000	7/27	12.48	10,500	<2
8	06853800	White Rock Creek near Burr Oak, KS	227	1957–93	1973	25.06	15,800	7/17	19.53	4,990	10–25
9	06854000	White Rock Creek at Lovewell, KS	345	1945–93	1950	21.62	23,300	7/22	19.16	4,780	5–10
10	06856000	Republican River at Concordia, KS	23,560	1945–93	1947	14.90	75,000	7/22	19.77	37,700	5–10
11	06856600	Republican River at Clay Center, KS	24,542	1917–93	1935	25.74	195,000	7/24	23.36	48,100	10–25
12	06857100	Republican River below Milford Dam, KS	24,890	1963–93	1964	22.10	17,200	7/26	21.52	33,700	--
13	06860000	Smoky Hill River at Elkader, KS	3,555	1939–93	1969	8.85	22,300	7/20	7.21	4,080	2–5
14	06861000	Smoky Hill River near Arnold, KS	5,220	1950–93	1951	12.57	23,800	7/20	7.95	4,890	2–5
15	06862700	Smoky Hill River near Schoenchen, KS	5,750	1964–93	1970	16.17	20,400	7/21	16.55	20,200	--
16	06862850	Smoky Hill River below Schoenchen, KS	5,810	1981–93	1987	13.57	3,740	7/21	17.60	20,500	--
17	06863500	Big Creek near Hays, KS	594	1946–93	1957	22.07	22,400	7/21	29.00	5,520	10–25
18	06864050	Smoky Hill River near Bunker Hill, KS	7,075	1939–93	1951	23.86	39,500	7/22	27.14	32,400	--
19	06867000	Saline River near Russell, KS	1,502	1945–53, 1959–93	1964	19.70	19,400	7/21	25.73	41,500	>100
20	06868200	Saline River at Wilson Dam, KS	1,917	1963–93	1973	18.84	3,320	7/22	14.17	1,870	--

Table 9. Maximum stages and discharges prior to and during March 1–September 30, 1993, in Kansas[1]—Continued

				Maximum prior to 1993				Maximum during March–September 1993			
Site no. (fig. 18)	Station no.	Stream and place of determination	Drainage area (mi^2)	Period	Year	Stage (ft)	Discharge (ft^3/s)	Date (month/day)	Stage (ft)	Discharge (ft^3/s)	Discharge recurrence interval (years)
21	06869500	Saline River at Tescott, KS	2,820	1919–93	1951	30.06	61,400	7/23	30.14	8,430	5–10
22	06870200	Smoky Hill River at New Cambria, KS	11,730	1962–93	1973	30.91	26,400	6/25	31.72	18,600	--
23	06871000	North Fork Solomon River at Glade, KS	849	1952–93	1957	18.55	23,300	8/15	11.88	2,520	2–5
24	06871500	Bow Creek near Stockton, KS	341	1950–93	1951	13.60	12,900	7/9	11.80	3,120	<2
25	06871800	North Fork Solomon River at Kirwin, KS	1,367	1919–25, 1928–32, 1941–93	1919	22.50	24,000	8/31	3.02	11	--
26	06872500	North Fork Solomon River at Portis, KS	2,315	1945–93	1951	30.41	35,700	7/21	23.43	9,180	--
27	06873000	South Fork Solomon River above Webster Reservoir, KS	1,040	1945–93	1951	14.90	55,200	8/31	11.41	5,940	2–5
28	06873200	South Fork Solomon River below Webster Reservoir, KS	1,150	1956–93	1962	--	2,070	9/1	4.41	53	--
29	06873460	South Fork Solomon River at Woodston, KS	1,502	1978–93	1987	19.82	4,380	7/21	22.89	8,710	--
30	06874000	South Fork Solomon River at Osborne, KS	2,012	1946–93	1951	27.65	81,200	7/21	28.33	59,100	>200
31	06875900	Solomon River near Glen Elder, KS	5,340	1964–93	1987	28.31	7,470	7/22	29.57	9,410	--
32	06876070	Solomon River near Simpson, KS	5,538	1990–93	--	--	--	7/8	32.69	10,700	--
33	06876700	Salt Creek near Ada, KS	384	1959–93	1961	23.25	16,000	7/19	22.43	9,170	10–25
34	06876900	Solomon River near Niles, KS	6,770	1897–03, 1917–93	1951	31.76	178,000	7/22	30.24	17,900	10
35	06877600	Smoky Hill River at Enterprise, KS	19,260	1934–93	1951	33.96	233,000	7/22	33.95	47,600	50
36	06879100	Kansas River at Fort Riley, KS	44,870	1963–93	1951	34.5	--	7/26	27.93	87,600	--
37	06879650	Kings Creek near Manhattan, KS	4.09	1979–93	1992	12.05	5,800	7/17	13.12	8,220	>500
38	06882510	Big Blue River at Marysville, KS	4,777	1984–93	1986	38.90	39,700	7/28	36.10	31,600	--
39	06884200	Mill Creek at Washington, KS	344	1959–93	1983	29.13	12,300	7/7	29.35	14,600	25
40	06884400	Little Blue River near Barnes, KS	3,324	1958–93	1973	27.70	53,700	7/15	19.99	23,800	5–10

Table 9. Maximum stages and discharges prior to and during March 1–September 30, 1993, in Kansas[1]—Continued

Site no. (fig. 18)	Station no.	Stream and place of determination	Drainage area (mi^2)	Maximum prior to 1993: Period	Year	Stage (ft)	Discharge (ft^3/s)	Maximum during March–September 1993: Date (month/day)	Stage (ft)	Discharge (ft^3/s)	Discharge recurrence interval (years)
41	06885500	Black Vermillion River near Frankfort, KS	410	1953	1959	32.28	38,300	7/5	--	18,700	5–10
42	06887000	Big Blue River near Manhattan, KS	9,640	1951, 1954–93	1951	36.04	93,400	7/23	32.71	58,800	--
43	06887500	Kansas River at Wamego, KS	55,280	1919–93	1951	30.56	400,000	7/26	27.33	199,000	25–50
44	06888350	Kansas River near Belvue, KS	55,870	1982–93	1984	20.34	65,000	7/26	26.00	170,000	--
45	06888500	Mill Creek near Paxico, KS	316	1953–93	1973	32.21	42,200	7/25	29.08	26,000	10
46	06889000	Kansas River at Topeka, KS	56,720	1917–93	1951	40.80	469,000	7/25	34.90	170,000	25–50
47	06889140	Soldier Creek near Soldier, KS	16.9	1964–93	1970	16.46	11,700	7/22	11.24	4,810	10–25
48	06889160	Soldier Creek near Circleville, KS	49.3	1964–93	1984	21.52	25,300	7/5	17.91	4,750	2–5
49	06889200	Soldier Creek near Delia, KS	157	1958–93	1982	23.95	29,400	7/10	21.10	8,520	5–10
50	06889500	Soldier Creek near Topeka, KS	290	1929–32, 1935–93	1882	27.44	30,400	7/10	23.42	18,900	10–25
51	06890100	Delaware River near Muscotah, KS	431	1969–93	1977	30.83	28,000	7/22	28.90	19,400	5–10
52	06890900	Delaware River below Perry Dam, KS	1,117	1969–93	1984	--	12,100	8/13	--	10,100	--
53	06891000	Kansas River at Lecompton, KS	58,460	1936–93	1951	30.23	483,000	7/27	24.65	175,000	25
54	06891500	Wakarusa River near Lawrence, KS	425	1929–93	1951	31.59	24,200	7/10	26.00	5,320	<2
55	06892000	Stranger Creek near Tonganoxie, KS	406	1929–93	1951	28.94	33,100	7/10	25.83	11,100	5
56	06892350	Kansas River at DeSoto, KS	59,756	1917–93	1951	37.30	510,000	7/27	26.91	172,000	25
57	06893080	Blue River near Stanley, KS	46	1970–93	1990	20.51	20,200	7/10	20.10	18,600	50–100
58	06893300	Indian Creek at Overland Park, KS	26.6	1963–93	1984	17.78	12,800	7/10	15.82	9,070	25
59	06910800	Marais des Cygnes River near Reading, KS	177	1969–93	1982	27.47	67,400	7/22	24.65	25,200	10–25
60	06911500	Salt Creek near Lyndon, KS	111	1939–93	1951	17.00	36,400	7/22	14.87	9,320	10–25

Table 9. Maximum stages and discharges prior to and during March 1–September 30, 1993, in Kansas[1]—Continued

Site no. (fig. 18)	Station no.	Stream and place of determination	Drainage area (mi^2)	Maximum prior to 1993 Period	Year	Stage (ft)	Discharge (ft^3/s)	Maximum during March–September 1993 Date (month/day)	Stage (ft)	Discharge (ft^3/s)	Discharge recurrence interval (years)
61	069 1900	Dragoon Creek near Burlingame, KS	114	1960–93	1982	22.64	34,400	7/22	21.60	17,600	10–25
62	069 2500	Hundred and Ten Mile Creek near Quenemo, KS	322	1939–73	1951	28.47	38,600	8/23	13.54	2,780	2
63	069 3000	Marais des Cygnes River near Pomona, KS	1,040	1922–38, 1968–93	1928	38.38	69,400	7/23	29.28	19,300	5–10
64	069 3500	Marais des Cygnes River near Ottawa, KS	1,250	1902–05, 1918–93	1951	42.50	142,000	7/22	33.87	17,100	2–5
65	069 4000	Pottawatomie Creek near Garnett, KS	334	1939–93	1961	35.38	57,000	7/22	29.93	17,200	2–5
66	069 5000	Big Bull Creek near Hillsdale, KS	147	1958–93	1961	20.85	39,600	7/7	11.18	1,190	--
67	069 5800	Marais des Cygnes River at La Cygne, KS	2,669	1984–93	1985	32.20	50,100	7/24	32.05	47,200	5–10
68	069 6600	Marais des Cygnes River near Kansas-Missouri State line	3,230	1958	1986	34.31	64,100	7/25	31.50	40,200	5
69	069 7000	Little Osage River at Fulton, KS	295	1949–93	1986	35.21	62,800	9/25	27.62	12,800	2–5
70	069 7380	Marmaton River near Marmaton, KS	292	1971–93	1986	42.87	106,000	9/25	35.58	28,700	5–10

[1]Arkansas River Basin floods listed in table 43.

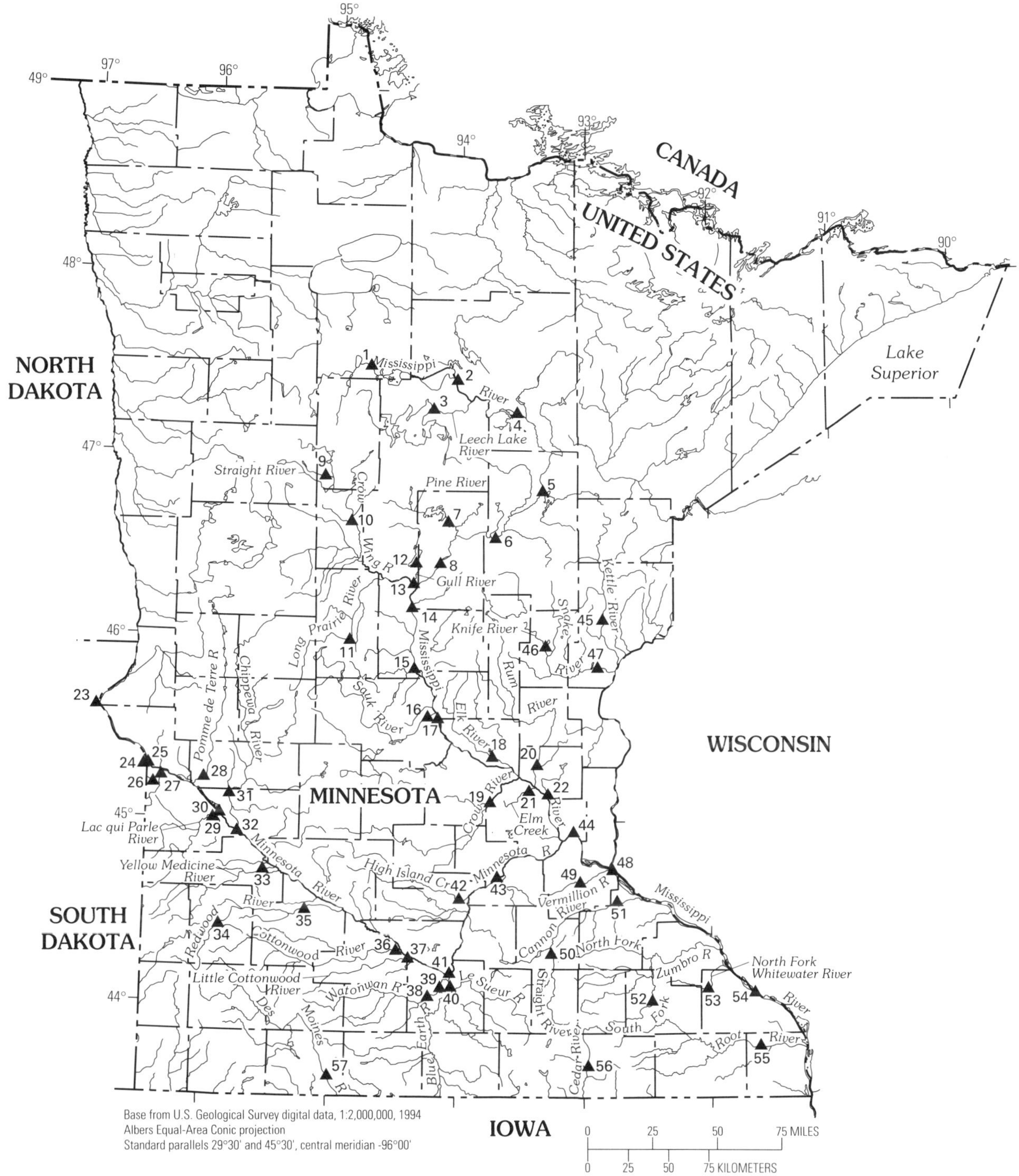

EXPLANATION

▲21 **Streamflow-gaging station and site number—** Number corresponds to that in table 10

Figure 19. Location of selected streamflow-gaging stations in Minnesota.

Table 10. Maximum stages and discharges prior to and during March 1–September 30, 1993, in Minnesota

[mi^2, square miles; ft, feet above an arbitrary datum; ft^3/s, cubic feet per second; --, not determined or not applicable; <, less than; >, greater than. Source: Recurrence intervals calculated from U.S. Geological Survey data. Other data from U.S. Geological Survey reports or data bases]

				Maximum prior to 1993				Maximum during March–September 1993			
Site no. (fig. 19)	Station no.	Stream and place of determination	Drainage area (mi^2)	Period	Year	Stage (ft)	Discharge (ft^3/s)	Date (month/day)	Stage (ft)	Discharge (ft^3/s)	Discharge recurrence interval (years)
1	05200510	Mississippi River near Bemidji, MN	610	1988–93	1989	4.87	887	05/07	4.93	938	--
2	05201500	Mississippi River at Winnibigoshish Dam near Deer River, MN	1,442	1885–93	1905	--	[1]4,370	08/06	--	[1]1,100	<2
3	05206500	Leech Lake River at Federal Dam, MN	1,163	1885–93	1957	--	[1,2]2,520	09/14	--	[1]840	<2
4	05211000	Mississippi River at Grand Rapids, MN	3,370	1884–93	1948	15.20	[2]12,500	07/20	10.26	3,030	4
5	05219000	Sandy River at Sandy Lake Dam near Deer River, MN	421	1896–93	1897	--	[1]3,740	06/25	--	[1]1,610	3
6	05227500	Mississippi River at Aitkin, MN	6,140	1945–93	1950	22.49	[1]19,900	07/10	14.20	[1]9,780	4
7	05231000	Pine River at Cross Lake Dam near Deer River, MN	562	1887–93	1896	--	[1]2,250	07/16	--	[1]1,250	3
8	05242300	Mississippi River at Brainerd, MN	7,320	1988–93	1989	12.40	10,800	07/10	13.65	12,200	10
9	05243725	Straight River near Park Rapids, MN	53.2	1987–93	1989	2.24	89	04/25	2.00	93	10
					1991	[3]2.71	--				
10	05244000	Crow Wing River at Nimrod, MN	1,010	1910–93	1950	[3]7.64	--	04/01	[3]5.60	--	
					1973	7.35	3,700	03/31	[3]5.55	[1]1,300	2
11	05245100	Long Prairie River at Long Prairie, MN	432	1972–93	1972	9.37	3,270	05/31	5.02	550	<2
								07/10	5.24	--	
12	05247000	Gull River at Gull Lake Dam near Brainerd, MN	287	1912–93	1938	--	1,120	06/01	--	584	4
13	05247500	Crow Wing River near Pillager, MN	3,520	1924–93	1965	--	[1]18,300	06/02	7.07	7,360	3
14	05261000	Mississippi River near Fort Ripley, MN	11,010	1971–93	1979	13.58	28,000	07/14	9.70	16,700	2
15	05267000	Mississippi River near Royalton, MN	11,600	1924–93	1965	--	[1]37,700	07/11	--	19,000	3
16	05270500	Sauk River near St. Cloud, MN	925	1909–93	1965	10.68	9,100	07/08	5.28	1,970	3
17	05270700	Mississippi River at St. Cloud, MN	13,320	1989–93	1991	7.71	19,700	07/11	8.08	21,700	2
18	05275000	Elk River near Big Lake, MN	615	1911–93	1965	10.86	7,360	06/30	2.97	716	<2
19	05280000	Crow River at Rockford, MN	2,520	1909–93	1965	19.27	22,400	07/08	13.23	10,000	12
20	05286000	Rum River near St. Francis, MN	1,360	1916–93	1965	11.57	10,100	07/02	5.64	2,410	<2

Table 10. Maximum stages and discharges prior to and during March 1–September 30, 1993, in Minnesota—Continued

Site no. (fig. 19)	Station no.	Stream and place of determination	Drainage area (mi^2)	Maximum prior to 1993 Period	Year	Stage (ft)	Discharge (ft^3/s)	Maximum during March–September 1993 Date (month/day)	Stage (ft)	Discharge (ft^3/s)	Discharge recurrence interval (years)
21	05287890	Elm Creek near Champlin, MN	84.9	1979–93	1986	9.93	597	06/22	8.41	315	2
22	05288500	Mississippi River near Anoka, MN	19,100	1916–93	1965	19.53	91,000	07/11	10.26	34,600	3
23	05290000	Little Minnesota River near Peever, SD	447	1920–93	1943 1952	[3]13.35 12.16	-- 4,370	07/25	13.58	8,900	>100
24	05291000	Whetstone River near Big Stone City, SD	389	1910–93	1919 1969	26.0 14.32	[2]29,000 6,870	07/18	11.25	3,890	6
25	05292000	Minnesota River at Ortonville, MN	1,160	1920–93	1952	12.92	3,060	07/28	9.99	2,950	33
26	05292704	North Fork Yellow Bank River near Odessa, MN	--	1991–93	1992	13.32	2,020	07/26	12.02	1,480	--
27	05293000	Yellow Bank River near Odessa, MN	398	1920–93	1969	19.07	6,970	07/26	10.36	1,910	3
28	05294000	Pomme de Terre River at Appleton, MN	905	1920–93	1969	14.58	5,520	07/10	9.56	2,370	13
29	05300000	Lac qui Parle River near Lac qui Parle, MN	983	1910–18, 1920–93	1969	[3]19.37	17,100	06/23	12.20	4,520	7
30	05301000	Minnesota River near Lac qui Parle, MN	4,050	1920–23	1969	39.75	29,400	08/03	35.95	10,200	10
31	05304500	Chippewa River near Milan, MN	1,870	1920–93	1969	15.45	11,400	08/01	9.56	4,790	8
32	05311000	Minnesota River at Montevideo, MN	6,180	1909–93	1969	21.68	35,100	08/04	16.46	11,500	11
33	05313500	Yellow Medicine River near Granite Falls, MN	653	1882–93	1919	17.5	25,200	06/21	11.46	8,380	22
34	05315000	Redwood River near Marshall, MN	259	1920–93	1969	--	5,590	05/09	17.00	6,380	>100
35	05316500	Redwood River near Redwood Falls, MN	629	1920–93	1957	15.92	19,700	06/18	15.73	12,600	53
36	05317000	Cottonwood River near New Ulm, MN	1,280	1920–93	1965 1969	[3]20.86 19.15	-- 28,700	06/19	18.90	24,300	83
37	05317200	Little Cottonwood River near Courtland, MN	230	1973–93	1985	8.96	1,340	06/20	10.45	3,520	50
38	05319500	Watonwan River near Garden City, MN	851	1920–93	1965	18.72	19,000	06/20	15.91	13,900	55
39	05320000	Blue Earth River near Rapidan, MN	2,430	1882–93	1965	21.36	43,100	06/20	13.32	20,300	18
40	05320500	Le Sueur River near Rapidan, MN	1,100	1920–93	1965	22.72	24,700	06/21	13.32	11,500	13

Table 10. Maximum stages and discharges prior to and during March 1–September 30, 1993, in Minnesota—Continued

Site no. (fig. 19)	Station no.	Stream and place of determination	Drainage area (mi²)	Maximum prior to 1993: Period	Year	Stage (ft)	Discharge (ft³/s)	Maximum during March–September 1993: Date (month/day)	Stage (ft)	Discharge (ft³/s)	Discharge recurrence interval (years)
41	05325000	Minnesota River at Mankato, MN	14,900	1881–93	1881	29.9	110,000	06/21	30.11	75,600	59
42	05327000	High Island Creek near Henderson, MN	237	1973–93	1981	9.09	--	06/17	9.72	2,750	22
					1992	8.32	1,770				
43	05330000	Minnesota River near Jordan, MN	16,200	1882–93	1965	35.07	117,000	06/24	33.52	92,200	71
44	05331000	Mississippi River at St. Paul, MN	36,800	1852–93	1965	26.01	171,000	06/26	19.15	104,000	30
45	05336700	Kettle River below Sandstone, MN	863	1951–93	1972	15.38	17,200	06/25	10.59	8,510	3
46	05337400	Knife River near Mora, MN	102	1972–93	1972	[2,3]14.0	--	06/25	4.21	489	<2
					1979	6.31	1,840				
47	05338500	Snake River near Pine City, MN	958	1914–93	1972	10.38	14,300	06/29	5.39	2,320	<2
48	05344500	Mississippi River at Prescott, WI	44,800	1852–93	1965	43.11	228,000	06/27	37.70	[1]130,000	27
49	05345000	Vermillion River near Empire, MN	110	1942–93	1992	10.00	6,570	06/18	8.37	1,780	7
50	05353800	Straight River near Faribault, MN	442	1966–93	1974	[3]12.74	--	06/17	11.16	5,730	12
					1990	11.31	6,030				
51	05355200	Cannon River at Welch, MN	1,320	1888–93	1965	14.01	36,100	06/17	13.19	17,200	15
52	05372995	South Fork Zumbro River at Rochester, MN	303	1855–93	1978	28.0	30,500	04/01	13.06	6,260	4
53	05376000	North Fork Whitewater River near Elba, MN	101	1966–93	1974	16.32	16,100	07/02	9.75	5,770	7
54	05378500	Mississippi River at Winona, MN	59,200	1879–93	1965	20.77	268,000	06/26	16.63	[1]168,000	16
55	05385000	Root River near Houston, MN	1,270	1909–93	1952	13.90	37,000	04/02	16.14	15,800	4
					1965	[3]18.32	--				
56	05457000	Cedar River near Austin, MN	425	1909–93	1978	20.35	12,400	08/15	19.43	10,800	25
57	05476000	Des Moines River at Jackson, MN	1,220	1909–93	1969	19.45	15,700	07/07	16.67	8,250	29

[1]Daily mean discharge.
[2]Stage and discharge resulted from dam failure.
[3]Backwater.

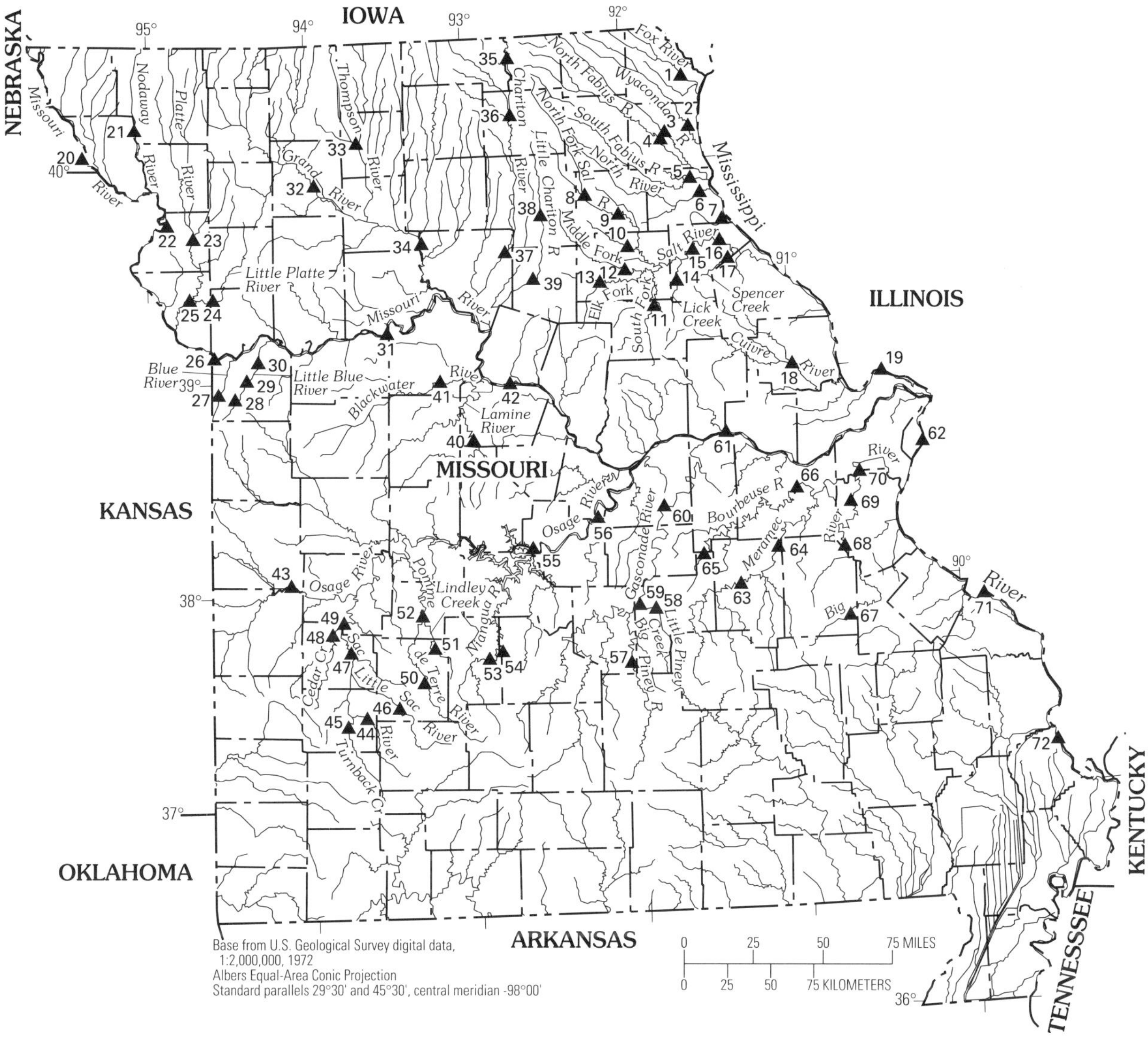

EXPLANATION

▲21 **Streamflow-gaging station and site number—**
Number corresponds to that in table 11

Figure 20. Location of selected streamflow-gaging stations in Missouri.

Table 11. Maximum stages and discharges prior to and during March 1–September 30, 1993, in Missouri

[mi^2, square miles; ft, feet above an arbitrary datum; ft^3/s, cubic feet per second; --, not determined or not applicable; >, greater than. Source: Recurrence intervals calculated from U.S. Geological Survey data. Other data from U.S. Geological Survey reports or data bases]

Site no. (fig. 20)	Station no.	Stream and place of determination	Drainage area (mi^2)	Maximum prior to 1993: Period	Year	Stage (ft)	Discharge (ft^3/s)	Maximum during March–September 1993: Date (month/day)	Stage (ft)	Discharge (ft^3/s)	Discharge recurrence interval (years)
1	05495000	Fox River at Wayland, MO	400	1922–93	1973	21.71	26,400	7/12	18.22	11,900	6
2	05496000	Wyaconda River above Canton, MO	393	1932–72,	1933	--	17,700	8/14	24.83	8,860	4
				1979–93	1986	31.33	--				
3	05497000	North Fabius River at Monticello, MO	452	1922–93	1973	33.03	20,700	8/13	27.76	13,600	8
4	05498000	Middle Fabius River near Monticello, MO	393	1945–93	1973	27.14	17,700	7/26	21.64	9,370	6
5	05500000	South Fabius River near Taylor, MO	620	1934–93	1947	19.5	19,700	7/1	14.41	12,700	6
6	05501000	North River at Palmyra, MO	373	1934–93	1973	29.7	57,500	7/1	29.36	28,600	15
7	05502000	Bear Creek at Hannibal, MO	31.0	1938–42, 1947–93	1957	14.05	6,500	9/22	10.05	2,080	--
8	05502300	North Fork Salt River at Hagers Grove, MO	365	1974–93	1947	19.7	26,900	7/2	19.14	20,300	12
9	05502500	North Fork Salt River near Shelbina, MO	481	1930–72, 1988–93	1947	27.4	23,000	7/3	23.78	16,000	23
10	05503800	Crooked Creek near Paris, MO	80.0	[1]1979–93	1973	15.53	12,100	7/1	12.68	8,100	27
11	05504800	South Fork Salt River above Santa Fe, MO	233	1940–93	1969	28.24	28,800	9/23	28.66	31,800	>100
12	05506500	Middle Fork Salt River at Paris, MO	356	1939–93	1973	33.5	45,000	9/24	19.03	11,500	6
13	05506800	Elk Fork Salt River near Madison, MO	200	1968–93	1973	33.4	42,300	9/23	26.81	17,500	9
14	05507600	Lick Creek near Perry, MO	104	1979–93	1983	21.30	9,360	9/23	21.96	10,900	19
15	05507800	Salt River near Center, MO	2,350	1979–93	1973	33.00	--	9/28	17.58	16,100	--
					1981	--	72,800				
16	05508000	Salt River near New London, MO	2,480	1922–93	1973	31.8	107,000	8/12	17.22	17,900	--
17	05508805	Spencer Creek below Plum Creek near Frankford, MO	206	1979–93	1981	16.86	16,200	9/22	18.54	20,300	>100

Table 11. Maximum stages and discharges prior to and during March 1–September 30, 1993, in Missouri—Continued

Site no. (fig. 20)	Station no.	Stream and place of determination	Drainage area (mi^2)	Maximum prior to 1993: Period	Year	Stage (ft)	Discharge (ft^3/s)	Maximum during March–September 1993: Date (month/ day)	Stage (ft)	Dis-charge (ft^3/s)	Discharge recurrence interval (years)
18	05514500	Cuivre River near Troy, MO	903	1922–72, 1979–93	1941	33.4	120,000	9/23	32.17	101,000	>100
19	05587450	Mississippi River at Grafton, IL	171,300	1879–92, 1929–93	1973	436.99	535,000	8/1	441.96	598,000	>100
20	06813500	Missouri River at Rulo, NE	414,900	1949–93	1952	25.60	358,000	7/24	25.37	307,000	>100
21	06817700	Nodaway River near Graham, MO	1,380	1982–93	1989	23.34	26,600	9/22	26.89	90,700	>100
22	06818000	Missouri River at St. Joseph, MO	420,300	1928–93	1952	26.82	397,000	7/26	32.07	335,000	>100
23	06820500	Platte River near Agency, MO	1,760	1924–30, 1932–93	1965	35.05	53,000	7/25	36.07	60,800	>100
24	06821150	Little Platte River at Smithville, MO	234	1965–93	1965	44.8	76,600	5/7	30.43	6,370	--
25	06821190	Platte River at Sharps Station, MO	2,380	1978–93	1984	34.55	29,000	7/26	36.43	37,800	--
26	06893000	Missouri River at Kansas City, MO	485,200	1897–1993	1951	36.2	573,000	7/27	48.87	541,000	>100
27	06893500	Blue River near Kansas City, MO	188	1939–93	1961	44.46	41,000	7/10	33.91	17,900	8
28	06893793	Little Blue River below Longview Dam at Kansas City, MO	50.3	1966–93	1982	21.24	18,700	7/6	12.96	1,080	--
29	06893890	East Fork Little Blue River near Blue Springs, MO	34.4	1974–93	1982	22.14	11,000	9/25	11.71	506	--
30	06894000	Little Blue River near Lake City, MO	184	1948–93	1961 1982	27.94 --	-- 42,300	9/25	17.93	6,600	4
31	06895500	Missouri River at Waverly, MO	487,200	1928–93	1984 1951	29.22 --	-- 549,000	7/27	31.15	633,000	>100
32	06897500	Grand River near Gallatin, MO	2,250	1921–93	1947	39.55	69,100	7/7	41.5	89,800	>100
33	06899500	Thompson River near Trenton, MO	1,670	1921–23, 1928–93	1947	25.7	95,000	7/6	22.8	54,000	17
34	06902000	Grand River near Sumner, MO	6,880	1923–93	1947	39.5	180,000	7/10 7/26	42.52	166,000	>100
35	06904050	Chariton River at Livonia, MO	864	1974–93	1982	28.33	9,200	7/11	25.97	8,280	--

Table 11. Maximum stages and discharges prior to and during March 1–September 30, 1993, in Missouri—Continued

Site no. (fig. 20)	Station no.	Stream and place of determination	Drainage area (mi^2)	Maximum prior to 1993: Period	Year	Stage (ft)	Discharge (ft^3/s)	Maximum during March–September 1993: Date (month/day)	Stage (ft)	Discharge (ft^3/s)	Discharge recurrence interval (years)
36	06904500	Chariton River at Novinger, MO	1,370	1930–52, 1954–93	1947	28.5	22,900	7/24	25.71	21,500	--
37	06905500	Chariton River near Prairie Hill, MO	1,870	1928–93	1973	21.96	31,900	7/1	21.93	31,500	>100
38	06906200	East Fork Little Chariton River near Macon, MO	112	1971–93	1973	20.60	8,700	7/7	14.12	1,360	--
39	06906300	East Fork Little Chariton River near Hunstville, MO	220	1962–93	1973	20.78	30,000	7/7	18.71	8,650	--
40	06906800	Lamine River near Otterville, MO	543	1987–93	1990	26.55	50,600	7/7	27.81	63,700	--
41	06908000	Blackwater River at Blue Lick, MO	1,120	1922–33, 1938–93	1986 1928	41.53 --	-- 54,000	7/9	33.07	19,100	5
42	06909000	Missouri River at Boonville, MO	501,700	1925–93	1951	32.82	550,000	7/29	37.10	755,000	>100
43	06918070	Osage River above Schell City, MO	5,410	1979–93	1986	--	133,000	11/26	--	49,000	--
44	06918440	Sac River near Dadeville, MO	257	1966–93	1986	20.83	13,600	9/25	27.56	36,100	>100
45	06918460	Turnback Creek above Greenfield, MO	252	1965–93	1986	23.74	44,000	9/25	26.34	42,700	35
46	06918740	Little Sac River near Morrisville, MO	237	1968–93	1972	21.95	22,300	9/25	23.33	29,100	>100
47	06919020	Sac River at Highway J below Stockton, MO	1,292	1973–93	1985 1986	24.91 --	-- 14,800	9/25	23.71	13,300	--
48	06919500	Cedar Creek near Pleasant View, MO	420	1923–26, 1948–93	1958	27.35	37,000	9/25	25.44	22,700	12
49	06919900	Sac River near Caplinger Mills, MO	1,810	1974–93	1986	30.00	60,000	9/25	29.93	51,200	--
50	06921070	Pomme de Terre River near Polk, MO	276	1968–93	1986	23.08	23,100	9/24	27.10	34,300	>100
51	06921200	Lindley Creek near Polk, MO	112	1957–93	1961 1986	23.60 --	-- 31,900	9/24	20.49	19,400	17
52	06921350	Pomme de Terre River near Hermitage, MO	615	1960–93	1961	15.02	9,000	9/27	9.51	3,630	--
53	06923250	Niangua River at Windyville, MO	377	1991–93	1992	10.44	4,700	9/24	24.36	44,700	>100

Table 11. Maximum stages and discharges prior to and during March 1–September 30, 1993, in Missouri—Continued

				Maximum prior to 1993				Maximum during March–September 1993			
Site no. (fig. 20)	Station no.	Stream and place of determination	Drainage area (mi²)	Period	Year	Stage (ft)	Discharge (ft³/s)	Date (month/day)	Stage (ft)	Discharge (ft³/s)	Discharge recurrence interval (years)
54	06923500	Bennett Spring at Bennett Springs, MO	--	1916–20, 1928–41, 1965–93	1986	11.1	14,400	9/25	7.00	5,000	--
55	06926000	Osage River near Bagnell, MO	14,000	1880–1993	1943	48.8	220,000	9/26	26.66	69,600	--
56	06926500	Osage River near St. Thomas, MO	14,500	1931–93	1943	43.8	216,000	9/26	25.55	82,100	--
57	0693000	Big Piney near Big Piney, MO	560	1921–82, 1988–93	1942	20.7	32,700	9/26	19.66	30,000	11
58	06932000	Little Piney Creek at Newburg, MO	200	1928–93	1985 1946	16.6 --	-- 32,500	9/24	11.78	9,700	3
59	06933500	Gasconade River at Jerome, MO	2,840	1903–06, 1923–93	1982	31.34	136,000	9/27	29.60	110,000	70
60	06934000	Gasconade River near Rich Fountain, MO	3,180	1921–59, 1986–93	1982	33.27	134,000	9/28	31.28	106,000	50
61	06934500	Missouri River at Hermann, MO	524,200	1897–1993	1986 1903	35.79 --	-- 676,000	7/31	36.97	750,000	>100
62	07010000	Mississippi River at St. Louis, MO	697,000	1861–1993	1973 1903	43.23 --	-- 1,019,000	8/1	49.58	1,080,000	>100
63	07013000	Meramec River near Steelville, MO	781	1922–93	1985	26.15	51,200	9/26	17.51	22,300	3
64	07014500	Meramec River near Sullivan, MO	1,475	1921–33, 1943–93	1945	32.0	77,300	9/26	22.14	34,200	4
65	07015720	Bourbeuse River near High Gate, MO	135	1965–93	1982	23.65	49,300	7/7	20.91	21,600	2
66	07016500	Bourbeuse River at Union, MO	808	1921–93	1982	33.8	73,300	7/9	26.32	36,700	13
67	07017200	Big River at Irondale, MO	175	1965–93	1972	27.92	43,200	6/25	18.08	20,900	4
68	07018100	Big River near Richwoods, MO	735	1942–93	1957	27.15	55,800	9/23	30.33	59,800	40

Table 11. Maximum stages and discharges prior to and during March 1–September 30, 1993, in Missouri—Continued

Site no. (fig. 20)	Station no.	Stream and place of determination	Drainage area (mi^2)	Maximum prior to 1993				Maximum during March–September 1993			
				Period	Year	Stage (ft)	Discharge (ft^3/s)	Date (month/ day)	Stage (ft)	Dis- charge (ft^3/s)	Discharge recurrence interval (years)
69	07018500	Big River at Byrnesville, MO	917	1921–93	1985	26.47	43,000	9/25	29.37	63,600	95
70	07019000	Meramec River near Eureka, MO	3,788	1903–06, 1921–93	1982	42.89	145,000	9/26	36.72	95,200	19
71	07020500	Mississippi River at Chester, IL	708,600	1927–93	1973 1844	43.32 --	-- [2]1,350,000	8/7	49.74	1,000,000	10–50
72	07022000	Mississippi River at Thebes, IL	713,200	1932–93	1973 1844	43.43 --	893,000 [2]1,375,000	8/7	45.51	996,000	10–50

[1]1966 to 1979 published by U.S. Army Corps of Engineers.
[2]Estimated.

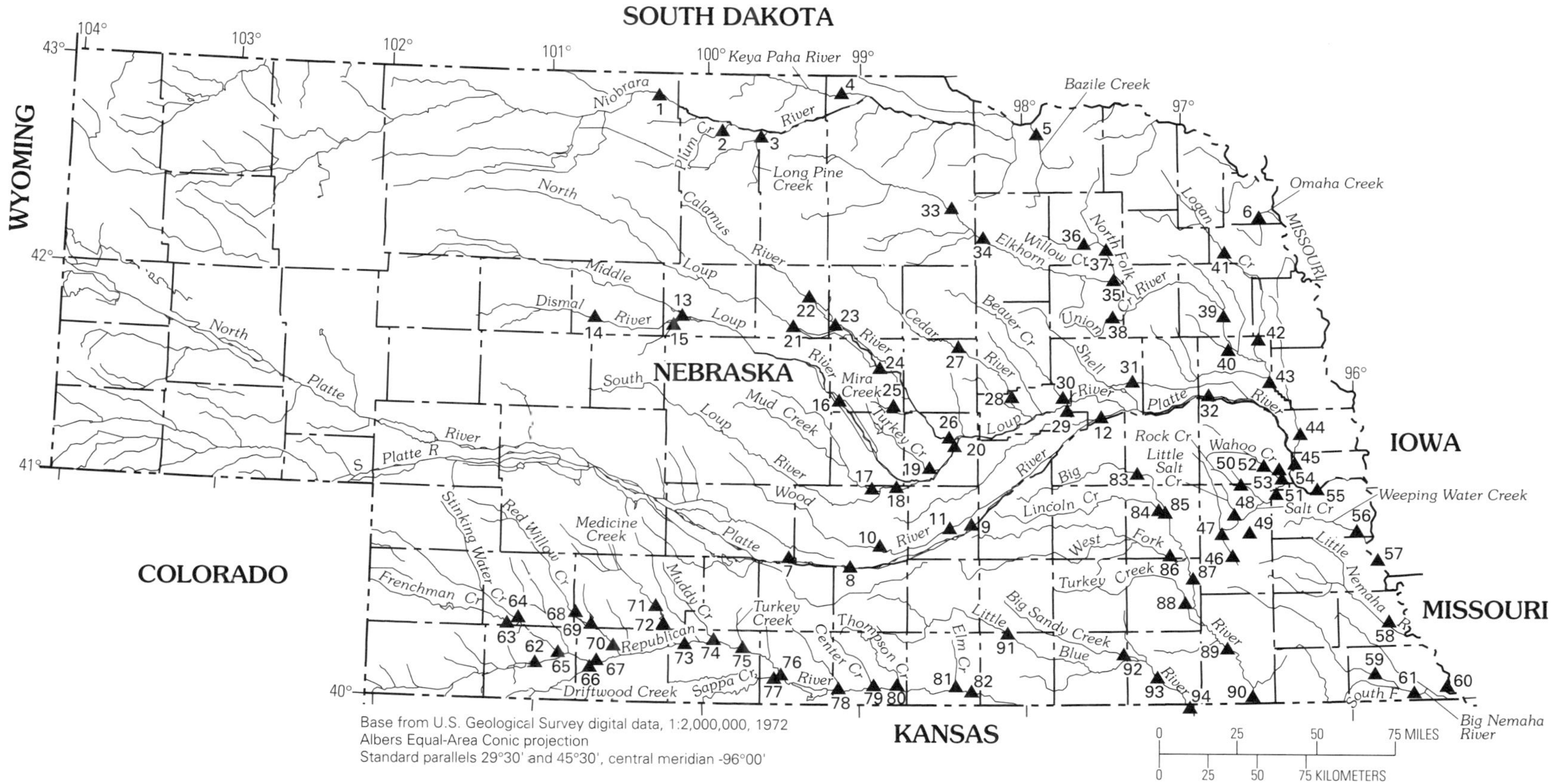

EXPLANATION

▲21 **Streamflow-gaging station and site number—**
Number corresponds to that in table 12

Figure 21. Location of selected streamflow-gaging stations in Nebraska.

Table 12. Maximum stages and discharges prior to and during March 1–September 30, 1993, in Nebraska

[mi^2, square miles; ft, feet above an arbitrary datum; ft^3/s, cubic feet per second; --, not determined or not applicable; <, less than; >, greater than. Source: Recurrence intervals calculated from U.S. Geological Survey data. Other data from U.S. Geological Survey reports or data bases]

				Maximum prior to 1993				Maximum during March–September 1993			
Site no. (fig. 21)	Station no.	Stream and place of determination	Drainage area (mi^2)	Period	Year	Stage (ft)	Discharge (ft^3/s)	Date (month/day)	Stage (ft)	Discharge (ft^3/s)	Discharge recurrence interval (years)
1	06461500	Niobrara River near Sparks, NE	8,090	1945–93	1949	6.73	10,200	8/16	3.60	1,710	<2
					1973	10.06	--	2/18	8.10	--	
2	06462500	Plum Creek at Meadville, NE	600	1947–93	1967	6.98	2,070	7/13	3.08	527	3
					1964	8.54	--				
3	06463500	Long Pine Creek near Riverview, NE	460	1948–93	1962	15.68	9,650	7/13	5.94	1,240	2
4	06464900	Keya Paha River near Naper, NE	1,630	1957–93	1962	10.91	9,280	3/10	8.06	2,260	3
					1960	13.34	--	3/8	11.22	--	
5	06466500	Bazile Creek near Niobrara, NE	440	1952–93	1957	19.96	68,600	3/9	18.82	4,820	3
6	06601000	Omaha Creek at Homer, NE	168	1945–93	1971	28.47	18,100	7/9	11.36	6,820	5
7	06768000	Platte River near Overton, NE	57,700	[1]1941–93	1983	6.38	22,900	3/9	4.46	4,930	<2
					1983	7.44	--	2/17	5.28	--	
8	06770200	Platte River near Kearney, NE	58,200	1982–93	1983	7.42	23,700	3/9	5.43	6,990	<2
9	06770500	Platte River near Grand Island, NE	58,800	[1]1941–93	1983	5.97	23,900	3/12	4.75	13,200	[2]8
					1960	6.16	--	3/10	5.54	--	
10	06771500	Wood River near Gibbon, NE	572	1949–76,	1967	16.79	4,050				
				1991–93				6/29	15.26	822	3
11	06772000	Wood River near Alda, NE	628	1953–93	1967	12.22	1,630	3/12	11.81	1,300	20
12	06774000	Platte River near Duncan, NE	60,900	[1]1942–93	1905	[3]6.50	44,100	3/11	7.86	18,000	10
13	06775500	Middle Loup River at Dunning, NE	1,850	1945–93	1989	3.55	2,160	7/20	3.96	707	2
					1993	7.09	--				
14	06775900	Dismal River near Thedford, NE	960	1966–93	1983	3.83	1,160	7/8	1.80	329	3
					1983	5.10	--				
15	06776500	Dismal River at Dunning, NE	2,040	1945–93	1983	2.40	1,290	7/20	1.50	701	7
					1947	5.21	--				

Table 12. Maximum stages and discharges prior to and during March 1–September 30, 1993, in Nebraska—Continued

Site no. (fig. 21)	Station no.	Stream and place of determination	Drainage area (mi²)	Maximum prior to 1993: Period	Year	Stage (ft)	Discharge (ft³/s)	Maximum during March–September 1993: Date (month/day)	Stage (ft)	Discharge (ft³/s)	Discharge recurrence interval (years)
16	06779000	Middle Loup River at Arcadia, NE	5,040	1937–93	1945	5.12	[2]9,700	3/10	5.64	7,200	[2]16
					1960	6.41	--	3/11	6.41	--	
17	06783500	Mud Creek near Sweetwater, NE	707	1946–93	1947	23.30	27,000	3/9	18.95	2,070	5
18	06784000	South Loup River at St. Michael, NE	2,350	1943–93	1947	2.00	50,000	3/9	10.32	9,970	7
19	06784800	Turkey Creek near Dannebrog, NE	66.2	1966–70,	1967	19.21	2,680				
				1978–93	1983	19.26	--	7/23	17.95	1,440	5
20	06785000	Middle Loup River at St. Paul, NE (since Farwell diversion)	8,090	1963–93	1947	12.69	72,000	7/27	6.00	24,100	[2]20
21	06786000	North Loup River at Taylor, NE	2,280	1936–93	1983	5.94	3,210	3/4	5.39	2,150	[2]8
					1957	9.50	--	2/6	6.48	--	
22	06787000	Calumus River near Harrop, NE	692	1978–93	1964	4.80	1,170	7/9	3.20	735	14
					1987	5.34	--	2/12	4.24	--	--
23	06787500	Calamus River near Burwell, NE (since Calamus Dam)	1,060	1985–93	1964	7.35	1,790	7/13	5.13	933	[2]--
					1967	8.90	--				
24	06788500	North Loup River at Ord, NE	3,750	1952–93	1962	5.52	10,100	7/9	4.27	3,290	[2]3
					1993	6.74	--				
25	06788988	Mira Creek near North Loup, NE	65.8	1979–93	1981	10.56	3,460	6/26	8.41	1,930	8
26	06790500	North Loup River near St. Paul, NE	4,290	1928–93	1896	14.90	90,000	7/23	5.92	11,100	[2]5
								3/7	8.85	--	
27	06791500	Cedar River near Spalding, NE	762	1944–53, 1957–93	1947	7.50	4,000	5/12	4.96	942	4
28	06792000	Cedar River near Fullerton, NE	1,220	1931–32, 1940–93	1966	16.90	64,700	7/23	9.43	8,160	9
29	06793000	Loup River near Genoa, NE	14,400	1943–93	1966	13.93	129,000	7/24	11.01	37,400	[2]10
30	06794000	Beaver Creek at Genoa, NE	647	1940–93	1950	18.70	21,200	8/31	16.13	6,620	9

Table 12. Maximum stages and discharges prior to and during March 1–September 30, 1993, in Nebraska—Continued

Site no. (fig. 21)	Station no.	Stream and place of determination	Drainage area (mi^2)	Maximum prior to 1993				Maximum during March–September 1993			
				Period	Year	Stage (ft)	Discharge (ft^3/s)	Date (month/day)	Stage (ft)	Discharge (ft^3/s)	Discharge recurrence interval (years)
31	06795500	Shell Creek near Columbus, NE	270	1947–75, 1977–93	1990	22.76	8,000	3/9	21.53	3,600	8
								7/10	21.73	--	
32	06796000	Platte River at North Bend, NE	77,100	1949–93	1960	10.04	112,000	3/10	9.71	97,800	50
								3/8	10.97	--	
33	06797500	Elkhorn River at Ewing, NE	1,400	1947–93	1962	10.60	7,500	5/11	8.73	4,310	8
								7/19	8.75	--	
34	06798500	Elkhorn River at Neligh, NE	2,200	1930–93	1987	11.99	14,100	5/11	9.32	6,440	10
					1947	12.53	--				
35	06799000	Elkhorn River at Norfolk, NE	2,790	1945–93	1967	8.52	16,900	7/8	7.70	8,250	5
					1949	15.63	--				
36	06799080	Willow Creek near Foster, NE	137	1975–93	1987	7.94	574	3/11	7.65	442	9
					1983	8.28	--				
37	06799100	North Fork Elkhorn River near Pierce, NE	700	1960–93	1971	15.10	15,200	3/10	13.64	5,760	9
38	06799230	Union Creek at Madison, NE	174	1978–93	1990	25.72	15,100	7/9	25.27	13,700	45
39	06799350	Elkhorn River at West Point, NE	5,100	1972–93	1969	13.21	33,000	7/9	14.02	28,800	10
					1978	16.09	--				
40	06799385	Pebble Creek at Scribner, NE	204	1978–93	1991	24.15	27,900	7/9	23.09	21,000	35
41	06799450	Logan Creek at Pender, NE	731	1965–93	1971	23.11	36,900	3/9	16.68	9,390	2
42	06799500	Logan Creek near Uehling, NE	1,030	1941–93	1971	20.15	25,200	7/9	18.57	11,300	5
43	06800000	Maple Creek near Nickerson, NE	450	1951–93	1990	[3]16.30	11,600	3/9	14.26	6,270	6
					1984	[3]17.65	--				
44	06800500	Elkhorn River at Waterloo, NE	6,900	1928–93	1944	[3]16.60	100,000	3/11	15.76	33,500	12
45	06801000	Platte River near Ashland, NE	84,200	1928–53, 1988–93	1944	[3]8.10	107,000	3/10	19.23	130,000	>100
								7/25	21.45	--	

Table 12. Maximum stages and discharges prior to and during March 1–September 30, 1993, in Nebraska—Continued

				Maximum prior to 1993				Maximum during March–September 1993			
Site no. (fig. 21)	Station no.	Stream and place of determination	Drainage area (mi²)	Period	Year	Stage (ft)	Discharge (ft³/s)	Date (month/ day)	Stage (ft)	Discharge (ft³/s)	Discharge recurrence interval (years)
46	06803000	Salt Creek at Roca, NE	167	1951–93	1950	26.00	67,000	7/25	20.10	5,140	9
47	06803500	Salt Creek at Lincoln, NE	684	1949–93	1951	26.15	28,200	7/24	26.52	28,400	30
48	06803510	Little Salt Creek near Lincoln, NE	43.6	1969–93	1985	18.24	8,000	7/24	20.58	8,480	50
					1887	20.02	--				
49	06803520	Stevens Creek near Lincoln, NE	47.8	1968–93	1989	19.42	12,900	7/24	18.59	9,760	45
					1984	19.57	--				
50	06803530	Rock Creek near Ceresco, NE	119	1970–93	1987	19.60	23,300	7/23	18.36	15,600	45
51	06803555	Salt Creek at Greenwood, NE	1,051	1951–93	1984	26.50	46,800	7/24	26.57	42,000	10
52	06804000	Wahoo Creek at Ithaca, NE	271	1949–93	1963	22.93	77,400	7/14	21.32	7,380	4
53	06804700	Wahoo Creek at Ashland, NE	416	1990–93	--	--	--	7/24	--	5,440	[5] NA
54	06804900	Johnson Creek near Memphis, NE	21.5	1990–93	--	--	--	7/23	--	240	[5] NA
55	06805500	Platte River at Louisville, NE	85,800	1953–93	1984	11.34	[4]144,000	7/25	11.90	160,000	100
					1960	12.45	--				
56	06806500	Weeping Water Creek at Union, NE	241	1950–93	1950	29.80	60,300	7/23	30.97	65,100	>100
57	06807000	Missouri River at Nebraska City, NE	410,000	1929–93	1952	27.66	414,000	7/23	27.19	196,000	50–100
58	06811500	Little Nemaha River at Auburn, NE	793	1949–93	1950	27.65	164,000	7/24	26.49	105,000	>50
59	06813500	Missouri River at Rulo, NE	414,900	1949–93	1952	25.60	358,000	7/24	25.37	307,000	>100
60	06814500	North Fork Big Nemaha River at Humboldt, NE	548	1952–93	1982	31.25	59,500	7/6	21.93	44,600	10
					1958	31.70	--				
61	06815000	Big Nemaha River at Falls City, NE	1,340	1944–93	1973	31.40	71,600	7/6	29.77	59,000	20
62	06829500	Republican River at Trenton, NE (since storage in Swanson Lake)	8,620	1953–93	1948	5.64	16,800	7/3	4.42	187	[2]1.1
63	06834000	Frenchman Creek at Palisade, NE	1,110	1950–93	1956	8.79	5,560	7/24	8.12	722	[2]3
64	06835000	Stinking Water Creek near Palisade, NE	1,500	1949–93	1956	11.30	3,030	7/27	8.47	458	3
65	06835500	Frenchman Creek at Culbertson, NE	2,770	1930–93	1935	14.80	15,000	9/19	7.57	717	[2]1.4

Table 12. Maximum stages and discharges prior to and during March 1–September 30, 1993, in Nebraska—Continued

Site no. (fig. 21)	Station no.	Stream and place of determination	Drainage area (mi^2)	Maximum prior to 1993: Period	Year	Stage (ft)	Discharge (ft^3/s)	Maximum during March–September 1993: Date (month/day)	Stage (ft)	Discharge (ft^3/s)	Discharge recurrence interval (years)
66	06836500	Driftwood Creek near McCook, NE	360	1946–93	1950	25.43	4,740	7/24	6.77	112	<2
67	06837000	Republican River at McCook, NE	12,310	1955–93	1960	9.14	5,890	7/26	5.20	516	<2
68	06837300	Red Willow Creek above Hugh Butler Lake, NE	600	1960–93	1972	13.27	4,020	7/23	12.34	687	4
69	06837500	Red Willow Creek near McCook, NE (since storage in Hugh Butler Lake)	740	1960–93	1947	31.95	30,000	7/26	10.01	127	<2
70	06838000	Red Willow Creek near Red Willow, NE (since storage in Hugh Butler Lake)	830	1961–93	1947	18.36	30,000	7/27	10.61	303	[2]2
71	06841000	Medicine Creek above Harry Strunk Lake, NE	770	1950–93	1967	20.05	11,600	7/12	16.35	1,820	2
72	06842500	Medicine Creek below Harry Strunk Lake, NE	880	1949–93	1960	5.97	1,300	7/27	3.27	570	[2]5
73	06843500	Republican River at Cambridge, NE	14,520	1945–93	1947	16.70	160,000	7/24	7.43	1,950	<2
74	06844000	Muddy Creek at Arapahoe, NE	246	1950–93	1986	28.90	10,800	7/24	20.10	2,440	4
75	06844210	Turkey Creek at Edison, NE	74.9	1977–93	1989	12.91	795	7/27	14.28	1,040	13
76	06844500	Republican River near Orleans, NE	15,640	1947–93	1948 1960	11.25 12.60	40,600 --	7/28 6/22	10.66 12.95	5,220 --	[2]4
77	06847500	Sappa Creek near Stamford, NE	3,741	1945–93	1966	22.13	43,400	3/9	8.50	440	2
78	06849500	Republican River below Harlan County Dam, NE	20,760	1952–93	1957	8.65	4,320	7/17	4.56	1,490	[2]2
79	06851000	Center Creek at Franklin, NE	177	1948–56, 1967–75, 1977–93	1950	6.80	3,150	7/24	7.36	1,860	10
80	06851500	Thompson Creek at Riverton, NE	279	1948–56, 1968-75, 1977-93	1950	13.22	12,200	7/17	14.70	7,000	15

Table 12. Maximum stages and discharges prior to and during March 1–September 30, 1993, in Nebraska—Continued

Site no. (fig. 21)	Station no.	Stream and place of determination	Drainage area (mi^2)	Maximum prior to 1993: Period	Year	Stage (ft)	Dis-charge (ft^3/s)	Maximum during March–September 1993: Date (month/day)	Stage (ft)	Dis-charge (ft^3/s)	Discharge recurrence interval (years)
81	06852000	Elm Creek at Amboy, NE	39.2	1948–93	1983	16.96	7,800	7/26	16.41	6,800	40
					1959	17.05	--				
82	06853020	Republican River at Guide Rock, NE	22,090	1950–93	1957	[3]20.73	29,200	7/19	15.47	9,860	10
83	06879900	Big Blue River at Surprise, NE	345	1964–93	1965	11.52	10,700	3/10	9.57	2,360	3
84	06880000	Lincoln Creek near Seward, NE	446	1953–93	1957	20.53	10,100	3/10	18.35	3,280	6
85	06880500	Big Blue River at Seward, NE	1,099	1953–93	1957	22.34	15,300	3/11	20.86	7,640	7
					1967	22.83	--				
86	06880800	West Fork Big Blue River near Dorchester, NE	1,206	1958–93	1986	22.62	11,800	3/11	21.71	12,400	20
87	06881000	Big Blue River near Crete, NE	2,716	1945–93	1950	28.74	27,600	3/12	28.80	10,000	4
					1986	29.86	--				
88	06881200	Turkey Creek near Wilber, NE	460	1959–93	1984	21.43	33,000	7/25	17.44	5,810	6
89	06881500	Big Blue River at Beatrice, NE	3,900	1974–93	1984	31.27	55,100	7/26	28.77	28,800	10
90	06882000	Big Blue River at Barneston, NE	4,447	1932–93	1941	34.30	57,700	7/27	27.87	32,100	12
91	06883000	Little Blue River near Deweese, NE	979	1953–72, 1974–93	1969	18.57	25,100	7/24	14.67	13,800	12
92	06883940	Big Sandy Creek at Alexandria, NE	607	1979–93	1984	16.71	21,900	7/26	12.80	5,700	3
93	06884000	Little Blue River near Fairbury, NE	2,350	1908–15, 1928–93	1992	24.53	54,000	7/27	21.17	24,100	10
94	06884025	Little Blue River at Hollenberg, KS	2,752	1974–93	1992	21.21	47,800	7/27	18.12	26,500	13

[1]Since Kingsley Dam.
[2]Regulated.
[3]Different site and datum.
[4]Date of maximum discharge. Maximum stage did not occur on the same day.
[5]Maximum average daily flow not available.

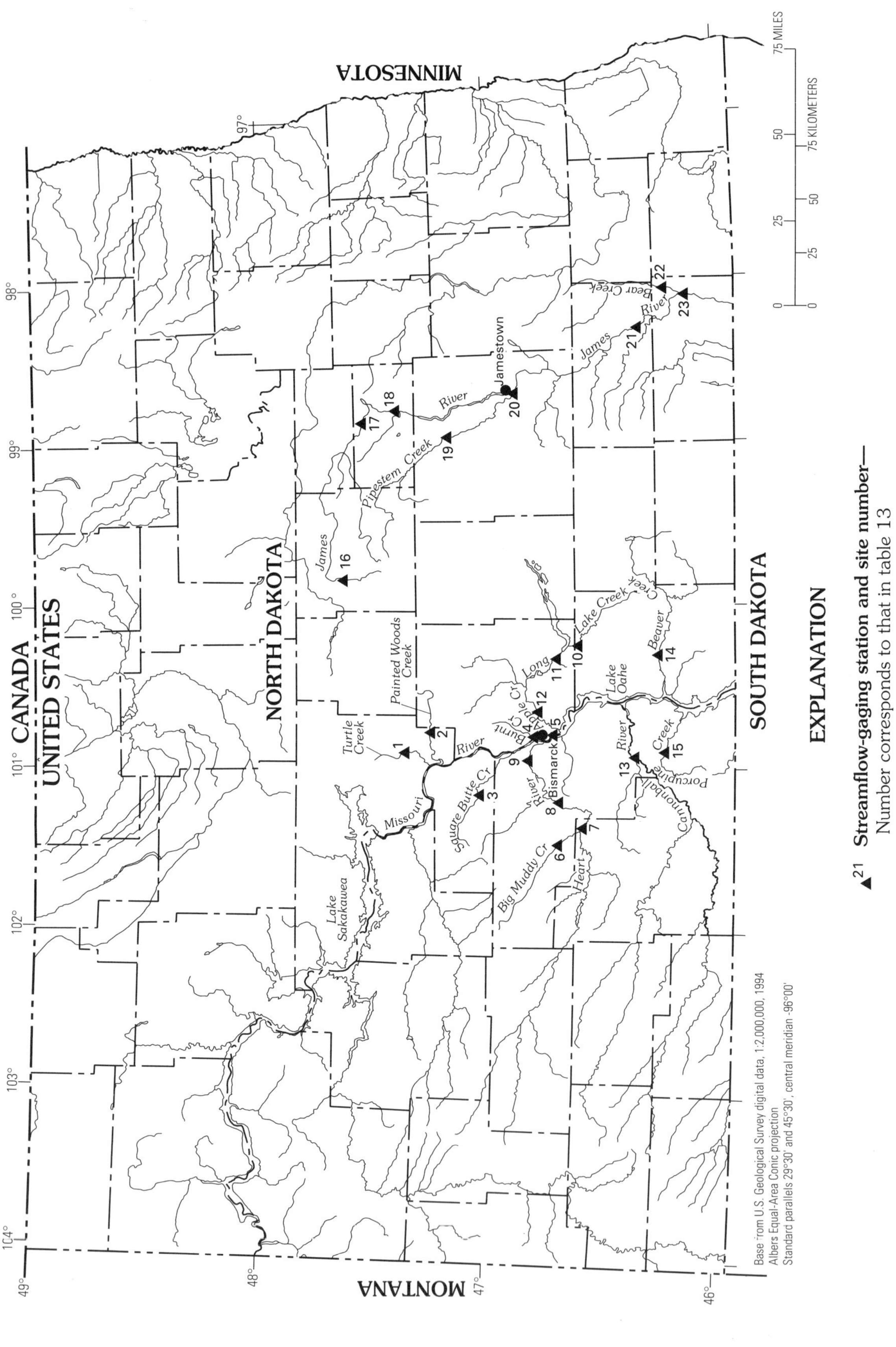

Figure 22. Location of selected streamflow-gaging stations in North Dakota.

Table 13. Maximum stages and discharges prior to and during March 1–September 30, 1993, in North Dakota

[mi², square miles; ft, feet above an arbitrary datum; ft³/s, cubic feet per second; --, not determined or not applicable; <, less than; >, greater than. Source: Recurrence intervals calculated from U.S. Geological Survey data. Other data from U.S. Geological Survey reports or data bases]

				Maximum prior to 1993				Maximum during March–September 1993			
Site no. (fig. 22)	Station no.	Stream and place of determination	Drainage area (mi²)	Period	Year	Stage (ft)	Discharge (ft³/s)	Date (month/day)	Stage (ft)	Discharge (ft³/s)	Discharge recurrence interval (years)
1	06341410	Turtle Creek above Washburn, ND	350	1987–93	1987	6.94	845	3/7	4.69	170	--
2	06341800	Painted Woods Creek, near Wilton, ND	427	1958–81, 1982–93	1979	9.64	4,050	7/23	8.13	1,580	10
3	06342260	Square Butte Creek below Center, ND	146	1965–93	1966	14.35	9,700	7/26	8.86	1,520	<10
4	06342450	Burnt Creek near Bismarck, ND	108	1968–93	1979	16.93	10,000	7/23	12.17	913	<10
5	06342500	Missouri River at Bismarck, ND	186,400	1928–93	1952	27.90	500,000	3/9	[1]9.01	--	[2]<10
								7/24	--	26,000	
6	06347500	Big Muddy Creek near Almont, ND	456	[3]1946–73, 1991–93	1950	30.70	20,200	7/23	30.99	8,390	25
7	06348000	Heart River near Lark, ND	2,750	1946–93	1950	20.70	29,200	7/23	16.85	12,100	10
8	06348300	Heart River at Stark Bridge near Judson, ND	2,930	1986–93	1987	16.70	9,500	7/23	15.47	9,390	--
9	06349000	Heart River near Mandan, ND	3,310	1924, 1928–33, 1937–93	1950	25.75	30,500	7/23	15.45	11,800	<10
10	06349215	Long Lake Creek above Long Lake near Moffit, ND	280	1989–93	1989	5.60	402	3/25	10.05	--	--
								7/16	--	1,760	
11	06349275	Long Lake Creek below Long Lake near Moffit, ND	700	1989–93	1989	1.81	12	7/18	3.54	271	--
12	06349500	Apple Creek near Menoken, ND	1,680	1945–93	1950	--	6,750	7/16	16.05	1,290	<10
					1979	17.46	--				
13	06354000	Cannonball River at Breien, ND	4,100	1934–93	1950	22.30	94,800	7/16	10.92	6,040	<10
14	06354580	Beaver Creek below Linton, ND	765	1990–93	1991	7.21	525	7/18	11.79	2,270	--
15	06354815	Porcupine Creek near Fort Yates, ND	220	1991–93	1991	10.35	505	7/16	13.62	1,200	--

Table 13. Maximum stages and discharges prior to and during March 1–September 30, 1993, in North Dakota—Continued

Site no. (fig. 22)	Station no.	Stream and place of determination	Drainage area (mi^2)	Maximum prior to 1993: Period	Year	Stage (ft)	Discharge (ft^3/s)	Maximum during March–September 1993: Date (month/day)	Stage (ft)	Discharge (ft^3/s)	Discharge recurrence interval (years)
16	06467600	James River near Manfred, ND	253	1955–93	1979	9.20	2,000	7/23	9.40	2,700	>100
17	06468170	James River near Grace City, ND	1,060	1968–93	1969	12.00	3,100	7/28	13.49	3,520	10
					1987	12.20	--				
18	06468250	James River above Arrowwood Lake near Kensal, ND	1,200	1986–93	1987	11.48	1,500	7/29	10.64	3,400	--
19	06469400	Pipestem Creek near Pingree, ND	700	1974–93	1979	11.60	2,520	7/26	10.64	1,150	<10
20	06470000	James River at Jamestown, ND	2,820	1928–33, 1935, 1937–39, 1943–93	1950	--	[4]6,390	7/16	13.58	1,300	>100
					1969	[4]16.94	--				
					1983	[5]8.76	[5]996				
21	06470500	James River at LaMoure, ND	4,390	1950–93	1969	[4]16.17	[4]6,800	7/21	14.64	2,830	[2]10
					1979	[5]14.09	[5]3,830				
22	06470800	Bear Creek near Oakes, ND	357	1977–93	1979	11.47	1,170	7/29	11.02	591	<10
23	06470875	James River at Dakota Lake Dam near Ludden, ND	5,480	1982–93	1987	13.76	2,300	8/3	13.53	1,660	[2]<10

[1]Ice affected.
[2]Discharge recurrence interval for regulated period.
[3]Annual maximum discharge only for 1971–73.
[4]Maximum before 1974 regulation.
[5]Maximum after 1974 regulation.

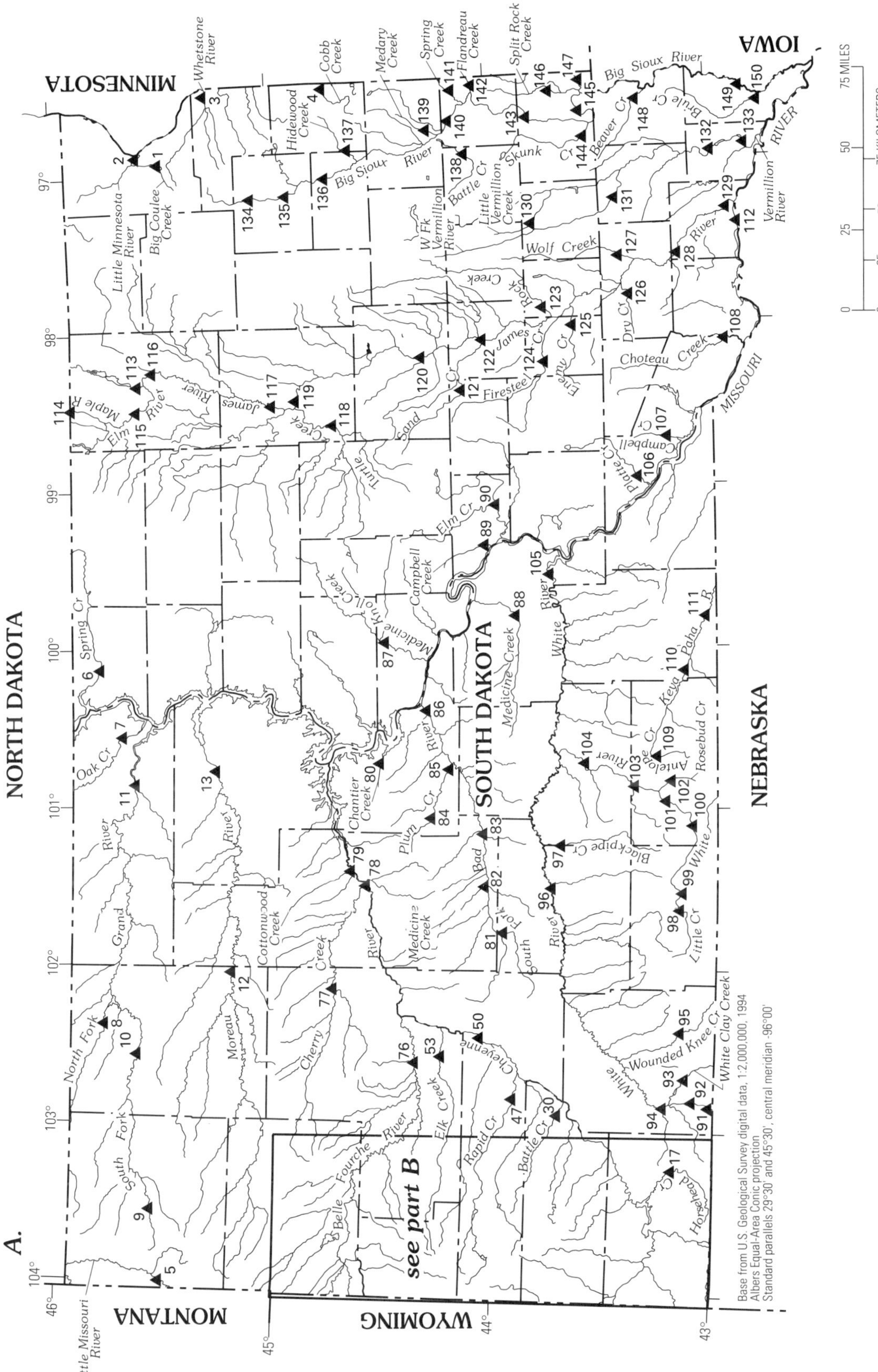

Figure 23. Location of selected streamflow-gaging stations in South Dakota.

B.

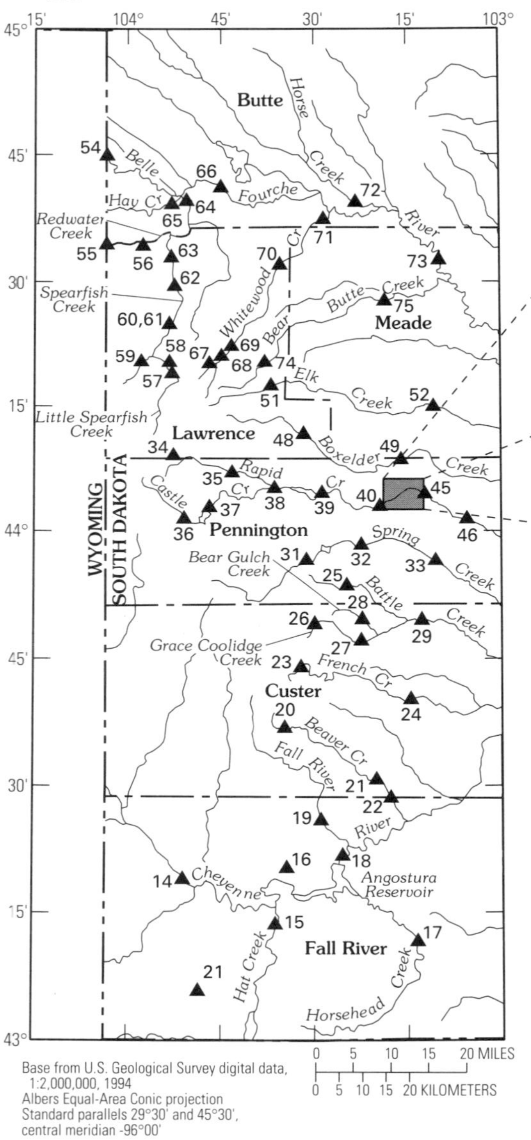

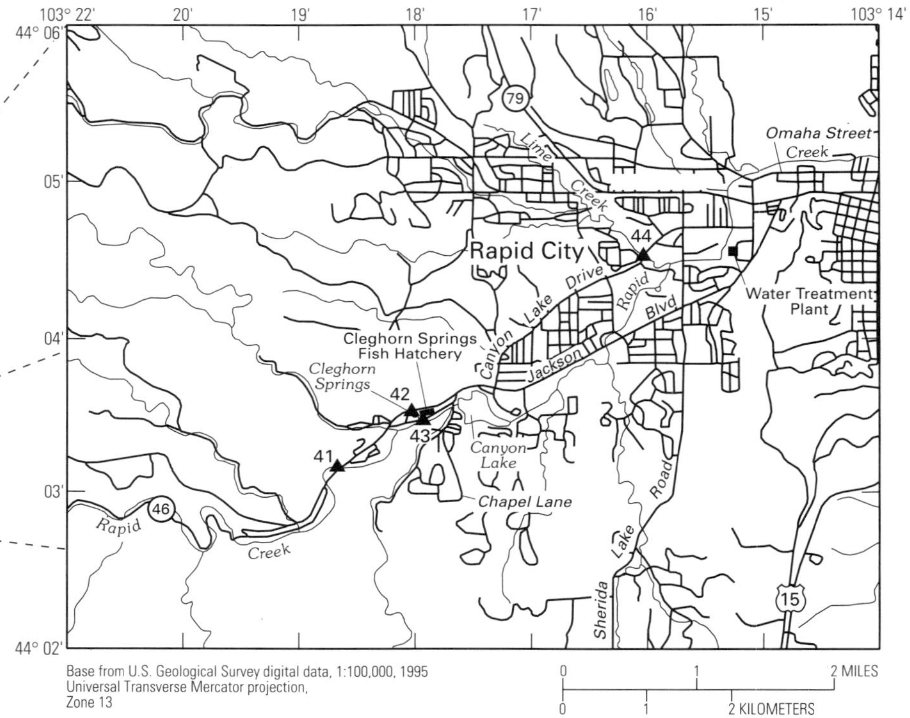

EXPLANATION

▲14 **Streamflow-gaging station and site number—** Number corresponds to that in table 14

Figure 23. Location of selected streamflow-gaging stations in South Dakota—Continued

Table 14. Maximum stages and discharges prior to and during March 1–September 30, 1993, in South Dakota

[mi^2, square miles; ft, feet above an arbitrary datum; ft^3/s, cubic feet per second; --, not determined or not applicable; <, less than. Source: Recurrence intervals calculated from U.S. Geological Survey data. Other data from U.S. Geological Survey reports or data bases]

Site no. (fig. 23)	Station no.	Stream and place of determination	Drainage area (mi^2)	Maximum prior to 1993: Period	Year	Stage (ft)	Discharge (ft^3/s)	Maximum during March–September 1993: Date (month/day)	Stage (ft)	Discharge (ft^3/s)	Discharge recurrence interval (years)
1	05289985	Big Coulee Creek near Peever, SD	12.1	1987–93	1991	8.21	456	7/25	8.13	446	--
2	05290000	Little Minnesota River near Peever, SD	447	1939–81, 1989–93	1952 1943	12.16 [1]13.35	4,730 --	7/25	13.58	8,900	50–100
3	05291000	Whetstone River near Big Stone City, SD	389	1910–12, 1919, 1931–93	1969	14.32	6,870	7/18	11.25	3,890	5–10
4	05299700	Cobb Creek near Gary, SD	70.3	1992–93				6/20 3/27	11.81 [1]11.86	829	--
5	06334500	Little Missouri River at Camp Crook, SD	1,970	1903–06, 1956–93	1978	16.90	9,420	6/10	13.31	3,970	2–5
6	06354860	Spring Creek near Herreid, SD	440 (220 non-contributing)	1962–86, 1989–93	1978	11.49	1,340	7/27	12.56	1,570	5–10
7	06354882	Oak Creek near Wakpala, SD	356	1984–93	1986 1987	17.73 [1]18.35	3,780 --	7/28	11.02	719	--
8	06355500	North Fork Grand River near White Butte, SD	1,190	1945–93, 1967–93	1978 1978	11.63 [1]12.08	6,710 --	3/6	8.64	2,750	5–10
9	06356000	South Fork Grand River at Buffalo, SD	148	1955–93	1963	9.01	2,780	7/28	8.68	2,160	10–25
10	06356500	South Fork Grand River near Cash, SD	1,350	1945–93	1950	15.40	27,000	7/27	7.97	3,030	2–5

Table 14. Maximum stages and discharges prior to and during March 1–September 30, 1993, in South Dakota—Continued

Site no. (fig. 23)	Station no.	Stream and place of determination	Drainage area (mi^2)	Maximum prior to 1993: Period	Year	Stage (ft)	Discharge (ft^3/s)	Maximum during March–September 1993: Date (month/day)	Stage (ft)	Discharge (ft^3/s)	Discharge recurrence interval (years)
11	06357800	Grand River at Little Eagle, SD	5,370	1958–93	1987	19.16	31,000	3/7	10.72	8,110	2–5
					1966	[1]21.76	--	7/28	11.20		
12	06359500	Moreau River near Faith, SD	2,660	1943–93	1944	20.90	26,000	7/27	12.96	6,590	2–5
13	06360500	Moreau River near Whitehorse, SD	4,880	1954–93	1982	26.00	27,700	7/22	18.02	10,700	2–5
					1972	[1]26.20	--				
14	06395000	Cheyenne River near Edgemont, SD	7,143	1903–06, 1928–33, 1946–93	1978	13.65	28,000	8/20	7.56	4,350	2–5
15	06400000	Hat Creek near Edgemont, SD	1,044	1905–06, 1950–93	1967	[2]13.35	13,300	7/16	14.94	3,230	5–10
16	06400497	Cascade Springs near Hot Springs, SD	.47	1976–93	1977	6.25	49	8/14	5.15	28	5
17	06400875	Horsehead Creek at Oelrichs, SD	187	1983–93	1991	18.57	8,270	7/17	8.48	548	--
18	06401500	Cheyenne River below Angostura Dam, SD	9,100	1945–93, 1950–93, 1978–93	1978	15.97	30,300	5/9	6.29	1,540	2–5
19	06402000	Fall River at Hot Springs, SD	137	1937–93, 1953–93	1977	3.32	486	8/19	4.42	354	2–5
					1988	4.62	--				
20	06402430	Beaver Creek near Pringle, SD	45.8	1990–93	1991	8.11	6.0	6/8	8.46	28	--
21	06402470	Beaver Creek above Buffalo Gap, SD	111	1990–93	1990	11.61	21	3/6	11.35	16	--
22	06402500	Beaver Creek near Buffalo Gap, SD	130	1937–93	1938	[2]16.46	11,700	8/20	5.01	49	<2
23	06402995	French Creek above Stockade Lake, near Custer, SD	68.7	1990–93	1991	7.31	320	6/7	6.50	173	--
24	06403300	French Creek above Fairburn, SD	105	1982–93	1987	2.73	329	6/8	2.58	266	--
25	06404000	Battle Creek near Keystone, SD	66	1945–47, 1961–93	1972	14.50	26,200	6/7	6.63	1,250	5–10

Table 14. Maximum stages and discharges prior to and during March 1–September 30, 1993, in South Dakota—Continued

Site no. (fig. 23)	Station no.	Stream and place of determination	Drainage area (mi^2)	Maximum prior to 1993: Period	Year	Stage (ft)	Discharge (ft^3/s)	Maximum during March–September 1993: Date (month/day)	Stage (ft)	Discharge (ft^3/s)	Discharge recurrence interval (years)
26	06404800	Grace Coolidge Creek near Hayward, SD	7.48	1989–93	1991	7.23	210	6/7	6.43	141	--
27	06404998	Grace Coolidge Creek near Game Lodge, near Custer, SD	25.2	1976–93	1989 1979	10.84 [1]12.76	1,030 --	6/7	9.26	298	2–5
28	06405800	Bear Gulch near Hayward, SD	4.23	1989–93	1989	10.68	1,250	6/7	7.46	115	--
29	06406000	Battle Creek at Hermosa, SD	178	1949–93	1972	17.72	21,400	6/7	10.63	1,130	5–10
30	06406500	Battle Creek below Hermosa, SD	285	1950–53, 1988–93	1952	[2]8.13	2,060	6/8	8.57	714	--
31	06406920	Spring Creek above Sheridan Lake, near Keystone, SD	127	1990–93	1991	10.77	455	6/8	10.45	362	--
32	06407500	Spring Creek near Keystone, SD	163	1945–47, 1986–93	1947	5.22	865	6/8	6.94	286	--
33	06408500	Spring Creek near Hermosa, SD	199	1949–93	1972	[2]13.12	13,400	8/16	5.40	366	5–10
34	06408700	Rhoads Fork near Rochford, SD	7.95	1981–93	1985 1982	[2]2.00 [2]2.19	9.7 --	7/20	3.83	6.5	<2
35	06408860	Rapid Creek near Rochford, SD	101	1988–93	1991 1990	5.58 [1]5.94	144 --	6/8	6.00	240	--
36	06409000	Castle Creek above Deerfield Reservoir, near Hill City, SD	79.2	1948–93	1952	5.81	1,120	6/8	3.45	113	5–10
37	06410000	Castle Creek below Deerfield Dam, SD	96.0	1946–93	1952 1991	-- 5.08	[3]200 --	6/10	--	[3]53	2–5
38	06410500	Rapid Creek above Pactola Reservoir, at Silver City, SD	292	1953–93	1965	10.44	2,060	5/10	7.91	867	10–25
39	06411500	Rapid Creek below Pactola Dam, SD	320	1928–32, 1946–93	1965	9.00	547	6/19	8.47	286	5–10
40	06412200	Rapid Creek above Victoria Creek, near Rapid City, SD	355	1988–93	1991	5.96	147	6/20	6.66	335	--

Table 14. Maximum stages and discharges prior to and during March 1–September 30, 1993, in South Dakota—Continued

Site no. (fig. 23)	Station no.	Stream and place of determination	Drainage area (mi^2)	Maximum prior to 1993: Period	Year	Stage (ft)	Discharge (ft^3/s)	Maximum during March–September 1993: Date (month/day)	Stage (ft)	Discharge (ft^3/s)	Discharge recurrence interval (years)
41	06412500	Rapid Creek above Canyon Lake, near Rapid City, SD	371	1946–93	1972	17.77	31,200	6/20, 21	3.62	328	2–5
42	06412810	Cleghorn Springs at Rapid City, SD	(7)	1992–93				3/21	--	[3]15	--
43	06412900	Rapid Creek below Cleghorn Springs, at Rapid City, SD	378	1987–93	1991	7.76	694	6/20	6.86	368	--
44	06413650	Lime Creek at mouth, at Rapid City, SD	10.0	1981–82, 1987–93	1991	[2]3.04	210	8/6	3.65	83	--
45	06414000	Rapid Creek at Rapid City, SD	410	1903–06, 1942–93	1972	19.66	50,000	7/21	5.64	500	2–5
46	06418900	Rapid Creek below Sewage Plant, near Rapid City, SD	452	1981–93	1982	9.12	1,680	7/21	6.62	879	2–5
47	06421500	Rapid Creek near Farmingdale, SD	602	1946–93	1972	11.85	7,320	5/8	8.95	826	2–5
48	06422500	Boxelder Creek near Nemo, SD	96.0	1945–47, 1966–93	1972	[2]20.40	30,100	6/7	3.70	293	2–5
49	06423010	Boxelder Creek near Rapid City, SD	128	1978–93	1978 1991	31.14 31.64	253 --	6/9 6/17	31.49 31.65	121 --	--
50	06423500	Cheyenne River near Wasta, SD	12,800	1914–15, 1928–32, 1934–93	1957 1982	[2]12.42 16.25	26,900 --	5/6	9.89	15,900	2–5
51	06424000	Elk Creek near Roubaix, SD	21.5	1991–93	1992	7.12	15	6/8	7.87	154	--
52	06425100	Elk Creek near Rapid City, SD	190	1978–93	1982 1986	10.79 [1]11.80	1,560 --	5/8	10.04	1,070	2–5
53	06425500	Elk Creek near Elm Springs, SD	540	1949–93	1952 1986	[2]10.61 13.25	8,540 --	5/8	12.80	3,660	5–10
54	06428500	Belle Fourche River at Wyoming-South Dakota State line	3,280	1946–93	1962	15.59	4,400	6/30	14.03	4,550	10–25
55	06430500	Redwater Creek at Wyoming-South Dakota State line	471	1929–31, 1936–37, 1954–93	1973	12.19	2,440	6/9	4.87	236	2–5

Table 14. Maximum stages and discharges prior to and during March 1–September 30, 1993, in South Dakota—Continued

Site no. (fig. 23)	Station no.	Stream and place of determination	Drainage area (mi^2)	Maximum prior to 1993				Maximum during March–September 1993			
				Period	Year	Stage (ft)	Discharge (ft^3/s)	Date (month/day)	Stage (ft)	Discharge (ft^3/s)	Discharge recurrence interval (years)
56	06430540	Cox Lake Outlet near Beulah, WY	0.07	1990–93	1991	--	[3]4.7	6/7	--	[3]4.9	--
57	06430770	Spearfish Creek near Lead, SD	63.5	1988–93	1991	7.53	34	6/8	7.64	51	--
					1991	[1]7.79	--				
58	06430800	Annie Creek near Lead, SD	3.55	1988–93	1989	4.27	12	6/8	4.96	19	--
					1991	4.75	--	3/17	[1]5.18		
59	06430850	Little Spearfish Creek near Lead, SD	25.8	1988–93	1991	--	[3]18	6/8	--	[3]18	--
60	06430898	Squaw Creek near Spearfish, SD	6.95	1988–93	1989	4.81	41	6/8	5.37	96	--
61	06430900	Spearfish Creek above Spearfish, SD	139	1988–93	1991	4.28	129	6/8	5.14	299	--
62	06431500	Spearfish Creek at Spearfish, SD	168	1946–93	1965	10.53	4,240	6/8	6.69	[6]100	<2
					1976	10.54	--				
63	06432020	Spearfish Creek below Spearfish, SD	204	1988–93	1991	5.31	163	6/29	5.35	127	--
					1989	[1]5.86	--				
64	06433000	Redwater River above Belle Fourche, SD	920	1945–93	1962	11.69	16,400	6/9	3.97	735	2–5
65	06433500	Hay Creek at Belle Fourche, SD	121	1953–93	1972	9.15	930	6/8	6.28	96	5–10
66	06436000	Belle Fourche River near Fruitdale, SD	4,540	1945–93	1982	14.32	12,700	6/30	9.72	3,890	2–5
67	06436156	Whitetail Creek at Lead, SD	6.15	1988–93	1991	3.60	33	7/20	3.56	39	--
68	06436170	Whitewood Creek at Deadwood, SD	40.6	1981–93	1982	7.54	2,660	7/20	6.09	2,120	10–25
69	06436180	Whitewood Creek above Whitewood, SD	56.3	1982–93	1991	5.68	2,080	7/21	4.41	722	5–10
70	06436190	Whitewood Creek near Whitewood, SD	77.4	1981–93	1982	4.52	3,050	5/5	3.42	1,050	2–5
71	06436198	Whitewood Creek above Vale, SD	102	1982–93	1986	4.32	3,680	5/5	5.06	2,520	10–25
72	06436760	Horse Creek above Vale, SD	464	1980–93	1982	24.80	17,700	7/2	13.76	2,670	2–5
73	06437000	Belle Fourche River near Sturgis, SD	5,870	1945–93	1982	19.10	36,400	5/9	12.18	9,980	5–10
74	06437020	Bear Butte Creek near Deadwood, SD	16.6	1988–93	1991	7.70	938	6/8	5.20	119	--
75	06437500	Bear Butte Creek near Sturgis, SD	192	1945–72, 1990–93	1962	12.45	12,700	5/6	9.48	1,300	2–5

Table 14. Maximum stages and discharges prior to and during March 1–September 30, 1993, in South Dakota—Continued

Site no. (fig. 23)	Station no.	Stream and place of determination	Drainage area (mi²)	Maximum prior to 1993: Period	Year	Stage (ft)	Discharge (ft³/s)	Maximum during March–September 1993: Date (month/day)	Stage (ft)	Discharge (ft³/s)	Discharge recurrence interval (years)
76	06438000	Belle Fourche River near Elm Springs, SD	7,210	1928–32, 1934–93	1964	15.90	45,100	5/9	12.85	21,000	5–10
					1982	18.22	--				
77	06439000	Cherry Creek near Plainview, SD	1,190	1945–93	1952	22.63	17,500	7/22	16.59	4,050	5–10
78	06439300	Cheyenne River at Cherry Creek, SD	23,900	1960–93	1982	15.77	55,900	5/9	14.63	39,200	10–25
79	06439430	Cottonwood Creek near Cherry Creek, SD	120	1982–93	1987	12.58	3,640	5/5	7.56	756	2–5
80	06439960	Chantier Creek near Hayes, SD	21.5	1990–93	1991	5.21	<10	7/13	14.81	8,000	--
81	06440200	South Fork Bad River near Cottonwood, SD	250	1988–93	1991	17.89	15,200	3/7	10.87	1,190	--
82	06440850	Medicine Creek near Philip, SD	56.5	1989–93	1991	6.82	--	6/17	5.50	168	--
83	06441000	Bad River near Midland, SD	1,460	1945–93	1967	[2]24.44	29,400	3/8	13.41	1,900	<2
								3/7	[1]15.22	--	
84	06441100	Plum Creek near Hayes, SD	24.5	1989–93	1991	2.72	49	6/7	2.73	67	--
85	06441110	Plum Creek below Hayes, SD	252	1989–93	1991	23.74	13,500	6/7	15.30	1,840	--
86	06441500	Bad River near Fort Pierre, SD	3,107	1928–93	1967	29.55	43,800	7/12	17.87	6,160	2–5
								3/6	[1]21.22	--	
87	06442000	Medicine Knoll Creek near Blunt, SD	317	1950–90, 1991–93	1991	12.98	6,370	3/30	11.05	661	5–10
88	06442500	Medicine Creek at Kennebec, SD	464	1954–90, 1991–93	1991	19.11	16,100	3/24	9.78	1,150	2–5
89	06442718	Campbell Creek near Lee's Corner, SD	54.1	1987–93	1988	14.19	2,920	7/24	15.01	3,260	--
90	06442900	Elm Creek near Gann Valley, SD	381	1987–93	1991	12.31	1,400	7/25	17.24	--	--

Table 14. Maximum stages and discharges prior to and during March 1–September 30, 1993, in South Dakota—Continued

Site no. (fig. 23)	Station no.	Stream and place of determination	Drainage area (mi^2)	Maximum prior to 1993				Maximum during March–September 1993			
				Period	Year	Stage (ft)	Dis-charge (ft^3/s)	Date (month/day)	Stage (ft)	Dis-charge (ft^3/s)	Discharge recurrence interval (years)
91	06445685	White River near Nebraska-South Dakota State line	1,440	1987–93	1991	19.07	3,820	6/8	12.88	1,110	--
92	06445700	White River at Slim Butte, SD	1,500	1962–70, 1990–93	1991	16.61	1,750	6/9 3/12	13.88 [1]14.38	926 --	<2
93	06445980	White Clay Creek near Oglala, SD	340	1965–81, 1987–93	1967 1966	14.74 [1]15.02	659 --	6/9	8.06	52	<2
94	06446000	White River near Oglala, SD	2,200	1943–93	1947 1967	23.50 23.61	5,200 --	6/8	19.48	2,190	5–10
95	06446100	Wounded Knee Creek at Wounded Knee, SD	82.5	1992–93				6/9	2.30	20	--
96	06447000	White River near Kadoka, SD	5,000	1942–93	1951 1982	13.83 16.18	21,700 --	3/21	--	10,000	2–5
97	06447230	Blackpipe Creek near Belvidere, SD	250	1992–93				6/29	6.34	1,760	--
98	06447500	Little White River near Martin, SD	310 (80 non-contributing)	1938–40, 1962–93	1965 1966	12.90 [1]13.21	1,190 --	3/14	[1]5.92	170	<2
99	06449000	Lake Creek below Refuge, near Tuthill, SD	120 (60 non-contributing)	1938–40, 1962–93	1987 1988	5.57 [2]6.46	594 --	3/24	4.93	86	<2
100	06449100	Little White River near Vetal, SD	590 (175 non-contributing)	1959–93	1991	12.53	3,540	3/27	5.20	244	<2
101	06449300	Little White River above Rosebud, SD	890 (260 non-contributing)	1981–93	1983 1988	3.51 [1]5.67	900 --	8/16	3.94	772	2–5
102	06449400	Rosebud Creek at Rosebud, SD	50.8	1974–93	1976	10.34	643	7/28	6.06	151	2–5
103	06449500	Little White River near Rosebud, SD	1,020 (260 non-contributing)	1943–93	1967	14.09	4,640	8/16	6.69	752	2–5

Table 14. Maximum stages and discharges prior to and during March 1–September 30, 1993, in South Dakota—Continued

				Maximum prior to 1993				Maximum during March–September 1993			
Site no. (fig. 23)	Station no.	Stream and place of determination	Drainage area (mi^2)	Period	Year	Stage (ft)	Discharge (ft^3/s)	Date (month/day)	Stage (ft)	Discharge (ft^3/s)	Discharge recurrence interval (years)
104	06450500	Little White River below White River, SD	1,570 (260 non-contributing)	1949–93	1967 1968	[2]10.02 [2]15.46	13,700 --	5/1 3/7	5.24 [1]7.99	1,650	<2
105	06452000	White River near Oacoma, SD	10,200	1928–93	1952 1978	[2]15.40 23.59	51,900 --	3/8 3/20	[1]17.14 [1]18.09	17,000 --	2–5
106	06452320	Platte Creek near Platte, SD	741	1988–93	1990	5.14	447	6/17	7.24	1,600	--
107	06452330	Campbell Creek near Geddes, SD	8.37	1989–93	1990	9.83	678	7/11	7.81	140	--
108	06453255	Choteau Creek near Avon, SD	602	1982–93	1984	13.93	7,280	5/8	13.11	5,120	5–10
109	06463900	Antelope Creek near Mission, SD	71.3	1990–93	1992	5.71	49	5/1	6.00	53	--
110	06464100	Keya Paha River near Keyapaha, SD	466	1981–93	1983 1982	7.95 [1]9.45	820 --	3/7	[1]8.58	500	2–5
111	06464500	Keya Paha River near Wewela, SD	1,070	1937–40, 1947–93	1952 1950	13.08 [1]13.50	5,430 --	3/9	[1]7.96	2,000	5–10
112	06467500	Missouri River at Yankton, SD	279,500	1930–93	1975 1975	23.07 23.17	63,700 --	5/21	15.14	27,100	<2
113	06471000	James River at Columbia, SD	5,857 (3,376 non-contributing)	1945–93	1979 1987	16.15 [1]17.11	2,340 --	8/19 8/12	[1]15.27 [1]15.31	1,160 --	2–5
114	06471200	Maple River at North Dakota-South Dakota State line	716 (332 non-contributing)	1956–93	1969	[1]16.05	5,930	7/21	9.46	708	2–5
115	06471500	Elm River at Westport, SD	1,493 (444 non-contributing)	1945–93	1969	22.11	12,600	7/22	8.91	923	2–5
116	06471550	James River below Columbia, SD	7,393 (3,820 non-contributing)	1988–93	1989	[1]15.26	1,700	8/20 8/12	[1]13.67 [1]13.76	1,390 --	--
117	06473000	James River at Ashton, SD	9,742 (4,069 non-contributing)	1945–93	1969 1969	20.63 [1]21.17	5,680 --	9/10 8/1	[1]12.92 [1]13.49	1,390 --	2–5

Table 14. Maximum stages and discharges prior to and during March 1–September 30, 1993, in South Dakota—Continued

Site no. (fig. 23)	Station no.	Stream and place of determination	Drainage area (mi²)	Maximum prior to 1993: Period	Year	Stage (ft)	Discharge (ft³/s)	Maximum during March–September 1993: Date (month/day)	Stage (ft)	Discharge (ft³/s)	Discharge recurrence interval (years)
118	06474000	Turtle Creek near Tulare, SD	1,124	1953–56, 1965–81, 1984–93	1969	[1]18.51	6,000	8/14	8.60	720	2–5
119	06475000	James River near Redfield, SD	13,911 (4,118 non-contributing)	1950–93	1969	[2]24.93	7,310	8/1	[1]14.64	2,100	2–5
120	06476000	James River at Huron, SD	15,869 (4,148 non-contributing)	1928–32, 1943–93	1969	16.70	9,000	7/29 7/31	[1]13.36 [1]13.41	3,440 --	5–10
121	06476500	Sand Creek near Alpena, SD	261	1950–93	1960 1950	13.35 [1]14.10	2,240 --	3/25	10.48	270	2–5
122	06477000	James River near Forestburg, SD	17,590 (4,148 non-contributing)	1950–93	1969	17.16	12,500	8/4	14.13	3,450	2–5
123	06477150	Rock Creek near Fulton, SD	240	1966–79, 1989–93	1969	[2]10.21	2,040	7/6	14.34	1,880	10–25
124	06477500	Firesteel Creek near Mount Vernon, SD	521	1955–93	1969 1969	15.34 [1]17.12	6,610 --	7/5	11.48	1,930	2–5
125	06478052	Enemy Creek near Mitchell, SD	163	1975–87, 1989–93	1984	15.15	4,280	7/6	14.97	4,050	10–25
126	06478300	Dry Creek near Parkston, SD	97.2	1955–80, 1989–93	1960	[2]12.70	4,210	5/8	8.92	2,240	10–25
127	06478390	Wolf Creek near Clayton, SD	396	1975–93	1984	18.01	6,520	7/5	17.39	5,390	10–25
128	06478500	James River near Scotland, SD	20,653 (4,148 non-contributing)	1928–93	1984	20.45	29,400	7/6	19.76	17,600	25–50
129	06478513	James River near Yankton, SD	20,942 (4,148 non-contributing)	1981–93	1984	24.34	26,400	7/8	21.15	15,800	5–10

Table 14. Maximum stages and discharges prior to and during March 1–September 30, 1993, in South Dakota—Continued

Site no. (fig. 23)	Station no.	Stream and place of determination	Drainage area (mi^2)	Maximum prior to 1993				Maximum during March–September 1993			
				Period	Year	Stage (ft)	Discharge (ft^3/s)	Date (month/day)	Stage (ft)	Discharge (ft^3/s)	Discharge recurrence interval (years)
130	06478540	Little Vermillion River near Salem, SD	78.6	1966–93	1984	[1]9.88	900	7/4	11.95	3,300	25–50
131	06478690	West Fork Vermillion River near Parker, SD	377	1961–93	1984	12.57	4,800	5/8	13.14	6,300	25–50
132	06479000	Vermillion River near Wakonda, SD	2,170 (494 non-contributing)	1945–93	1984	17.62	17,000	7/7	17.33	13,000	25–50
133	06479010	Vermillion River near Vermillion, SD	2,302 (494 non-contributing)	1983–93	1984	31.77	21,400	7/8	[1]26.68	10,200	10–25
134	06479215	Big Sioux River near Florence, SD	638 (570 non-contributing)	1984–93	1986	9.08	1,810	7/25	9.18	1,280	--
135	06479438	Big Sioux River near Watertown, SD	1,007 (779 non-contributing)	1972–93	1986 1991	11.08 [1]11.13	4,970 --	3/28	[1]11.07	4,000	10–25
136	06479525	Big Sioux River near Castlewood, SD	1,997 (1,427 non-contributing)	1976–93	1986	11.73	2,250	3/28	[1]11.39	1,500	5–10
137	06479640	Hidewood Creek near Estelline, SD	164	1968–85, 1990–93	1992	13.10	17,300	3/28	[1]11.44	3,000	5–10
138	06479928	Battle Creek near Nunda, SD	163 (4.8 non-contributing)	1987–93	1990	9.64	578	7/4	[1]12.99	3,400	--
139	06479980	Medary Creek near Brookings, SD	200	1981–93	1984	11.27	2,590	7/4	11.78	3,710	10–25
140	06480000	Big Sioux River near Brookings, SD	3,898 (1,479 non-contributing)	1953–93	1969	14.77	33,900	7/4	13.50	13,300	10–25

Table 14. Maximum stages and discharges prior to and during March 1–September 30, 1993, in South Dakota—Continued

Site no. (fig. 23)	Station no.	Stream and place of determination	Drainage area (mi^2)	Maximum prior to 1993: Period	Year	Stage (ft)	Discharge (ft^3/s)	Maximum during March–September 1993: Date (month/day)	Stage (ft)	Discharge (ft^3/s)	Discharge recurrence interval (years)
141	06480400	Spring Creek near Flandreau, SD	63.2	1982–93	1984	15.72	2,030	7/3	16.84	4,480	25–50
142	06480650	Flandreau Creek above Flandreau, SD	100	1981–93	1984	11.02	2,650	7/4	11.28	2,410	5–10
143	06481000	Big Sioux River near Dell Rapids, SD	4,483 (1,479 non-contributing)	1948–93	1969	16.47	41,300	7/4	15.56	16,400	10–25
144	06481500	Skunk Creek at Sioux Falls, SD	622 (8.51 non-contributing)	1948–93	1957	[2]17.78	29,400	7/11	10.81	8,640	10–25
145	06482020	Big Sioux River at North Cliff Avenue at Sioux Falls, SD	5,216 (1,487 non-contributing)	1971–93	1984	25.40	21,600	7/07	23.84	18,000	25–50
146	06482610	Split Rock Creek at Corson, SD	464	1951–93	1969	15.00	17,800	5/08	17.58	18,900	25–50
147	06482745	Beaver Creek at Valley Springs, SD	104	1986–93	1992	24.69	2,200	5/08	24.21	2,000	--
148	06482848	Beaver Creek at Canton, SD	124	1982–93	1984	13.72	2,570	6/07	12.93	3,680	10–25
					1983	[1]14.61	--				
149	06485500	Big Sioux River at Akron, SD	8,424 (1,487 non-contributing)	1928–93	1969	22.99	80,800	5/10	23.05	66,700	25–50
150	06485696	Brule Creek near Elk Point, SD	204	1982–93	1983	22.39	6,290	3/28	15.32	2,210	2–5
								7/15	15.34	--	

[1]Backwater from ice or other sources.

[2]Site and datum then in use.

[3]Discharge is maximum daily.

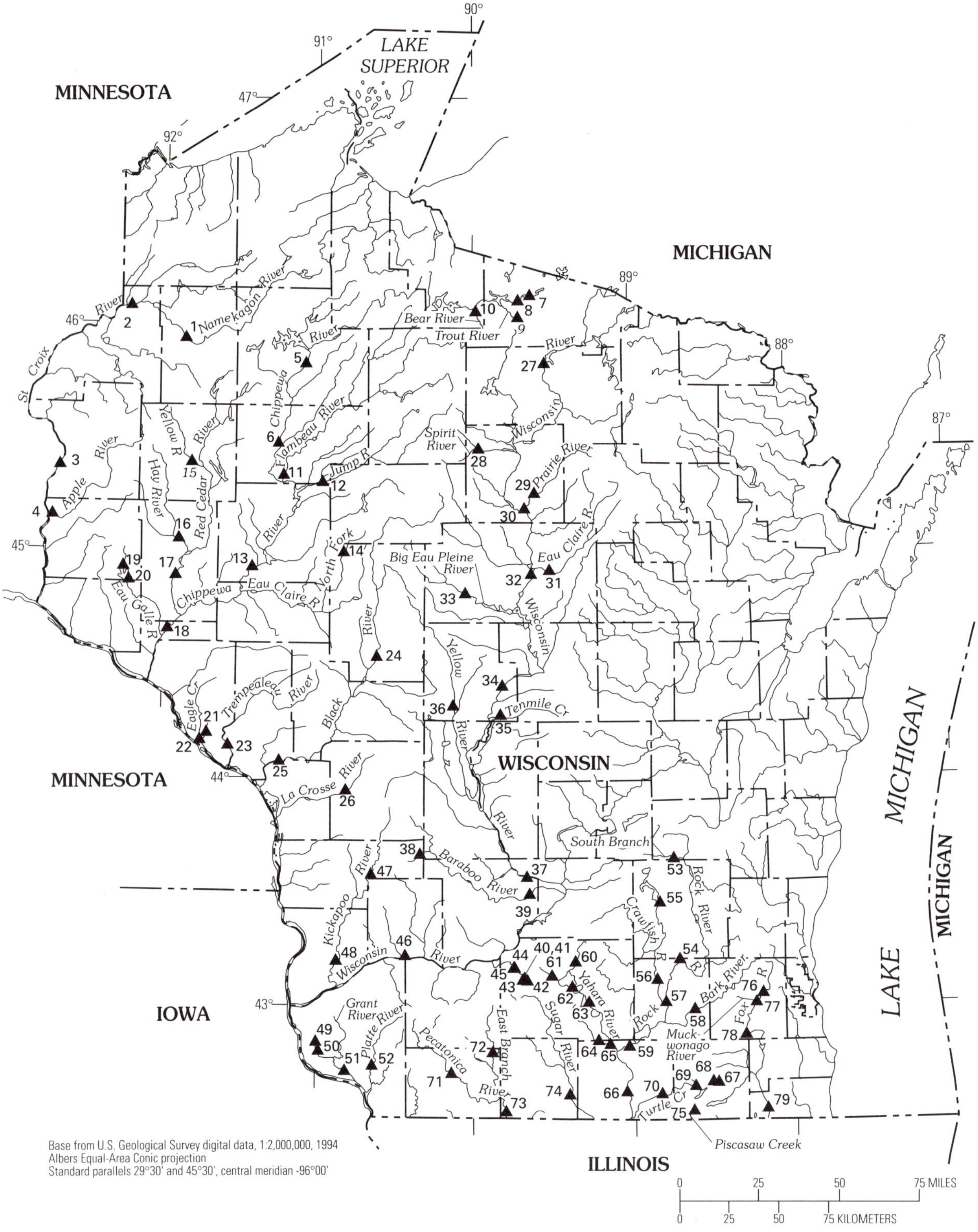

EXPLANATION

▲66 **Streamflow-gaging station and site number**—
Number corresponds to that in table 15

Figure 24. Location of selected streamflow-gaging stations in Wisconsin.

Table 15. Maximum stages and discharges prior to and during March 1–September 30, 1993, in Wisconsin

[mi^2, square miles; ft, feet above an arbitrary datum; ft^3/s, cubic feet per second; --, not determined or not applicable; <, less than; >, greater than. Source: Recurrence intervals calculated from U.S. Geological Survey data. Other data from U.S. Geological Survey reports or data bases]

				Maximum prior to 1993				Maximum during March–September 1993			
Site no. (fig. 24)	Station no.	Stream and place of determination	Drainage area (mi^2)	Period	Year	Stage (ft)	Discharge (ft^3/s)	Date (month/day)	Stage (ft)	Discharge (ft^3/s)	Discharge recurrence interval (years)
1	05332500	Namekagon River near Trego, WI	488	1927–70, 1987–93	1941	--	5,200	5/3,6/1	--	1,080	<2
2	05333500	St. Croix River near Danbury, WI	1,580	1914–81, 1984–93	1950	8.22	10,200	5/28	3.33	3,260	<2
3	05340500	St. Croix River at St. Croix Falls, WI	6,240	1902–93	1950	25.19	54,900	6/26, 27	10.08	20,700	<2
4	05341500	Apple River near Somerset, WI	579	1914–70, 1986–93	1965	--	2,510	6/26	--	1,290	[1]2
5	05356000	Chippewa River, at Bishops Bridge, near Winter, WI	790	1912–93	1941	11.05	7,520	6/24	8.53	4,280	4
6	05356500	Chippewa River near Bruce, WI	1,650	1913–93	1941	20.46	25,800	6/21	12.34	12,700	[2]4
7	05357215	Allequash Creek (head of Trout River) at CTH M, near Boulder Junction, WI	8.43	1991–93	1991	2.36	79	6/20	1.85	36	--
8	05357225	Stevenson Creek, at County Highway M, near Boulder Junction, WI	7.96	1991–93	1991	9.62	39	6/20	8.58	12	--
9	05357245	Trout River, at Trout Lake, near Boulder Junction, WI	46.2	1991–93	1991	1.90	77	6/21	1.82	69	--
10	05357335	Bear River near Manitowish Waters, WI	81.3	1991–93	1992	2.66	331	6/21	2.66	248	--
11	05360500	Flambeau River near Bruce, WI	1,860	1951–93	1986	10.45	17,600	6/21	10.11	16,500	10
12	05362000	Jump River at Sheldon, WI	576	1915–93	1941	18.80	46,000	6/21	13.20	16,400	13
13	05365500	Chippewa River at Chippewa Falls, WI	5,650	1888–1983, 1986–93	1941	24.80	102,000	6/21	19.84	60,300	8
14	05365707	North Fork Eau Claire River near Thorp, WI	51.0	1986–93	1986	10.13	9,050	6/20	8.10	4,210	--
15	05367446	Yellow River at Barron, WI	153	1991–93	1992	5.89	1,160	6/21	5.73	1,080	--

Table 15. Maximum stages and discharges prior to and during March 1–September 30, 1993, in Wisconsin—Continued

Site no. (fig. 24)	Station no.	Stream and place of determination	Drainage area (mi²)	Maximum prior to 1993: Period	Year	Stage (ft)	Discharge (ft³/s)	Maximum during March–September 1993: Date (month/day)	Stage (ft)	Discharge (ft³/s)	Discharge recurrence interval (years)
16	05368000	Hay River at Wheeler, WI	418	1950–93	1967	15.04	13,600	6/18	11.01	3,710	2
17	05369000	Red Cedar River at Menomonie, WI	1,770	1907–08, 1913–93	1934	16.00	40,000	6/19	5.29	7,380	<2
18	05369500	Chippewa River at Durand, WI	9,010	1928–93	1967	16.93	123,000	6/23	15.76	90,100	21
19	05369945	Eau Galle River, at low-water bridge, at Spring Valley, WI	47.9	1981–83, 1986–93	1986	8.80	6,000	8/9	8.06	5,050	--
20	05370000	Eau Galle River at Spring Valley, WI	64.1	1944–93	1980	19.90	3,030	8/9	18.65	2,170	4
21	05378183	Joos Valley Creek near Fountain City, WI	5.89	1990–93	1990	9.46	574	7/3	8.89	302	--
22	05378185	Eagle Creek, at County Highway G, near Fountain City, WI	14.3	1990–93	1990	9.02	919	7/3	7.59	388	--
23	05379500	Trempealeau River at Dodge, WI	643	1913–19, 1934–93	1956	[1]10.35	17,400	6/22	11.15	5,880	4
24	05381000	Black River at Neillsville, WI	749	1905–09, 1913–93	1938	23.80	48,800	6/20	19.30	30,400	24
25	05382000	Black River near Galesville, WI	2,080	1931–93	1967	[1]14.63	65,500	6/21	16.64	64,000	>100
26	05382325	La Crosse River at Sparta, WI	167	1992–93	1993	8.78	1,100	6/20	8.78	1,100	--
27	05391000	Wisconsin River, at Rainbow Lake, near Lake Tomahawk, WI	757	1936–93	1941	7.59	3,570	6/21	4.58	1,650	<2
28	05393500	Spirit River at Spirit Falls, WI	81.6	1942–93	1942	10.00	4,180	6/20	7.43	2,730	10
29	05394500	Prairie River near Merrill, WI	184	1914–31, 1939–93	1941	9.45	5,800	6/21	6.77	2,000	4
30	05395000	Wisconsin River at Merrill, WI	2,760	1902–93	1941	18.26	49,400	6/21	12.25	20,600	[4]8
31	05397500	Eau Claire River at Kelly, WI	375	1914–26, 1939–93	1926	8.40	8,300	6/21	7.15	3,900	3
32	05398000	Wisconsin River at Rothschild, WI	4,020	1944–93	1965	[1]18.46	49,200	6/21	27.48	44,400	10
33	05399500	Big Eau Pleine River near Stratford, WI	224	1914–25, 1937–93	1938	24.50	41,000	6/20	16.78	11,100	3

Table 15. Maximum stages and discharges prior to and during March 1–September 30, 1993, in Wisconsin—Continued

Site no. (fig. 24)	Station no.	Stream and place of determination	Drainage area (mi^2)	Maximum prior to 1993: Period	Year	Stage (ft)	Discharge (ft^3/s)	Maximum during March–September 1993: Date (month/day)	Stage (ft)	Discharge (ft^3/s)	Discharge recurrence interval (years)
34	05400760	Wisconsin River at Wisconsin Rapids, WI	5,420	1914–50, 1957–93	1938	--	70,400	6/21	--	64,600	34
35	05401050	Tenmile Creek near Nekoosa, WI	73.3	1963–79, 1987–93	1979	6.62	456	6/21	6.34	249	3
36	05402000	Yellow River at Babcock, WI	215	1944–93	1952	17.38	11,600	6/9	15.63	8,180	8
37	05404000	Wisconsin River near Wisconsin Dells, WI	8,090	1934–93	1938	23.83	72,200	6/24	18.16	59,100	17
38	05404116	South Branch Baraboo River at Hillsboro, WI	39.1	1988–93	1990	15.60	4,010	3/29	10.35	424	--
39	05405000	Baraboo River near Baraboo, WI	609	1913–22, 1942–93	1917	[1]17.50	7,900	7/18	22.78	6,340	19
40	05406460	Black Earth Creek at Cross Plains, WI	12.8	1984–86, 1989–93	1993	13.32	230	7/5	13.32	230	--
41	05406470	Brewery Creek at Cross Plains, WI	[2]10.5	1984–86, 1989–93	1993	15.05	420	7/6	15.05	420	--
42	05406476	Black Earth Creek, at Mills Street, at Cross Plains, WI	[2]25.5	1989–93	1993	--	[3]250	7/6	--	250	--
43	05406491	Garfoot Creek near Cross Plains, WI	5.39	1984–86, 1989–93	1985	5.84	128	7/5	7.57	111	--
44	05406497	Black Earth Creek, at South Valley Road, near Black Earth, WI	[2]40.6	1989–93	1993	8.60	1,100	7/6	8.60	1,100	--
45	05406500	Black Earth Creek at Black Earth, WI	[2]45.6	1954–93	1954	6.58	1,750	7/6	6.13	1,320	31
46	05407000	Wisconsin River at Muscoda, WI	10,400	1913–93	1938	11.48	80,800	6/26	10.34	59,600	12
47	05408000	Kickapoo River at La Farge, WI	266	1938–93	1978	14.92	14,300	5/3	11.14	2,390	<2
48	05410490	Kickapoo River at Steuben, WI	687	1933–93	1978	[1]14.81	16,500	5/5	12.96	2,960	<2
49	054134435	Kuenster Creek, at Muskellunge Road, near North Andover, WI	9.59	1991–93	1993	8.74	834	7/10	8.74	834	--
50	05413449	Rattlesnake Creek near North Andover, WI	42.4	1987–93	1991	11.20	7,000	7/10	8.40	3,640	--

Table 15. Maximum stages and discharges prior to and during March 1–September 30, 1993, in Wisconsin—Continued

Site no. (fig. 24)	Station no.	Stream and place of determination	Drainage area (mi^2)	Maximum prior to 1993				Maximum during March–September 1993			
				Period	Year	Stage (ft)	Discharge (ft^3/s)	Date (month/day)	Stage (ft)	Discharge (ft^3/s)	Discharge recurrence interval (years)
51	05413500	Grant River at Burton, WI	269	1934–93	1950	24.82	25,000	7/9	21.15	6,950	2
52	05414000	Platte River near Rockville, WI	142	1934–93	1950	17.26	43,500	7/9	11.17	3,980	<2
53	05423500	South Branch Rock River at Waupun, WI	63.6	1948–69, 1987–93	1959	7.97	1,500	7/6	7.53	920	6
54	05425500	Rock River at Watertown, WI	969	1931–70, 1976–93	1979	6.19	5,080	4/20	6.03	4,620	25
55	05425912	Beaverdam River at Beaver Dam, WI	157	1985–93	1993	9.32	758	7/9	9.32	758	--
56	05426000	Crawfish River at Milford, WI	762	1931–93	1959	11.15	6,140	4/23	9.36	4,140	12
57	05426031	Rock River at Jefferson, WI	1,850	1978–93	1979	10.84	10,300	4/22	10.29	8,660	--
58	05426250	Bark River near Rome, WI	122	1979–93	1993	2.56	476	4/20	2.56	476	--
59	05427570	Rock River at Indianford, WI	2,630	1975–93	1979	[1]16.23	11,900	4/25	15.85	10,200	--
60	05427718	Yahara River at Windsor, WI	73.6	1976–81, 1989–93	1993	6.58	2,050	7/6	6.58	2,050	--
61	05427948	Pheasant Branch at Middleton, WI	[4]18.3	1974–93	1993	8.92	746	7/6	8.92	746	18
62	05427965	Spring Harbor storm sewer at Madison, WI	3.29	1976–93	1993	4.16	754	7/5	4.16	754	21
63	05429500	Yahara River near McFarland, WI	327	1930–93	1959	[1]5.82	867	4/21	5.93	681	26
64	05430150	Badfish Creek near Cooksville, WI	82.6	1977–93	1981	8.11	870	3/24	7.61	734	4
65	05430175	Yahara River near Fulton, WI	517	1977–93	1981	8.36	3,040	4/20	6.62	1,820	4
66	05430500	Rock River at Afton, WI	3,340	1914–93	1929	11.81	13,000	4/23	11.41	10,700	10
67	05431014	Jackson Creek, at Petrie Road, near Elkhorn, WI	8.96	1983–93	1993	9.13	448	4/20	9.13	448	--
68	054310157	Jackson Creek tributary near Elkhorn, WI	4.34	1983–93	1993	10.00	210	4/19	10.00	210	--
69	05431022	Delavan Lake outlet, at Borg Road, near Delavan, WI	[5]42.1	1983–93	1993	7.99	372	4/21	7.99	372	--
70	05431486	Turtle Creek, at Carvers Rock Road, near Clinton, WI	[6]199	1939–93	1973	[5]12.85	16,500	6/30	10.38	5,580	14

Table 15. Maximum stages and discharges prior to and during March 1–September 30, 1993, in Wisconsin—Continued

				Maximum prior to 1993				Maximum during March–September 1993			
Site no. (fig. 24)	Station no.	Stream and place of determination	Drainage area (mi^2)	Period	Year	Stage (ft)	Discharge (ft^3/s)	Date (month/day)	Stage (ft)	Discharge (ft^3/s)	Discharge recurrence interval (years)
71	05432500	Pecatonica River at Darlington, WI	273	1939–93	1950	20.71	22,000	7/6	18.22	12,400	24
72	05433000	East Branch Pecatonica River near Blanchardville, WI	221	1939–86, 1987–93	1948	15.74	11,700	7/6	16.54	5,650	9
73	05434500	Pecatonica River at Martintown, WI	1,034	1939–93	1969	21.46	15,100	7/8	20.36	9,600	6
74	05436500	Sugar River near Brodhead, WI	523	1914–93	1915	11.40	14,800	3/25	8.01	4,310	3
75	05438283	Piscasaw Creek near Walworth, WI	9.58	1992–93	1993	10.05	322	6/30	10.05	322	--
76	05543800	Fox River, at Watertown Road, near Waukesha, WI	77.4	1993	--	--	--	4/21	11.55	1,170	--
77	05543830	Fox River at Waukesha, WI	126	1963–93	1973	7.42	2,260	4/21	6.92	1,520	9
78	05544200	Mukwonago River at Mukwonago, WI	74.1	1973–93	1976	[5]2.50	300	4/21	3.44	268	9
79	05546500	Fox River at Wilmot, WI	868	1940–93	1960	9.25	7,520	4/22	7.67	5,060	14

[1]Site and datum then in use.
[2]2.80 square miles noncontributing.
[3]Mean daily discharge.
[4]1.22 square miles noncontributing.
[5]2.30 square miles noncontributing.
[6]2.33 square miles noncontributing.

REFERENCES

Flood Insurance Administration, 1981, Flood insurance study, City of Kansas City, Kansas: U.S. Department of Housing and Urban Development, February 1981, 31 p.

Krug, W.R., and House, L.B., 1980, Streamflow model of Wisconsin River for estimating flood frequency and volume: U.S. Geological Survey Open-File Report 80–1103, 44 p.

National Oceanic and Atmospheric Administration, 1992, Daily normals of temperature, precipitation, and heating and cooling degree days, 1961–90—By State: National Oceanic and Atmospheric Administration Climatography of the United States [variously paged].

National Weather Service, 1993a, Midwestern floods—Heat and drought in the East, special climate summary: Asheville, N.C., v. 93, no. 1, 4 p.

———1993b, Update on Midwestern floods—Heat and drought in the East, special climate summary: Asheville, N.C., v. 93, no. 2, 8 p.

———1993c, Growing season summary, special climate summary: Asheville, N.C., v. 93, no. 3, 9 p.

———1994, Climatological data for Illinois (v. 98), Iowa (v. 104), Kansas (v. 107), Minnesota (v. 99), Missouri (v. 97), Nebraska (v. 98), North Dakota (v. 102), South Dakota (v. 98), and Wisconsin (v. 98): Asheville, N.C., National Climatic Data Center [variously paged].

Parrett, Charles, Melcher, N.B., and James, R.W., Jr., 1993, Flood discharges in the upper Mississippi River Basin, 1993, *in* Floods in the upper Mississippi River Basin, 1993: U.S. Geological Survey Circular 1120–A, 14 p.

Perry, C.A., 1994, Effects of reservoirs on flood discharges in the Kansas and Missouri River Basins, 1993, *in* Floods in the upper Mississippi River Basin, 1993: U.S. Geological Survey Circular 1120–E, 20 p.

U.S. Geological Survey, 1994, Water resources data for Colorado, Iowa, Illinois, Kansas, Minnesota, Missouri, Montana, Nebraska, North Dakota, South Dakota, and Wisconsin, water year 1993: U.S. Geological Survey Water-Data Reports CO–93–1, IA–93–1, IL–93–1, KS–93–1, MN–93–1, MO–93–1, MT–93–1, NE–93–1, ND–93–1, SD–93–1, and WI–93–1 [variously paged].

U.S. Interagency Advisory Committee on Water Data, 1982, Guidelines for determining flood flow frequency: U.S. Geological Survey, Hydrology Subcommittee Bulletin 17B, 183 p.

Viessman, Warren, Jr., Harbaugh, T.E., and Knapp, J.W., 1972, Introduction to hydrology: New York, Intext Educational Publications, p. 173–221.

Wahl, K.L., Vining, K.C., and Wiche, G.J., 1993, Precipitation in the upper Mississippi River Basin, January 1 through July 31, 1993, *in* Floods in the upper Mississippi River Basin, 1993: U.S. Geological Survey Circular 1120–B, 13 p.

———1994, Precipitation in the upper Mississippi River Basin during the great floods of 1993: Hydrological Science and Technology, v. 10, no. 1–4, p. 140–152.

SUMMARY OF FLOODS OF 1992

January 5–6, 1992, in Puerto Rico

By Heriberto Torres-Sierra

During January 5–6, 1992, most of the mountainous interior and several areas along the north, east, and south coasts of Puerto Rico experienced moderate to severe floods (fig. 25). More than one-half of the island's towns were affected (40 municipalities). This flood was a result of intense rainfall generated by the unusual combination of a cold front and an upper-level pressure trough. As much as 20 inches of rain fell on the interior of Puerto Rico. Damage to houses, businesses, farmlands, livestock, highways, bridges, and other public and private properties was $155 million. This value was estimated from reports submitted by different agencies, which included the Federal Emergency Management Agency, the Puerto Rico Department of Agriculture, and the Puerto Rico Department of Transportation and Public Works.

The U.S. Geological Survey collected and analyzed data on the magnitude and frequency of maximum discharge recorded at streamflow-gaging stations throughout Puerto Rico. At streamflow-gaging stations where recording instruments failed or were damaged during the flood, high-water marks were surveyed to determine the maximum stage and discharge. Maximum discharges at the Río de La Plata at Proyecto La Plata and at the Río Grande de Patillas near Patillas gaging stations were estimated using indirect-discharge measurement techniques. This article provides a general description of the storm and flood, flood damage, and maximum stages and discharges for the January 5–6, 1992, flood and for previous maximum floods at selected streamflow-gaging stations.

Information related to rainfall quantities and intensities throughout the island was provided by the National Weather Service, National Oceanic and Atmospheric Administration, U.S. Department of Commerce. The Federal Emergency Management Administration, the Puerto Rico Department of Agriculture, and the Puerto Rico Department of Transportation and Public Works provided information on damage caused by the rain, floods, and landslides.

THE STORM

On January 4, 1992, an extensive area of low pressure at the surface and aloft was located off the east coast of the United States. An associated cold front extended from the Atlantic Ocean east of the Bahama Islands southwest to the Caribbean Sea. At 0800 on January 5, the front was positioned northeast to southwest across the Dominican Republic. As the day progressed, convective activity ahead of the cold front became more intense. By 1400, a surface pressure trough had developed over Puerto Rico ahead of the nearly stationary cold front. A weak surface low pressure also developed near the Gulf of Venezuela reinforcing the moist airflow into the storm system (fig. 26).

The unusual combination of these weather systems produced nearly stationary thunderstorms and intense rain on the mountainous interior of Puerto Rico. By the evening of January 5, the showers and isolated thunderstorms had spread over the island. Strong thunderstorm activity and intense rain continued throughout that night and into the early morning hours of January 6 as the stationary cold front remained over the Dominican Republic. The surface trough ahead of the cold front and the weak low-pressure system to the southwest of Puerto Rico had dissipated. Although scattered showers and thunderstorms persisted, the intense rains had diminished.

Rainfall totals ranged from about 2 inches in the northern part of Puerto Rico to 20 inches in its interior (fig. 27). The largest 24-hour rainfall total recorded was 20.3 inches at Toro Negro Forest (table 16). As much as 19.6 inches of rain fell in Cayey, exceeding the previous 24-hour record of 12.8 inches on August 30, 1979. Rainfall intensities for 1-, 2-, 3-, and 6-hour durations of 5.9, 8.4, 11.2, and 18.4 inches exceeded previous islandwide maximums of 4.3, 6.7, 7.8, and 10.4 inches, respectively.

FLOOD DESCRIPTION AND DAMAGES

During January 5–6, 1992, at least 40 of the 78 municipalities in Puerto Rico were affected by the

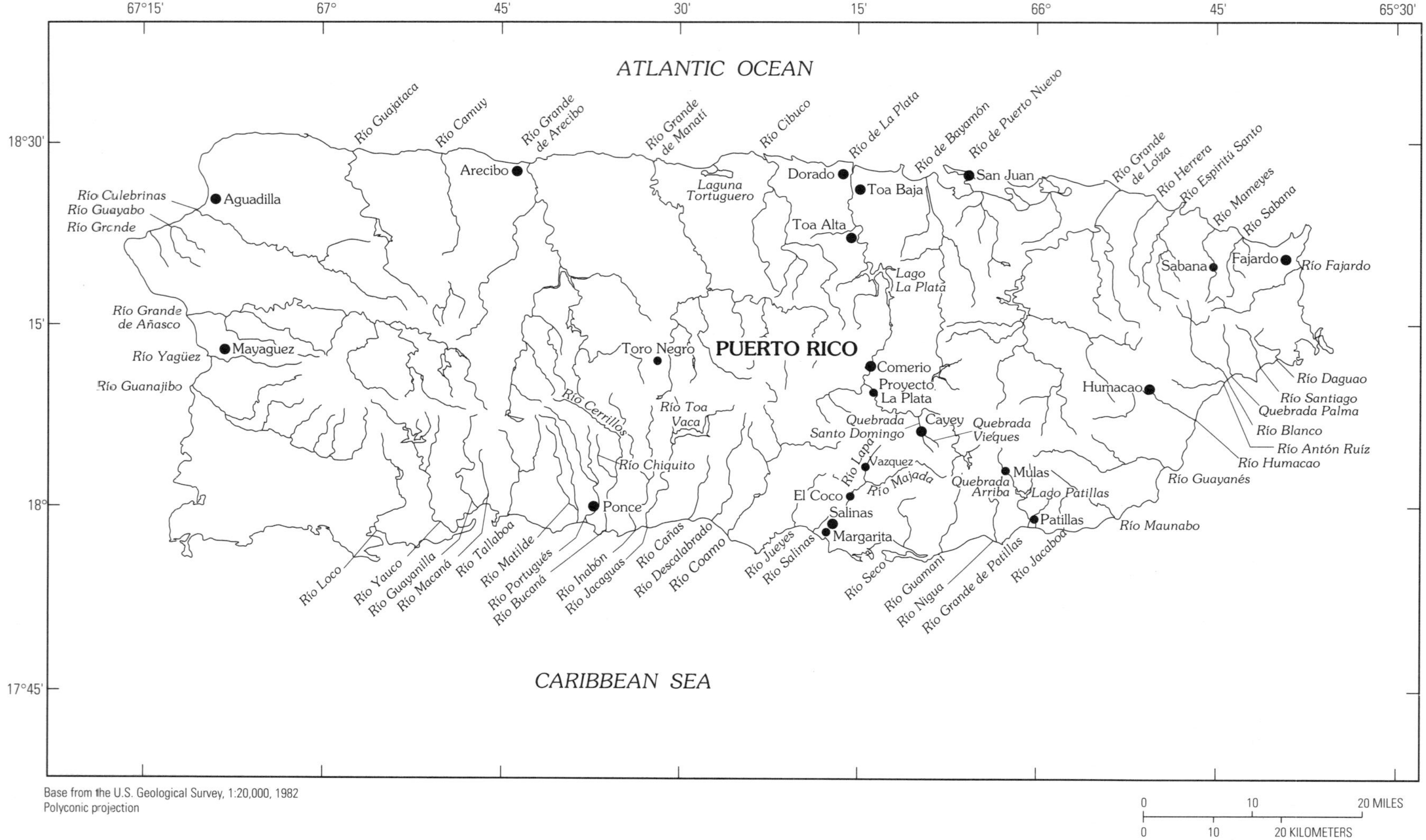

Figure 25. Location of municipalities in Puerto Rico where flooding was most severe during January 5–6, 1992, and other locations mentioned in this article.

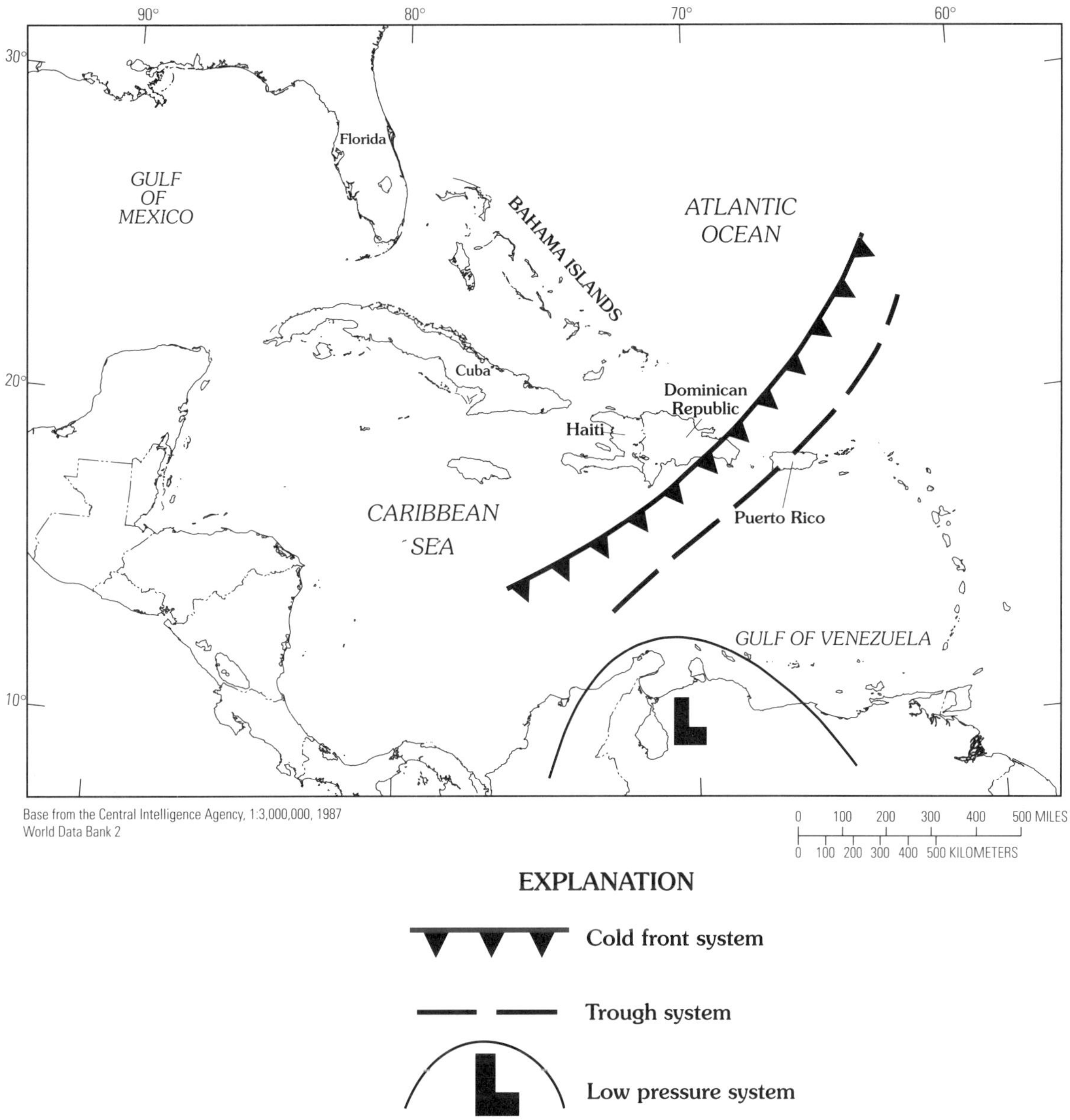

Figure 26. Location of weather systems that produced intense rain over Puerto Rico during evening of January 5, 1992.

worst flooding since the October 7, 1985, flood. Floodwaters inundated downtown areas and neighborhoods, washing away homes, businesses, vehicles, bridges, and portions of roads. Twenty persons were drowned as floodwaters swept away their vehicles as they attempted to cross flooded bridges and roads. The most affected municipalities were Patillas, Cayey, Comerío, Toa Baja, Dorado, Salinas, and Ponce. Many landslides occurred on the slopes of the island's mountainous interior. About 15 main thoroughfares were obstructed or destroyed by floodwaters and landslides.

At Patillas, the floodwaters of the Río Grande de Patillas and its tributaries destroyed 15 houses and severely damaged another 141 houses. Most of these houses were affected by the Quebrada Arriba, a major tributary of the Río Grande de Patillas upstream from the Lago Patillas. The residents of Barrio Mulas were isolated when the Río Grande de Patillas destroyed the bridge on Highway 754. The streamflow-gaging station on the Río Grande de Patillas was extensively damaged. Several older citizens living along the Quebrada Arriba and the Río Grande de Patillas indicated that this flood was the largest they had ever experienced.

Floods along the Río de La Plata and its upper reach tributaries were particularly noteworthy. At Cayey, the Quebrada Santo Domingo and the

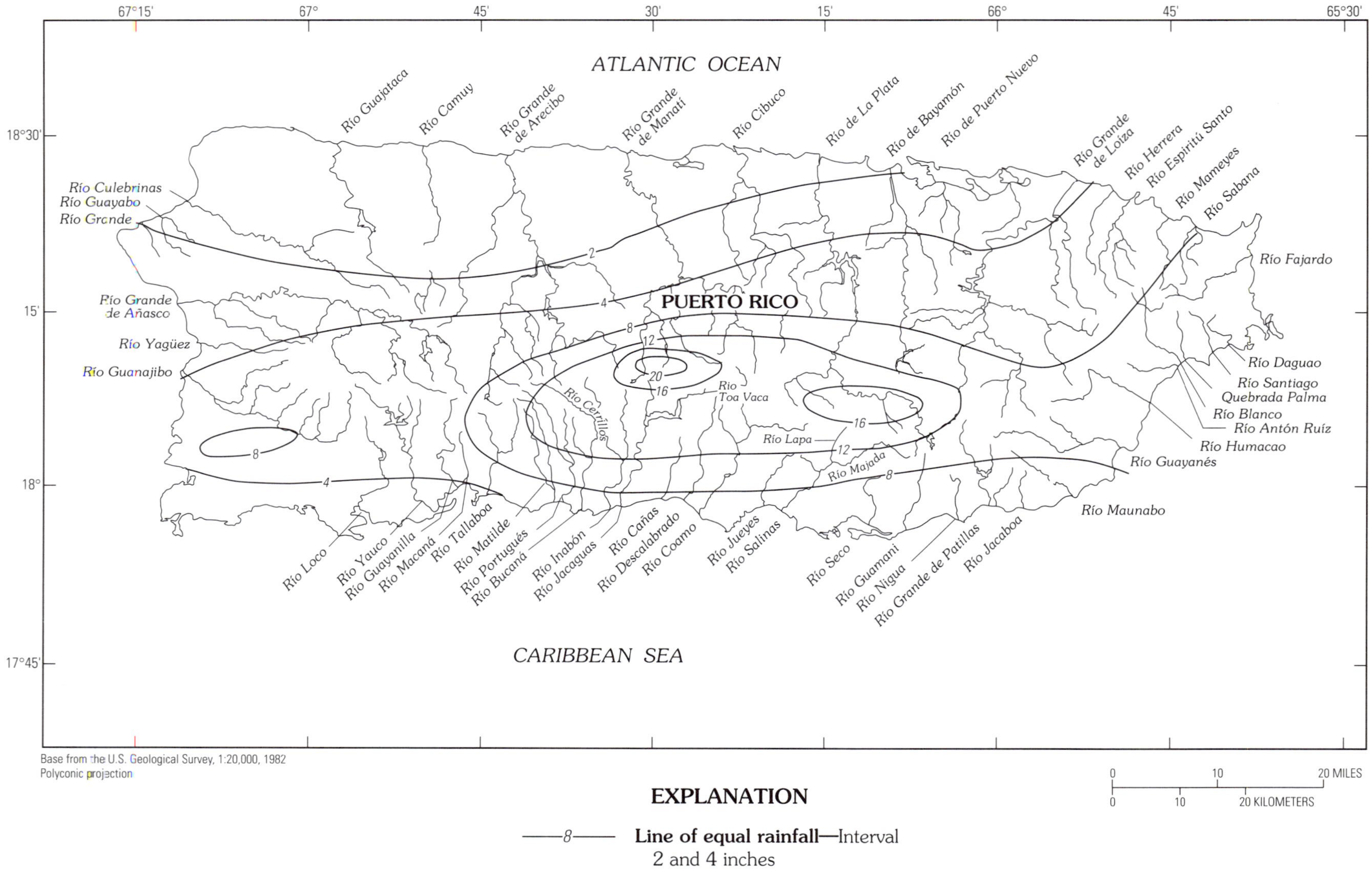

Figure 27. Cumulative rainfall over Puerto Rico during January 5–6, 1992 (source: rainfall data compiled by National Weather Service).

Table 16. Daily rainfall reported for January 5–6, 1992, at selected National Weather Service-National Oceanic and Atmospheric Administration (NWS-NOAA) stations throughout Puerto Rico (data from National Oceanic and Atmospheric Administration, 1992)

[Data recorded at the NWS–NOAA stations are reported for a 24-hour period starting at 0800 of one day and ending at 0800 of the date reported. --, daily rainfall value not available]

Climatic subdivision and name of NWS-NOAA station	Rainfall (inches)		
	January 5	January 6	Total
WESTERN INTERIOR			
Adjuntas substation	0.31	6.63	6.94
Arecibo observatory	0	1.70	1.70
Cerro Maravilla	2.00	12.00	14.00
Coloso	0	1.40	1.40
Corozal substation	0	6.10	6.10
Dos Bocas	0	1.22	1.22
Jayuya	.14	7.20	7.34
Las Marías	.02	1.14	1.16
Maricao 2 SSW	--	--	6.40
Morovis 1 N	0	2.96	2.96
Negro-Corozal	0	8.00	8.00
San Sebastián 2 WNW	0	1.37	1.37
Toro Negro Forest	.75	20.30	21.05
Utuado	0	3.55	3.55
EASTERN INTERIOR			
Aibonito	0.30	5.00	5.30
Caguas 1 W	.05	6.95	7.00
Cayey 1 E	.10	19.56	19.66
Cidra 1 E	.05	10.36	10.41
Gurabo substation	.07	7.11	7.18
Juncos 1 NNE	0	6.57	6.57
Paraiso	.40	8.20	8.60
Pico del Este	1.60	10.10	11.70
Río Blanco Lower	5.14	2.11	7.25
San Lorenzo 3 S	.54	5.60	6.14
NORTHERN SLOPES			
Canóvanas	0.15	4.95	5.10
Fajardo	0	8.00	8.00
Isabela substation	.02	1.92	1.94
La Muda-Caguas	0	4.93	4.93
Manatí 2 E	0	1.43	1.43
Rincón Powerplant	--	--	2.25
Trujillo Alto 2 SSW	.05	4.01	4.06

Climatic subdivision and name of NWS-NOAA station	Rainfall (inches)		
	January 5	January 6	Total
SOUTHERN SLOPES			
Benavente-Hormigueros	0	3.05	3.05
Coamo 2 SW	0	9.00	9.00
Corral Viejo	.32	12.15	12.47
Guayama	0	4.82	4.82
Humacao 2 SSE	0	12.00	12.00
Juana Díaz Camp	0	10.15	10.15
Maunabo	0	7.60	7.60
Patillas	0	2.10	2.10
Puerto Real	0	4.80	4.80
Roosevelt Roads	4.65	1.64	6.29
Sabana Grande 2 ENE	0	5.00	5.00
Yabucoa 1 NNE	0	10.90	10.90
NORTH COAST			
Borinquen Airport	0	0.40	0.40
Quebradillas	--	--	3.66
Río Piedras AES	.05	2.00	2.05
Río Piedras Heights	0	4.60	4.60
San Juan WSFO	0	2.24	2.24
Toa Baja 1 SSW	0	3.30	3.30
SOUTH COAST			
Central Aguirre	0	2.05	2.05
Central San Francisco	.02	2.92	2.94
Ensenada	.05	4.16	4.21
Lajas substation	.01	10.00	10.01
Ponce 4 E	.20	8.30	8.50
Ponce City	1.66	5.49	7.15
Santa Isabel 2 ENE	0	5.00	5.00

Quebrada Vieques rose quickly, flooding homes and businesses in downtown Cayey. Floodwaters washed away 10 municipal vehicles, including 3 new garbage trucks worth $55,000 each. At the Proyecto La Plata streamflow-gaging station (site 2, table 17), the river crested at a stage of 36.39 feet, exceeding the previous maximum of record by about 4 feet. The bridge on Highway 173, downstream from the streamflow-gaging station, was washed away by the floodwaters. The Río de La Plata flooded homes and businesses in the Comerío downtown area and partially destroyed the bridge on Highway 775, a newly built sewage pipeline, and the streamflow-gaging station at Comerío (site 3, table 17).

Farther downstream at Lago La Plata, the water level increased 19 feet in 6 hours. The streamflow-gaging station at Highway 2 (site 6, table 17) recorded a maximum discharge of 110,000 cubic feet per second, which exceeded the maximum previously recorded discharge of 95,000 cubic feet per second, produced during the flood of September 6, 1960 (Barnes and Bogart, 1961). Downtown Toa Baja and nearby low-lying areas were hard hit by the floodwaters of the Río de La Plata. In many homes the floodwaters reached a depth of 7 feet. Water hyacinths carried from the Lago La Plata by the Río de La Plata proved to be a menace to bridges and culverts. At Dorado, water hyacinths and bamboo blocked the opening of the river bridge, forcing floodwaters over and around the bridge, resulting in severe damage to the bridge deck, abutments, and piers.

On the south coast, the towns of Salinas and Ponce were the most affected by the January 5–6, 1992, flood. At Salinas, the Río Lapa flooded the rural area of Vázquez, and the Río Nigua flooded the communities of El Coco and Margarita, destroying a bridge and 10 houses and damaging 164 other houses. At Ponce, an extensive area near the confluence of the Río Chiquito and the Río Portugués was inundated when floodwaters from both streams combined.

The northeastern area of Puerto Rico also experienced notable floods when the Río Sabana and the Río Fajardo overflowed their banks. At the streamflow-gaging station on the Río Sabana at Sabana (site 18, table 17) the maximum stage exceeded the previous maximum of record of 19.35 feet by 0.39 feet, while at the Río Fajardo near Fajardo streamflow-gaging station (site 19, table 17) the maximum stage almost equalled the maximum of record of 20.00 feet. The Río Fajardo gage house was washed away and found destroyed about 2 miles downstream.

The rain, floods, and landslides caused extensive damage to private and public property. Total damages were estimated at $155 million by the Federal Emergency Management Administration. The flood of January 5–6, 1992, resulted in 23 deaths, 20 of which involved motor vehicles. There also were 167 persons injured; 17 required hospitalization (Torres-Sierra, 1995).

Emergency housing, medical attention, food, and clothing were provided for thousands of people as 78 houses were destroyed and 4,241 others damaged. Damage to homes, including those destroyed, was estimated at $20.5 million. As a result of the flooding about 550 persons were left homeless. Damage to businesses was estimated at $11.5 million. Damages occurred mostly in the towns of Toa Baja, Cayey, and Patillas (Torres-Sierra, 1995).

The Puerto Rico Department of Agriculture reported considerable damage, with losses of approximately $5.0 million. The damage to vegetable crops alone was $1.6 million. The plantain and banana crops suffered a loss of $0.5 million. Damage to the agricultural infrastructure in Puerto Rico was about $1.0 million (Torres-Sierra, 1995).

Public facilities, roads, and bridges sustained more than $24 million of damage. At least 5 bridges were destroyed, and 20 were damaged. Many water-filtration plants, sewage-treatment plants, pumping stations, and aqueduct systems were severely damaged (Torres-Sierra, 1995).

SUMMARY OF FLOOD STAGES AND DISCHARGES

The U.S. Geological Survey collected maximum stage and discharge information from its network of streamflow-gaging stations throughout Puerto Rico during the January 5–6, 1992, flood. High-water marks were surveyed shortly after the flood to determine the maximum stages and discharges by indirect methods at sites where recording instruments failed or were damaged during the flood. The streamflow-gaging stations on the Río Grande de Patillas near Patillas and on the Río Fajardo (sites 20 and 19, fig. 28) near Fajardo were washed away by the floodwaters. Twelve others were rendered inoperable because of the accumulation of heavy debris and partial washouts. Discharges at the Río Grande de Patillas near Patillas and at the Río de La Plata at Proyecto La Plata were calculated using

Table 17. Maximum stages and discharges prior to and during January 5–6, 1992, at selected U.S. Geological Survey streamflow-gaging stations in Puerto Rico

[mi^2, square miles; ft, feet above an arbitrary datum; ft^3/s, cubic feet per second; --, not determined or not applicable; >, greater than. Source: Recurrence intervals calculated from U.S. Geological Survey data. Other data from U.S. Geological Survey reports or data bases]

Site no. (fig. 28)	Station no.	Stream and place of determination	Drainage area (mi^2)	Maximum prior to January 1992				Maximum during January 5–6, 1992			
				Period	Year	Stage (ft)	Discharge (ft^3/s)	Day	Stage (ft)	Discharge (ft^3/s)	Discharge recurrence interval (years)
1	50038320	Río Cibuco below Corozal, PR	15.1	1969–92	1979	19.80	13,600	5	19.32	11,900	6
2	50043000	Río de La Plata at Proyecto La Plata, PR	54.8	1960–92	1961	32.20	59,600	5	36.39	73,600	40
3	50043800	Río de La Plata at Comerío, PR	109	1988–92	1989	17.36	32,000	5	29.22	127,000	80
4	50044830	Río Guadiana at Guadiana, PR	9.19	1990–92	1991	11.54	4,600	5	13.36	6,670	10
5	50045010	Río de La Plata below La Plata Dam, PR	173	1989–92	1989	22.98	48,800	5	34.76	127,000	25
6	50046000	Río de La Plata at Highway 2 near Toa Alta, PR	208	1960–92	1960	[1]36.35	95,500	5	26.39	118,000	25
7	50050900	Río Grande de Loíza at Quebrada Arenas, PR	6.00	1977–92	1983	14.78	11,700	5	17.52	18,200	60
8	50051150	Quebrada Blanca at El Jagual, PR	3.25	1984–92	1985	14.58	7,400	5	14.30	7,180	12
9	50051800	Río Grande de Loíza at Highway 183 near San Lorenzo, PR	25.0	1990–92	1990	16.60	7,510	5	31.37	40,700	70
10	50053025	Río Turabo above Borinquen, PR	7.14	1990–92	1990	--	[2]2,400	5	21.07	12,000	35
11	50055000	Río Grande de Loíza at Caguas, PR	89.8	1960–92	1960	31.17	71,500	5	24.32	43,300	6
12	50055100	Río Cagüitas near Aguas Buenas, PR	5.30	1990–92	1990	15.24	1,460	5	15.93	1,760	5
13	50055225	Río Cagüitas at Villa Blanca at Caguas, PR	16.9	1990–92	1991	13.16	2,060	5	19.91	13,400	20
14	50055390	Río Bairoa at Bairoa, PR	5.08	1991–92	1991	10.24	742	5	12.32	1,580	4
15	50058350	Río Cañas at Río Cañas, PR	7.53	1990–92	1990	20.55	3,830	5	20.15	3,580	7
16	50059050	Río Grande de Loíza below Damsite, PR	209	1987–92	1987	39.57	124,300	5	33.79	79,500	7
17	50065500	Río Mameyes near Sabana, PR	6.88	1969–73, 1983–92	1989	13.19	20,500	5	10.27	10,300	2
18	50067000	Río Sabana at Sabana, PR	3.96	1980–92	1983	19.35	9,010	5	19.74	9,600	13
19	50071000	Río Fajardo near Fajardo, PR	14.9	1962–92	1989	20.00	23,500	5	19.88	23,300	15
20	50092000	Río Grande de Patillas near Patillas, PR	18.3	1966–92	1975	12.45	14,800	5	--	30,900	>100
21	50100200	Río Lapa near Rabo del Buey, PR	9.92	1989–92	1990	10.23	1,750	5	17.82	15,700	60
22	50100450	Río Majada at La Plena, PR	16.7	1989–92	1990	10.34	3,820	5	17.19	15,200	15
23	50110900	Río Toa Vaca above Lago Toa Vaca, PR	7.64	1989–92	1989	9.62	3,740	5	13.24	8,700	25
24	50111500	Río Jacaguas at Juana Díaz, PR	49.8	1984–92	1985	29.42	40,000	5	22.81	20,500	4

Table 17. Maximum stages and discharges prior to and during January 5–6, 1992, at selected U.S. Geological Survey streamflow-gaging stations in Puerto Rico—Continued

				Maximum prior to January 1992				Maximum during January 5–6, 1992			
Site no. (fig. 28)	Station no.	Stream and place of determination	Drainage area (mi^2)	Period	Year	Stage (ft)	Discharge (ft^3/s)	Day	Stage (ft)	Discharge (ft^3/s)	Discharge recurrence interval (years)
25	50113800	Río Cerrillos above Lago Cerrillos near Ponce, PR	15.4	1989–92	1990	--	[3]2,500	5	9.65	8,140	4

[1]Datum then in use.
[2]Discharge estimated based on a mean daily discharge correlation with station 50055000.
[3]Discharge estimated based on a mean daily discharge correlation with station 50115000.

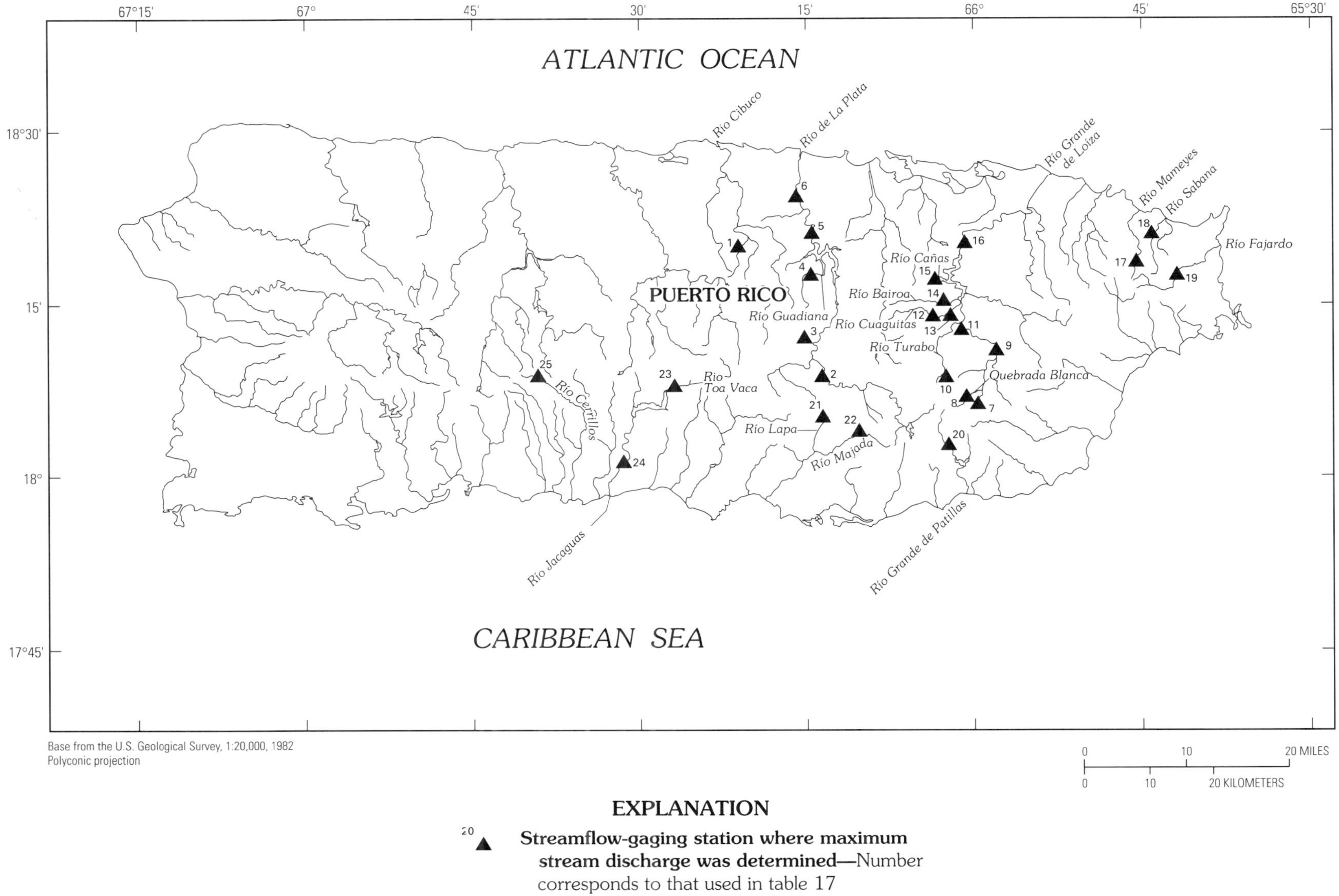

Figure 28. Location of selected streamflow-gaging stations that recorded flood data in most affected areas during January 5–6, 1992, flood.

indirect methods and were found to exceed the historical maximum discharges.

Hydrologic information indicates that floods of moderate to severe intensity occurred in the basins of the Río de La Plata, the Río Grande de Loíza, the Río Sabana, the Río Fajardo, the Río Grande de Patillas, the Río Lapa, the Río Majada, the Río Toa Vaca, and the Río Cerrillos (fig. 28). Flood stages, discharges, recurrence intervals, and other pertinent information for selected streamflow-gaging stations within these basins are summarized in table 17.

Previous maximum discharges were exceeded at 18 streamflow-gaging stations, and 11 streams had maximum flows in excess of 1,000 cubic feet per second per square mile. The largest flows during the flood were recorded at the streamflow-gaging stations along the Río de La Plata. The Río de La Plata at Comerío had a maximum discharge of 127,000 cubic feet per second (1,170 cubic feet per second per square mile), the Río de La Plata below La Plata Dam had a maximum discharge of 127,000 cubic feet per second (734 cubic feet per second per square mile), and the Río de La Plata at Highway 2 near Toa Alta had a maximum flow of 118,000 cubic feet per second (590 cubic feet per second per square mile). The streamflow-gaging station with the largest flow per square mile during the flood was the Río Grande de Loíza at Quebrada Arenas station with 3,030 cubic feet per second per square mile (18,200 cubic feet per second). Discharge hydrographs for selected streamflow-gaging stations are shown in figure 29.

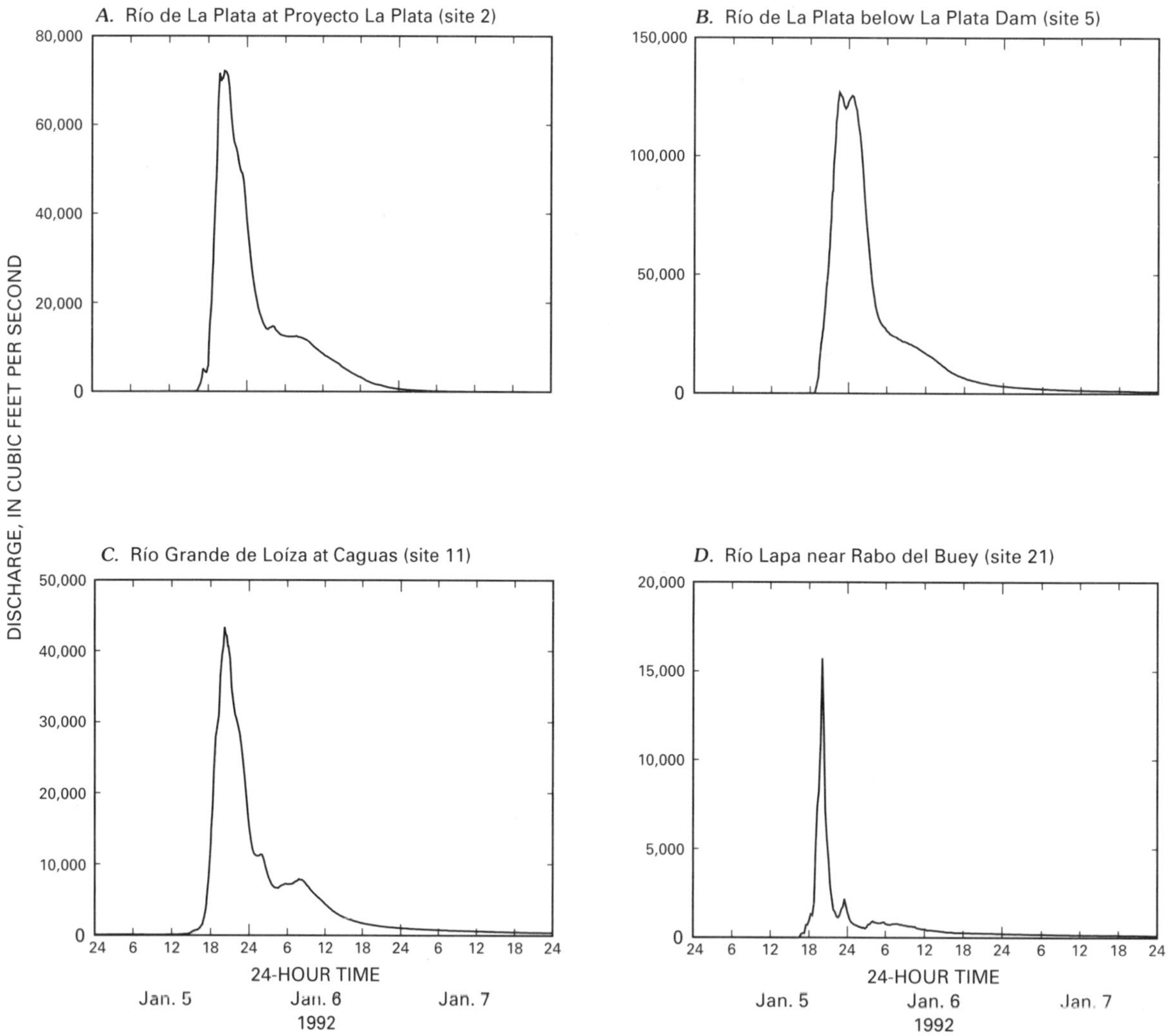

Figure 29. Discharge hydrographs for selected streamflow-gaging stations during January 5–7, 1992.

REFERENCES

Barnes, H.H., Jr., and Bogart, D.B., 1961, Floods of September 6, 1960, in eastern Puerto Rico: U.S. Geological Survey Circular 451, 13 p.

López, M.A., Colón-Dieppz, Eloy, and Cobb, E.D., 1979, Floods in Puerto Rico, magnitude and frequency: U.S. Geological Survey Water-Resources Investigations Report 78–141, 68 p.

National Oceanic and Atmospheric Administration, 1992, National disaster survey report—Puerto Rico flash floods, January 5–6, 1992: National Oceanic and Atmospheric Administration, National Weather Service, 68 p.

Torres-Sierra, Heriberto, 1995, Flood of January 5–6, 1992, in Puerto Rico: U.S. Geological Survey Open-File Report 95–374, 13 p.

U.S. Interagency Advisory Committee on Water Data, 1982, Guidelines for determining flood flow frequency: U.S. Geological Survey, Hydrology Subcommittee Bulletin 17B, 183 p.

March 11, 1992, Ice-Jam Flood in Montpelier, Vermont

By Jon C. Denner *and* Robert O. Brown

The ice-jam flood on March 11, 1992, caused the largest inundation of Vermont's State capital since the flood of 1927. Unlike the 1927 flood, which occurred because of intense rainfall and excessive runoff, the high water levels of March 1992 were generated by ice-induced backwater of the Winooski River. Although the areal extent of an ice-jam flood is small, the consequences to a community may be severe. That was the case in Montpelier when the Winooski River overflowed its banks and inundated the downtown area.

PRE-BREAKUP CONDITIONS IN THE WINOOSKI RIVER BASIN

In mid-February, the Winooski River and North Branch Winooski River in the Montpelier area were covered with solid ice and snow. Streamflow on unregulated streams was low because of consistently cold temperatures. On February 18, a discharge of 232 cubic feet per second was measured in the reach upstream from the streamflow-gaging station on the Winooski River at Montpelier (fig. 30). The discharge measurement was made through ice cover using a current meter. Backwater attributed to ice effect, as determined by the discharge measurement, was 1.21 feet. The average ice thickness in the cross section was about 1.7 feet. After a brief thaw and light rains on February 19, temperatures remained seasonably cold through the first week of March.

A moderating trend began as a storm system developed over the mid-Atlantic Coast and moved in a northeasterly direction along a cold front bringing rain and above-freezing temperatures to Vermont. When ice-covered rivers, such as the Winooski River, are subject to mild weather, the following two processes are likely to occur: (1) increased runoff by snowmelt and rains can result in increased uplift and frictional forces applied to the ice cover, and (2) increased heat input to the ice can reduce its strength (Beltaos and others, 1990, p. 39).

Although snow depth had diminished to 2–4 inches in the Montpelier area, a considerable snowpack remained in the higher altitudes on March 10. Snow-course data, collected on March 3 at an elevation of 1,300 feet in the headwaters of the Winooski River Basin, indicated an average snow depth of 18 inches and a water depth equivalent of 5 inches (unpublished data on file with the Bow, New Hampshire, office of the U.S. Geological Survey).

Late in the evening on March 10, a low-intensity rainfall commenced over the Winooski River Basin and continued overnight. By 0600 on March 11, the storm dropped about 0.60 inch of rain. An additional 0.20 inch accumulated throughout the day on March 11. About 0650, local police reported an ice jam near the Washington County Railroad Bridge, west of Pioneer Street Bridge (fig. 30). The jam subsequently released, and the ice and water surged downstream. At 0700 another police report described an ice jam near the Bailey Avenue Bridge, and flooding was reported on State Street (Times Argus, March 15, 1992). Within less than 1 hour, downtown Montpelier was inundated to a depth of 2 to 5 feet. A formal state of emergency was declared by the Governor of Vermont at 0900.

Ice from the 1.5-mile section of the Winooski River downstream from the confluence with Stevens Branch probably caused the ice jam in Montpelier (Federal Emergency Management Agency, 1992). A surge of water released from the upstream ice jam may have triggered ice breakup along the Winooski River in Montpelier. The ice run stalled at a bend in the channel about 300 feet downstream from the Bailey Avenue Bridge. Stage data recorded at the streamflow-gaging station downstream from the major ice jam indicated substantial backwater in this reach prior to the formation of the jam (fig. 31). Another ice jam downstream from the streamflow-gaging station, possibly in Middlesex, may have caused the backwater. Channels affected by backwater typically have a marked reduction in water-surface slope and a decrease in flow velocities. These factors may have contributed to the stalling of the ice run and its subsequent ice-jam formation on the Winooski River.

After the initial ice jam formed near the Bailey Avenue Bridge, additional ice fragments continued to arrive from upstream. The jam thickened primarily because of underturning of ice blocks. Rising water levels caused large ice fragments to become entangled with the low steel of the Bailey Avenue and Taylor

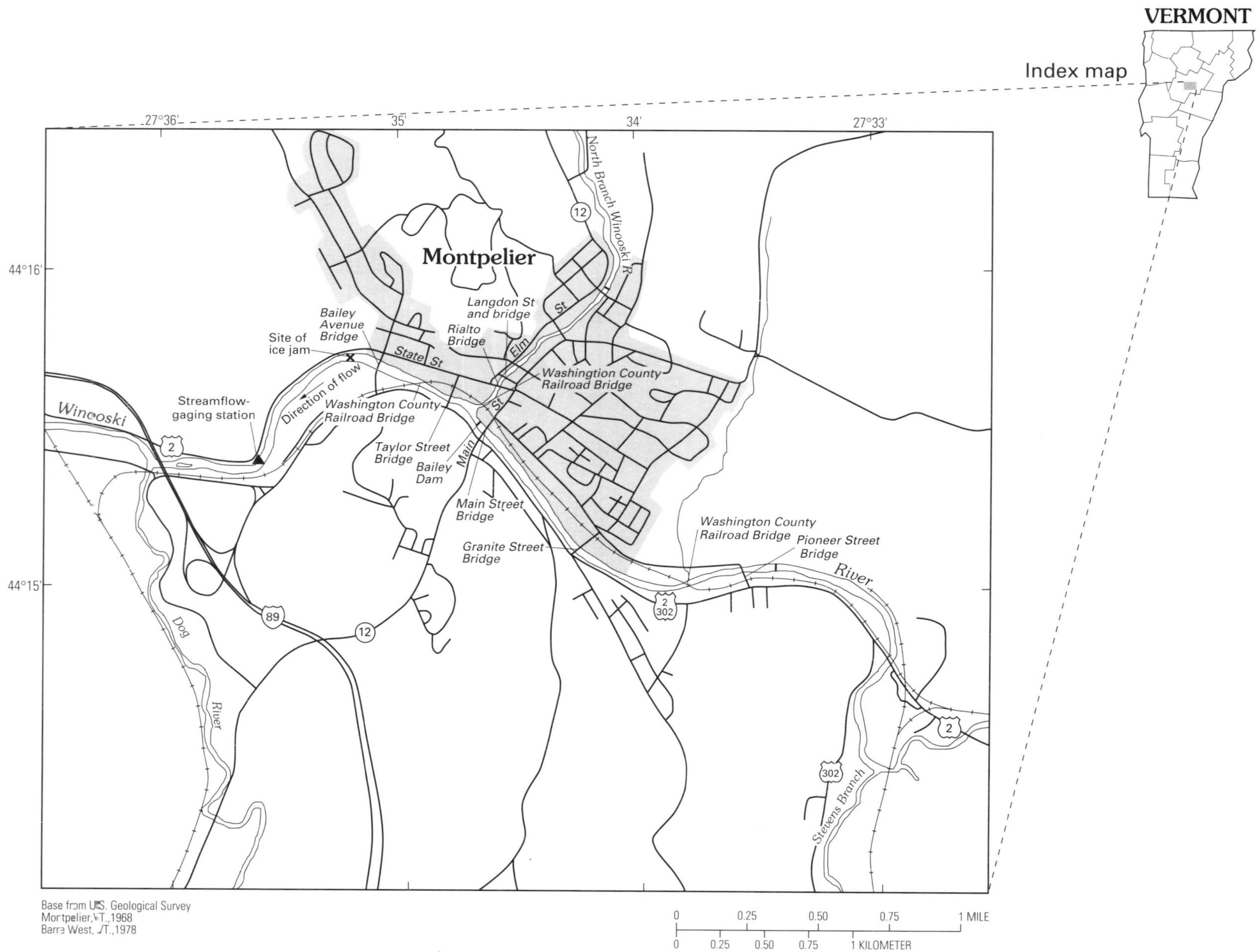

Figure 30. Location of the Winooski River, North Branch Winooski River, and Stevens Branch in Montpelier, Vermont.

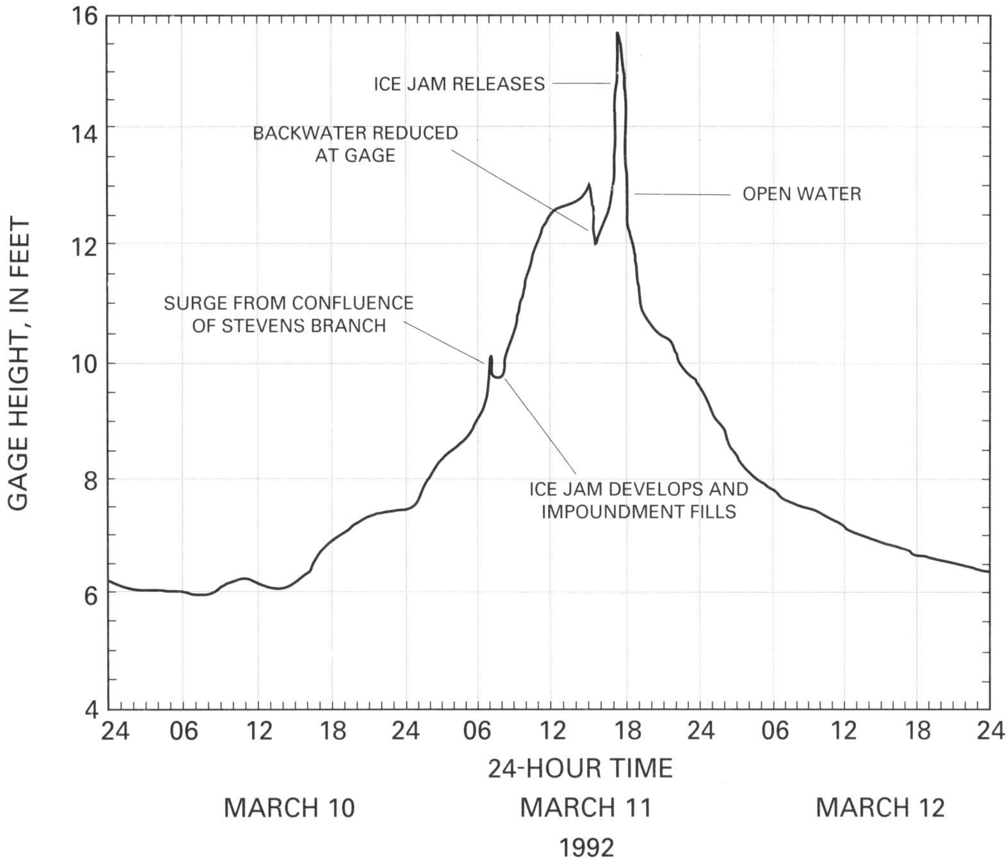

Figure 31. Gage height, Winooski River at Montpelier, Vermont, March 10–12, 1992 (source: U.S. Geological Survey files, Bow, N.H.).

Street Bridges. Ice was stacked against the upstream side of the Washington County Railroad Bridge, but about 1.5 feet separated the Main Street Bridge from the ice pack. Bridges and other structures can contribute support to ice jams. The ice jam extended about 1 mile along the Winooski River, its toe (downstream end) was located near the Bailey Avenue Bridge, and its head (upstream end) was upstream from the Granite Street Bridge (Federal Emergency Management Agency, 1992).

MONTPELIER FLOODING

Flooding of the downtown area was rapid. The ice jam formed about 0700, and by 0800, Main, Elm, and State Streets were inundated. Most office workers, merchants, and residents had little warning of the impending flood. Some waded through thigh-deep water in parking lots only to find their vehicles stranded. Hundreds of people were evacuated by local and State police, fire departments, and private citizens using an array of small watercraft. Fortunately, the flood occurred during daylight; otherwise, the rescue operations would have been more difficult.

Flooding was observed first on State Street in a low-lying area near the confluence of the North Branch and Winooski River. The high water levels on the Winooski River, resulting from the ice jam, created backwater on the North Branch Winooski River. The North Branch overflowed sending floodwater onto State Street. Ice cover along the North Branch was uplifted but remained intact as the water level increased. Downstream movement of ice on the North Branch was prevented because the ice pack on the Winooski River blocked its outlet. Furthermore, ice cover on the North Branch lodged against the Langdon, Rialto, and Washington County Railroad Bridges.

Flow on the North Branch Winooski River is regulated by the Wrightsville Detention Reservoir (site 3, fig. 32), located 4.2 miles upstream from the confluence with the Winooski River. The earthfill reservoir, constructed for flood-control storage, was completed in

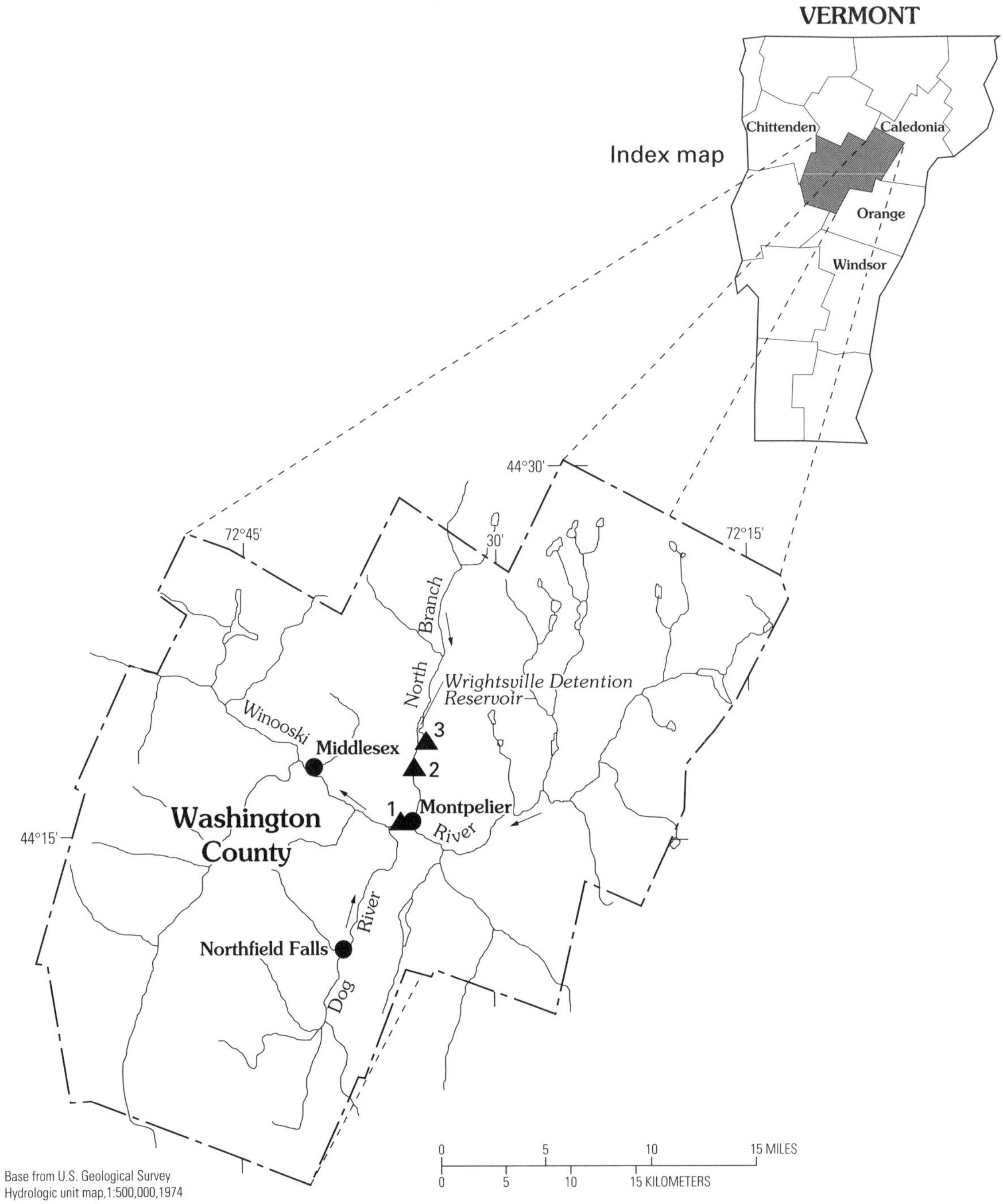

EXPLANATION

▲1 Streamflow-gaging station and site number

→ Direction of flow

Figure 32. Location of U.S. Geological Survey streamflow-gaging stations in vicinity of Montpelier, Vermont.

1935. The effectiveness of the structure was documented during the 1936 flood; the reservoir contributed to reducing the flood crest and potential for flood damage in Montpelier (Denner, 1991, p. 539). As originally designed, the reservoir was an uncontrolled, self-regulating detention basin. Outflow was dependent on the capacity of the outlet opening near the base of the dam. Since 1985, a hydroelectric-generating station has operated at the reservoir outlet. When the reservoir stage is below 635 feet, discharge is through a conduit leading to the generating units. Water levels higher than 635 feet flow out an uncontrolled conduit; discharge at high stages, then, is a direct function of the reservoir stage.

The streamflow-gaging station on the North Branch Winooski River (site 2, fig. 32) is 0.8 mile downstream from Wrightsville Detention Reservoir (site 3, fig. 32). Recorded data showed discharge on March 11, from midnight to 0500, at about the minimum-flow rate of 30 cubic feet per second (fig. 33). At 0600, discharge increased to 215 cubic feet per second as a result of powerplant operation. Powerplant operation continued until shutdown at about 1015. Meanwhile, the water level at Wrightsville Detention Reservoir was increasing (fig. 34). Outflow to the uncontrolled conduit began at about 1100 when the reservoir stage exceeded 635 feet. Discharge on the North Branch increased during the afternoon; a discharge of 563 cubic feet per second was recorded at 1715, approximately when the ice jam released on the Winooski River. Discharge probably was slightly higher at the mouth because of additional inflow from streams between the streamflow-gaging station and Montpelier. A maximum discharge of 842 cubic feet per second occurred at the North Branch Winooski River streamflow-gaging station on March 12, at 0530.

Discharge on the North Branch Winooski River alone was too small to account for the rapid flooding of Montpelier. Estimated streamflow on the Winnooski River was about 3,000 cubic feet per second during the period of inundation; thus, the Winooski River was

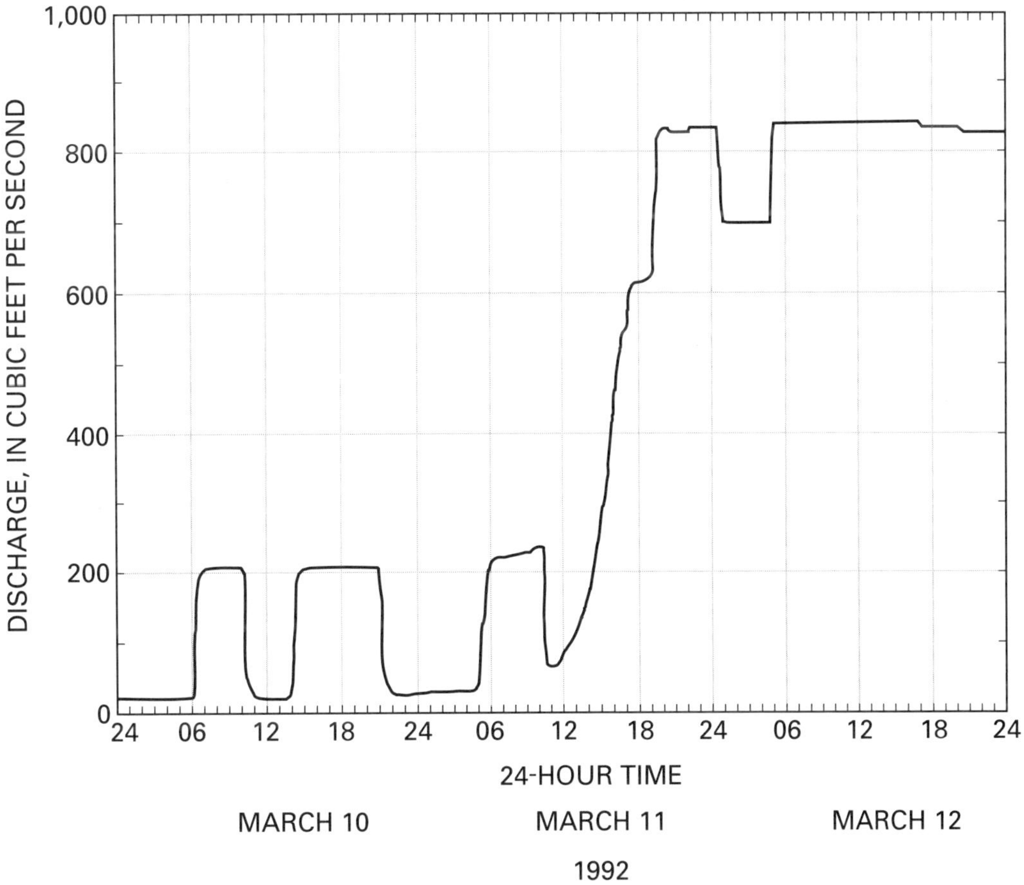

Figure 33. Discharge, North Branch Winooski River at Wrightsville, Vermont, March 10–12, 1992 (source: U.S. Geological Survey files, Bow, N.H.).

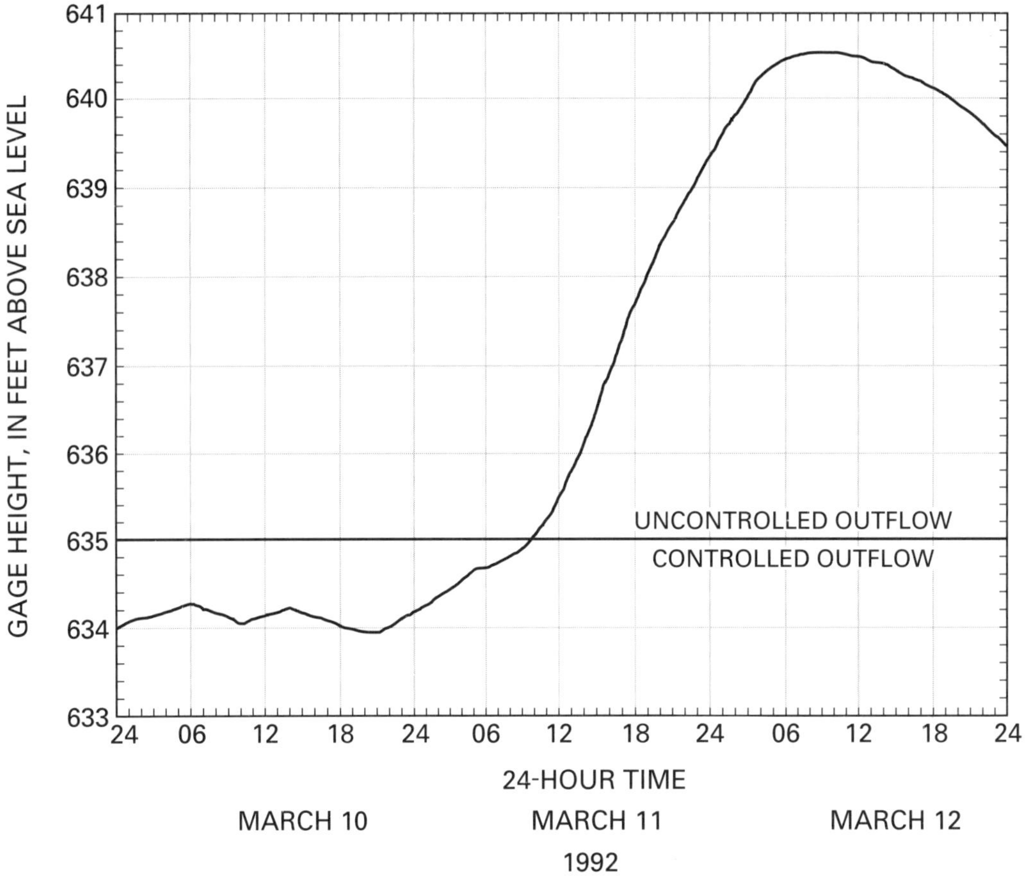

Figure 34. Gage height, Wrightsville Detention Reservoir at Wrightsville, Vermont, March 10–12, 1992 (source: U.S. Geological Survey files, N.H.).

most likely the major source of floodwater in the downtown area.

HIGH WATER LEVELS RESULTING FROM ICE JAM

A major consequence to communities during ice-jam flooding is the high water levels attained behind the ice dam. An important constraint to the size of an ice jam and thus the maximum water level is flow diversion around the ice jam (Beltaos and others, 1990, p. 77). After the Winooski River and North Branch Winooski River overflowed onto the flood plain, water was free to move around the ice jam. The ice dam at the Bailey Avenue Bridge was bypassed on the north bank. Lower State Street, in effect, became a spillway for the ice dam. Overbank flow on the flood plain reconverged with the main channel about 200 feet downstream from the Bailey Avenue Bridge (fig. 30).

The water-surface elevation in the impoundment was relatively stable during most of the flood. Water levels, based on onsite inspections, ranged from 523.4 feet at 1055 to 524.3 feet at 1630. A maximum water level of 525.1 feet (0.9 foot below the current Federal Emergency Management Agency 100-year flood elevation) was determined from high-water marks found at the Federal building on State Street. The maximum elevation probably occurred during the surge of ice and water at about 1700. In comparison, the 1927 flood crest of 533.9 feet exceeded the 100-year flood elevation by 7.9 feet.

ICE-JAM RELEASE

Between 1430 and 1500, a section of the toe of the ice jam dislodged as a result of high flows and intervention by construction equipment; a crane operating on the left bank dropped a steel beam on the ice fragments while excavators pushed blocks downstream. The ice jam redeveloped, however, when upstream ice fragments moved downstream. Ultimately, the ice jam was pushed out by a major surge of ice and water. The surge originated in the steep section of the channel between the Stevens Branch confluence and the upstream dam at Levesque Station. An ice jam at the confluence broke up at about 1615. The flood surge traveled downstream to the Bailey Avenue Bridge, causing breakup of the ice jam there at about 1710.

As the jam moved out, ice damaged the right truss of the Washington County Railroad Bridge, thereby causing the bridge to fail. The bridge was driven off its center pier by a large mass of ice and snow that had accumulated over the winter as snow was removed from the city streets and dumped into the Winooski River. The mass remained lodged against the bridge after the water receded (Federal Emergency Management Agency, 1992).

When ice jams release, water in storage discharges, and sudden increases in water levels and velocities are generated downstream. The maximum gage height and discharge recorded at the Winooski River streamflow-gaging station downstream from the Bailey Avenue Bridge were 15.71 feet and 11,500 cubic feet per second, respectively, at 1730. The maximum discharge had a recurrence interval of about 10 years (10-percent chance in a given year). The 1992 flood maximum was much smaller than the 1927 flood maximum. A maximum gage height of 27.1 feet and a maximum discharge of 57,000 cubic feet per second occurred during the 1927 flood; the recurrence interval was greater than 100 years (1-percent chance in a given year).

The duration of flooding in the downtown area was about 11 hours. After the ice jam released, floodwater quickly receded from the streets of Montpelier. The surge of ice and water traveled downstream causing overbank flooding in the fields between Montpelier and Middlesex. The arrival of sharply colder weather later in the day on March 11 reduced runoff and thus lessened the potential for more flooding in the Montpelier area.

Despite backwater from ice-affected streamflows at the streamflow-gaging station on the Winooski River, recorded stage data provided valuable information on the ice-jam flood. The gage-height plot (fig. 31) illustrates streamflow trends. Stage and discharge increased during the early morning on March 11. The sharp spike at 0715 shows the surge following the release of the ice jam upstream from Montpelier. Discharge decreased after the ice jam formed downstream from the Bailey Avenue Bridge. Between 0730 and 0800, the flow by the streamflow-gaging station was relatively stable, probably because of water being retained by the ice dam. The upward trend after the inundation of the downtown resulted from flow bypassing the jam and increased runoff in the basin. About 1500, heavy equipment dislodged some ice in mid-channel. The rapid drop during this period probably was not caused by a reduction in discharge but instead may indicate a reduction in backwater. By 1600, the ice jam redeveloped, and flows continued to increase. The flood wave that ultimately caused the ice jam to fail arrived at about 1700. High flows, related to the dynamic breakup, are represented by the sharp upward trend. The recession after the maximum at 1730 shows decreasing discharges under mostly open-water conditions.

FLOOD DAMAGE

The downtown commercial district of Montpelier received severe damage from the flooding. Water levels were 2 to 3 feet above the main-level floors in many businesses. Flood damage consisted primarily of destroyed inventory, machinery and equipment, and records and utilities housed in basements and on main-level floors. Buildings, streets, sidewalks, and a railroad bridge were damaged (Federal Emergency Management Agency, 1992).

Martial law was declared in Montpelier on March 11, and only business owners and displaced residents were allowed in the city. Cleanup efforts were hampered by extremely cold weather and light snows. The first priority for many property owners was to pump out basements and to repair heating and utility units because subfreezing temperatures could have further damaged properties. More than 200 automobiles were damaged or totally destroyed by floodwater. Some vehicles, not towed to heated garages, sustained more damage because engine blocks and transmission cases were cracked by expanding ice.

Petroleum spills caused pollution and safety hazards. An estimated 8,000 gallons of fuel oil were discharged into the floodwater (Federal Emergency Management Agency, 1992). In addition, gasoline leaked from automobile-service stations and vehicles. These contaminants either evaporated or were flushed downstream with the high flows after the ice jam was released. However, some petroleum residue remained in buildings and soils and created potentially hazardous conditions for emergency crews. In Montpelier, the ice-jam flood caused an estimated $4 million in damage. Other flood damage totaling $1.1 million occurred in Caledonia, Orange, Washington, Windsor, and Chittenden Counties (Federal Emergency Management Agency, 1992). The President of the United States declared the flood-affected counties a disaster area (Times Argus, March 15, 1992). No deaths or serious injuries were reported.

REFERENCES

Beltaos, Spyridon, Gerard, R.E., Petryk, S., and Prowse, T.D., 1990, Working group on river ice jams, field studies and research needs: Saskatoon, Canada, Nation Hydrology Research Institute, 121 p.

Denner, J.C., 1991, Vermont floods and droughts, *in* U.S. Geological Survey, National Water Summary 1988–89: U.S. Geological Survey Water-Supply Paper 2375, p. 535–542.

Federal Emergency Management Agency, 1992, Interagency hazard mitigation team report for Vermont: FEMA–936DR–UT.

April 1992, in Central and Southwestern Virginia

By Byron J. Prugh, Jr.

A strong frontal system moved eastward across Virginia during the third week in April 1992. The combination of a low-pressure system over the Midwest coupled with a high-pressure system over the Atlantic Ocean off the eastern seaboard helped bring moisture northward from the Gulf of Mexico. The resultant rainfall amounts varied from 1 inch in the eastern part of the State to as much as 8 inches at several locations along the crest of the Blue Ridge Mountains (National Oceanic and Atmospheric Administration, 1992). Runoff from the rain produced widespread flooding, with the most severe damage noted along streams draining the western slopes of the Blue Ridge Mountains and in the valley to the west of the mountains.

There were two large areas where the maximum flood discharges had a recurrence interval in excess of 10 years (fig. 35). The largest area, encompassing more than 5,000 square miles, was along the Blue Ridge Mountains from the Virginia-North Carolina State line northward for about 150 miles. A second, smaller area of nearly 400 square miles was located at the northern end of the Shenandoah River Valley and on adjacent tributaries to the Potomac River.

Additionally, streamflow-gaging stations in two smaller areas—one in the New River Basin on Big Reed Island Creek (site 15, fig. 35) and Beaverdam Creek at Hillsville (site 16, fig. 35) and a second in the upper James River Basin on the Maury River (site 5, fig. 35)—recorded flood discharges with a recurrence interval in excess of 50 years.

Flooding was less severe along the eastern slopes of the Blue Ridge Mountains in the headwaters of the upper Rappahannock, Rapidan, and Rivanna River Basins where flood discharges with 2- to 10-year recurrence intervals were recorded. Along the eastern slope of the Blue Ridge Mountains in the headwaters of the Roanoke and Dan Rivers, there was considerable variation in response to runoff from the rainfall, and flood discharges from less than 2 years to more than 30 years were recorded.

The maximum stages and discharges recorded at selected streamflow-gaging stations are summarized in table 18. New maximum discharges for the period of record were recorded at three stations in the New River Basin during the April 1992 flooding. A new maximum of record—257 cubic feet per second—was recorded on Mira Fork tributary near Dugspur, Virginia (site 14, fig. 35), on April 21. The previous maximum for this 0.62-square-mile watershed was 184 cubic feet per second during May 1989. At the streamflow-gaging station on Big Reed Island Creek near Allisonia, Virginia (site 15, fig. 35), a discharge of 17,900 cubic feet per second was recorded on April 21. The previous record of 14,500 cubic feet per second was established during September 1959. A third peak of record was measured at a partial-record station on Beaverdam Creek at Hillsville, Virginia (site 16, fig. 35), and was determined to be 876 cubic feet per second. The previous record maximum discharge of 834 cubic feet per second was recorded during September 1979.

Damages from the flooding were in the millions of dollars with more than 300 private residences inundated and numerous roads, bridges, and public utilities damaged (Federal Emergency Management Agency, 1992). Preliminary estimates of damages in the vicinity of Roanoke and surrounding counties was $7.5 million. Effects of the flooding were felt as far downstream as the James River at Richmond (fig. 35) where sandbags were used to fill gaps in a new floodwall that was being constructed to protect low-lying areas. One death was attributed to the flooding.

REFERENCES

Federal Emergency Management Agency, 1992, Interagency hazard mitigation team report for Virginia: FEMA–944DR–VA.

National Oceanic and Atmospheric Administration, 1992, Climatological data, Virginia: Asheville, N.C., National Climatic Data Center, v. 102, no. 4, 23 p.

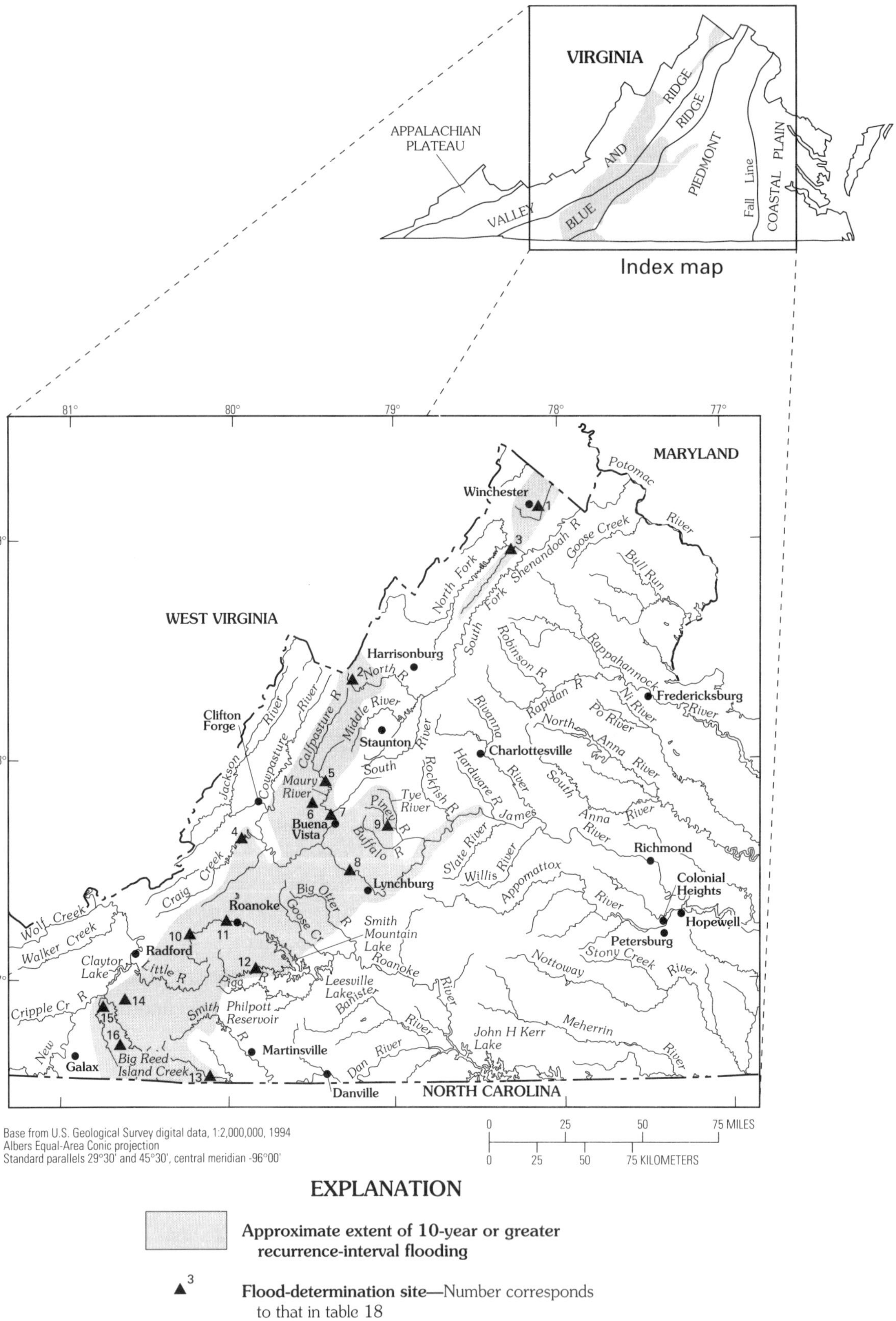

Figure 35. Approximate extent of 10-year or greater recurrence-interval flooding during April 1992, location of selected streamflow-gaging stations, and physiographic provinces in central and southwestern Virginia.

Table 18. Maximum stages and discharges prior to and during April 21–22, 1992, in central and southwestern Virginia

[mi^2, square miles; ft, feet above an arbitrary datum; ft^3/s, cubic feet per second; --, not determined or not applicable. Source: Recurrence intervals calculated from U.S. Geological Survey data. Other data from U.S. Geological Survey reports or data bases]

				Maximum prior to 1992				Maximum during April 21–22, 1992			
Site no. (fig. 35)	Station no.	Stream and place of determination	Drainage area (mi^2)	Period	Year	Stage (ft)	Dis-charge (ft^3/s)	Day	Stage (ft)	Dis-charge (ft^3/s)	Discharge recurrence interval (years)
1	01615000	Opequon Creek near Berryville, VA	57.4	1942,	1942	18.4	--	21	12.00	8,100	35
				1944–92	1988	13.49	12,600				
2	01620500	North River near Stokesville, VA	17.2	1947–92	1949	10.9	9,530	21	7.22	3,740	35
					1985	[1]19.8	7,600				
3	01635500	Passage Creek near Buckton, VA	87.8	1933–92	1942	15.5	21,000	22	12.66	9,860	30
4	02018000	Craig Creek at Parr, VA	329	1926–92	1985	24.76	58,500	22	17.29	23,600	50
5	02021500	Maury River at Rockbridge Baths, VA	329	1929–92	1985	19.19	87,700	22	13.28	34,500	55
6	02022500	Kerrs Creek near Lexington, VA	35.0	1927–92	1950	--	23,000	21	11.26	9,230	25
7	02024000	Maury River near Buena Vista, VA	646	1939–92	1969	31.23	105,000	22	18.99	35,100	30
8	02025500	James River at Holcomb Rock, VA	3,260	1900–17, 1927–92	1985	42.15	207,000	22	29.48	105,000	25
9	02027500	Piney River at Piney River, VA	47.6	1949–92	1969	13.80	38,000	21	9.76	9,330	30
10	02054500	Roanoke River at Lafayette, VA	257	1944–92	1972	15.60	24,500	21	13.09	16,400	25
11	02055000	Roanoke River at Roanoke, VA	395	1899–92	1985	23.35	32,300	22	18.09	22,000	25
12	02056900	Blackwater River near Rocky Mount, VA	115	1977–92	1985	21.92	20,800	21	19.93	13,200	25
13	02069700	South Mayo River near Nettleridge, VA	84.6	1963–92	1979	22.00	20,600	21	17.10	10,600	30
14	03167300	Mira Fork tributary near Dugspur, VA	.62	1967–92	1989	5.98	184	21	7.20	257	45
15	03167500	Big Reed Island Creek near Allisonia, VA	278	1909–15, 1939–92	1959	12.54	14,500	21	14.06	17,900	75
16	03167700	Beaverdam Creek at Hillsville, VA	4.75	1962–92	1990	[1]8.50	--	21	7.63	876	55
					1979	7.42	834				

[1]Stage affected by backwater.

May–August 1992, in Alaska

By Bruce B. Bigelow

Several floods occurred in Alaska during 1992 as spring streamflow was above normal. The flooding during May and June was intensified by backwater from ice jams. Some flooding was caused by excessive rainfall during August.

YUKON RIVER

The ice breakup in 1992 started with an ice-jam flood May 12 on the Yukon River near the Canadian border. The stage at the streamflow-gaging station at Eagle (site 1, fig. 36) was almost equal to the maximum observed stage caused by an ice jam during May 1962 (table 19). The ice-jam flooding dissipated within the next 100 miles downstream from Eagle. The spring streamflows throughout much of Alaska were above normal. In the Yukon River Basin, the Chena River near Two Rivers (site 6, fig. 36) and the streamflow-gaging station on the Yukon River near Stevens Village (site 3, fig. 36) recorded 100-year floods on June 3 and 11, respectively. Other stations (sites 1 and 2, fig. 36) had peak flows with recurrence intervals from 10 to 30 years during early June. The Kobuk River near Kiana (site 9, fig. 36) in northwest Alaska also had a significant spring-melt discharge during early June.

FAIRBANKS

During the last week in May and the first week in June, flooding occurred in the upper Chena River Basin near Fairbanks. The streamflow-gaging station on the Chena River near Two Rivers (site 6, fig. 36) recorded a flood with a recurrence interval of 100 years. This discharge produced the highest stage since the flood that occurred in 1967. The discharge of the August 13, 1967, flood was not determined, but its stage was 4.56 feet higher than that recorded during the flood of June 3, 1992. In contrast, the 1992 maximum discharge at the streamflow-gaging station on the Salcha River (site 4, fig. 36) and at the crest-stage gage on Monument Creek (site 5, fig. 36, a small tributary in the Chena River headwater) had recurrence intervals of only 4 years. Maximum discharge on the Chena River occurred about a week later than the maximum discharges on the Salcha River and Monument Creek. The relatively high discharge on the Chena River was the result of local rainfall during June 1–3 on parts of the headwaters at the time of maximum snowmelt runoff.

For the first time since the Moose Creek flood-control dam on the Chena River became operational in 1981, water was diverted (May 30–June 6) into the Tanana River. The impoundment of floodwater in the Moose Creek Reservoir and the diversion of some of this water prevented major flooding in Fairbanks. The impoundment and diversion elevated ground-water levels in the area. These high ground-water levels along with poor drainage patterns and high water in the upper Chena River Basin caused damage estimated at about $500,000. However, damages that were prevented by the Chena Lakes Flood Control Project (of which Moose Creek Reservoir is a part) were estimated by the U.S. Army Corps of Engineers at about $10 million (U.S. Army Corps of Engineers, 1994).

ARCTIC SLOPE

A flood occurred along the Sagavanirktok River extending from Atigun Pass in the Brooks Range to the North Slope of Alaska during the last week of August 1992. On August 24–26 where the road between Fairbanks and Prudhoe Bay climbs to Atigun Pass, precipitation fell as snow. However, where the road descends to the lower elevations of the Arctic Slope, the precipitation fell as rain and caused flooding at both gaging stations along the Sagavanirktok River.

The August 27 maximum discharge at the gage on Sagavanirktok River tributary near Pump Station 3 (site 11, fig. 36) 20 miles north of Atigun Pass had a recurrence interval of 50 years. At the Sagavanirktok River near Pump Station 3 (site 12, fig. 36), about 140 miles north of Atigun Pass, the August 27 discharge had a recurrence interval greater than 100 years.

At the Kuparuk River near Deadhorse (site 10, fig. 36), 10 miles upstream from the Arctic Ocean (Prudhoe Bay), the August 28 maximum discharge had a recurrence interval less than 2 years. Usually the spring-melt discharge at this station far exceeds any summer discharge.

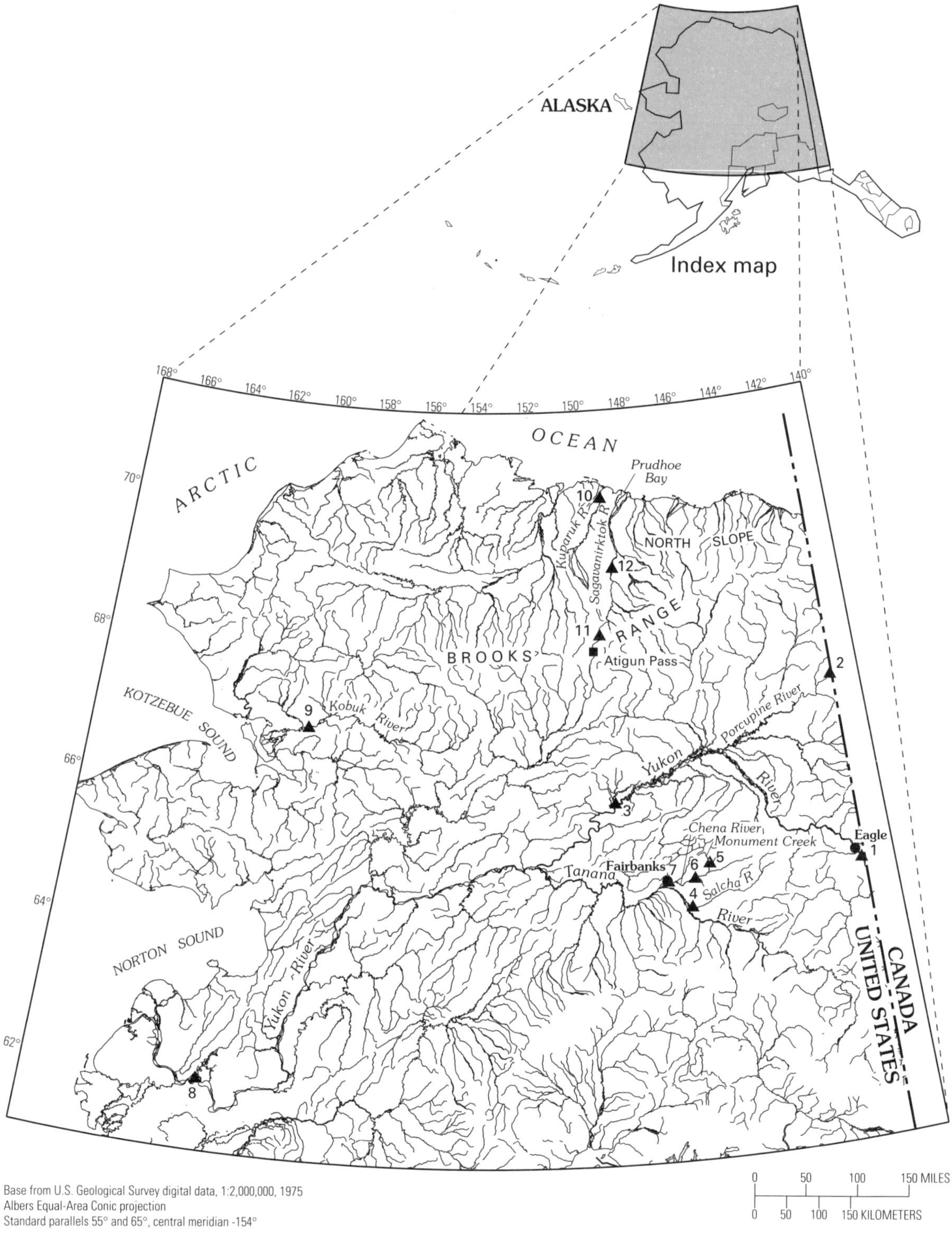

EXPLANATION

▲11 **Streamflow-gaging station and site number—** Number corresponds to that in table 19

Figure 36. Location of selected U.S. Geological Survey streamflow-gaging stations in Alaska used to determine maximum stages and discharges for May–August 1992.

Table 19. Maximum stages and discharges prior to and during May–August 1992, in Alaska

[mi^2, square miles; ft, feet above an arbitrary datum; ft^3/s, cubic feet per second; --, not determined or not applicable; <, less than; >, greater than. Source: Recurrence intervals calculated from U.S. Geological Survey data. Other data from U.S. Geological Survey reports or data bases]

				Maximum prior to 1992				Maximum during May–August 1992			
Site no. (fig. 36)	Station no.	Stream and place of determination	Drainage area (mi^2)	Period	Year	Stage (ft)	Discharge (ft^3/s)	Date (month/day)	Stage (ft)	Discharge (ft^3/s)	Discharge recurrence interval (years)
1	15356000	Yukon River near Eagle, AK	113,500	1911–13,	1962	[1]35.94	--	5/12	[1]35.90	--	--
				1950–92	1964	33.85	545,000	6/25	29.96	428,000	15
2	15388960	Porcupine River near International Boundary, AK	23,100	1987–92	1990	46.58	179,000	6/25	50.76	250,000	25
3	15453500	Yukon River near Stevens Village, AK	196,300	1976–92	1977	54.49	670,000	6/11	59.60	827,000	100
4	15484000	Salcha River near Salchaket, AK	2,170	1948–92	1967	21.78	97,000	5/28	15.65	22,600	4
5	15490000	Monument Creek at Chena Hot Springs, AK	26.7	1967–92	1967	29.10	1,490	5/26	25.91	694	4
6	15493000	Chena River near Two Rivers, AK	937	1967–92	1967	26.60	--	6/3	22.04	20,000	100
7	15514000	Chena River at Fairbanks, AK	1,995	1947–92	1967	[2]18.82	74,400	5/27	10.08	[3]11,400	--
8	15565447	Yukon River near Pilot Station, AK	321,000	1975–92	1985	27.50	1,070,000	6/21	27.38	788,000	5
9	15744500	Kobuk River near Kiana, AK	9,520	1976–92	1982	59.46	152,000	6/4	62.94	161,000	15
10	15896000	Kuparuk River near Deadhorse, AK	3,130	1971–92	1978	37.60	118,000	8/28	32.94	30,800	<2
11	15906000	Sagavanirktok River tributary near Pump Station 3, AK	28.4	1979–92	1979	19.99	700	6/11	21.20	1,080	50
12	15908000	Sagavanirktok River near Pump Station 3, AK	1,860	1982–92	1983	21.07	23,000	8/27	20.67	42,900	>100

[1]Backwater from ice jam.
[2]Site and datum then in use.
[3]Regulated by Moose Creek Dam.

Damage to roads and other public facilities was minor. A culvert on the Fairbanks to Prudhoe Bay road was washed out and delayed traffic briefly. At Prudhoe Bay, the Sagavanirktok River flooded some oil-field areas, which halted oil production for several days.

REFERENCE

U.S. Army Corps of Engineers, 1994, U.S. Army Corps of Engineers annual flood damage report to Congress for fiscal year 1993: Washington, D.C., 17 p.

June 1992, in Western Virginia

By Byron J. Prugh, Jr.

During the first week of June 1992, a strong frontal system moved across Virginia. The front was accompanied by widespread rainshowers, some of which were locally intense. Total rainfall amounts for June 4–6 generally varied from 1 to 4 inches; however, some areas in the southwestern part of the State received more than 6 inches of rain (National Oceanic and Atmospheric Administration, 1992).

Runoff from the rains produced high water across much of the western one-half of the State; however, the area of major flooding was confined to the New River Basin (fig. 37). Maximum discharges at streamflow-gaging stations on the main stem New River had recurrence intervals of approximately 7 to 20 years (table 20). At streamflow-gaging stations on tributaries to the New River, the recurrence interval of the flood peaks varied from less than 2 to 22 years, except on Walker Creek where the recurrence interval was estimated to be in excess of 100 years (site 10, table 20).

The maximum discharge of 25,000 cubic feet per second at the streamflow-gaging station on Walker Creek at Bane, Virginia (site 10, fig. 37), was the largest recorded since the beginning of gaging-station operation in 1938. Information on a historical flood that occurred in 1878 was provided by local residents when the station was established in 1938. The flood of September 1878 had a maximum discharge estimated at 40,000 cubic feet per second on the basis of historical high-water marks and channel geometry measured after the June 1992 flood. Table 20 summarizes the maximum stages and discharges for the June 1992 flood at selected streamflow-gaging stations in the New River Basin.

REFERENCE

National Oceanic and Atmospheric Administration, 1992, Climatological data, Virginia: Asheville, N.C., National Climatic Data Center, v. 102, no. 6, 23 p.

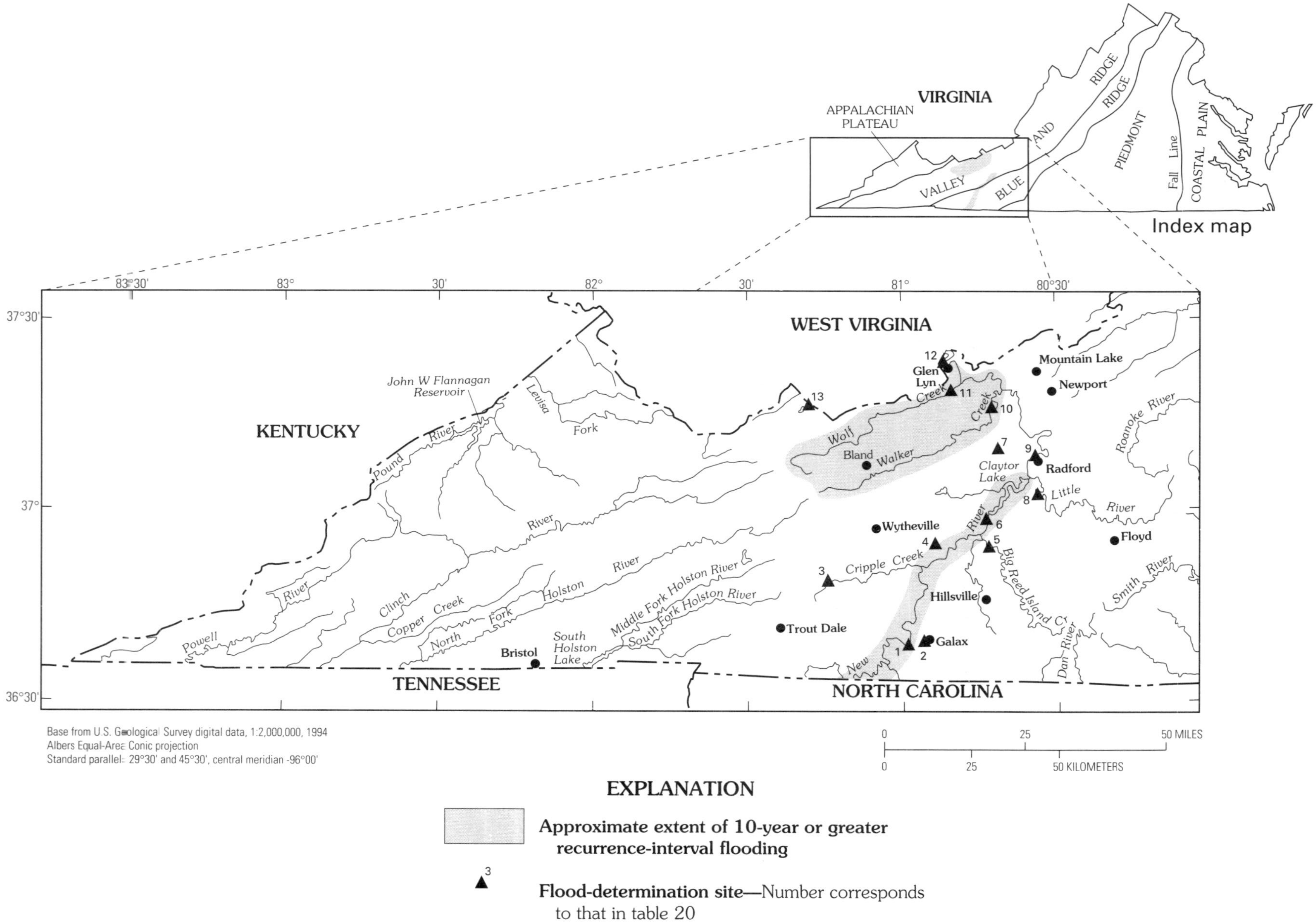

Figure 37. Approximate extent of 10-year or greater recurrence-interval flooding and location of selected streamflow-gaging stations in western Virginia, June 1992.

Table 20. Maximum stages and discharges prior to and during June 4–5, 1992, in western Virginia

[mi^2, square miles; ft, feet above an arbitrary datum; ft^3/s, cubic feet per second; <, less than; >, greater than. Source: Recurrence intervals calculated from U.S. Geological Survey data. Other data from U.S. Geological Survey reports or data bases]

				Maximum prior to June 1992				Maximum during June 4–5, 1992			
Site no. (fig. 37)	Station no.	Stream and place of determination	Drainage area (mi^2)	Period	Year	Stage (ft)	Discharge (ft^3/s)	Day	Stage (ft)	Discharge (ft^3/s)	Discharge recurrence interval (years)
1	03164000	New River near Galax, VA	1,131	1930–92	1940	25.7	141,000	5	12.00	49,500	17
2	03165000	Chestnut Creek at Galax, VA	39.4	1940,	1940	[1]17.4	11,000	4	6.15	2,250	3
				1945–92	1947	14.4	6,980				
3	03165700	Cripple Creek at Cedar Springs, VA	11.3	1967–92	1977	20.37	1,860	5	14.52	484	<2
4	03167000	Reed Creek at Grahams Forge, VA	247	1909–16,	1916	11.4	17,500	5	9.07	8,560	15
				1927–92	1977	10.01	14,000				
5	03167500	Big Reed Island Creek near Allisonia, VA	278	1909–15,	1912	6.6	7,500	5	6.47	4,300	<2
				1939–92	1992	14.06	17,900				
6	03168000	New River at Allisonia, VA	2,202	1930–92	1940	23.42	185,000	5	13.72	80,300	20
7	03168750	Thorne Springs Branch near Dublin, VA	4.77	1957–92	1973	8.01	2,200	5	.93	11	<2
8	03170000	Little River at Graysontown, VA	300	1929–92	1972	13.40	22,800	5	4.98	4,370	<2
9	03171000	New River at Radford, VA	2,748	1878,	1878	37.4	217,000	5	18.81	74,100	7
				1908–30,	1916	35.7	200,000				
				[2]1939–92	1940	35.96	[2]218,000				
10	03173000	Walker Creek at Bane, VA	305	1878,	1878	23.5	40,000	5	19.28	25,000	>100
				1938–92	1977	16.69	16,900				
11	03175500	Wolf Creek near Narrows, VA	223	1909–16,	1916	13.0	11,000	5	11.91	11,200	20
				1938–92	1957	13.8	12,900				
12	03176500	New River at Glen Lyn, VA	3,768	1878,	1878	33.1	240,000	5	17.47	100,000	10
				1915–38,	1916	31.1	210,000				
				1939-92	1940	27.5	[2]226,000				
13	03177710	Bluestone River at Falls Mills, VA	44.2	1981–92	1984	8.37	1,050	5	6.09	713	<2

[1]Datum then in use.

[2]Flows in the lower New River downstream from Clayton Lake have been regulated since 1939.

June 14–15, 1992, Along the North Fork Powder River Below Pass Creek, Near Mayoworth, Wyoming

By Jeff C. Vigil

Intense rain, associated with strong convection, fell over much of north-central Wyoming on the night of June 14 and morning of June 15, 1992, and caused flooding on the North Fork of the Powder River and other tributaries of the Powder River (Evans, 1993). The National Oceanic and Atmospheric Administration (1992) reported 4.06 inches of rain at Buffalo and 5.60 inches, 15 miles south of Buffalo, on June 14th and 15th. Unofficial reports indicated that more than 6 inches of rain fell southwest of Buffalo (fig. 38). Most of the rainfall occurred in an 18-hour period and was equivalent to a 24-hour design storm having a recurrence interval of 100 years (Evans, 1993).

The streamflow-gaging station on North Fork Powder River below Pass Creek, near Mayoworth, Wyoming (site 1, fig. 38), had a maximum stage of 7.47 feet and a maximum discharge of 1,090 cubic feet per second on June 15. This discharge has a recurrence interval of about 20 years, but it did not exceed the 1984 maximum discharge of 1,590 cubic feet per second (table 21). At a bridge about 7 miles downstream from site 1, floodwaters overtopped the road embankment.

Stream stages in the affected drainage areas increased downstream from the area of greatest rainfall. Wyoming State Engineer Office personnel reported an increase in stage of more than 7 feet at the streamflow-gaging station on Crazy Woman Creek at Trabing Bridge (site 2, fig. 38) (Carmine LoGuidice, Wyoming State Engineer's Office, written commun., 1993).

Johnson County Road and Bridge Department officials (oral commun., 1993) reported minor damage to roads and bridges in Johnson County. The greatest damage occurred to road surfaces and shoulders of roads as a result of overflowing water, and some roads were closed for as long as a week. Rip-rap structures around bridges and culverts also were reported to have been damaged. The area affected is sparsely populated, so that monetary damage resulting from this flood was small.

REFERENCES

Evans, J.S., 1993, Flash flooding in central Johnson County, Wyoming: National Weather Service Central Region Technical Attachment 93–19, 7 p.

National Oceanic and Atmospheric Administration, 1992, Climatological data, Wyoming, June 1992: U.S. Department of Commerce, v. 101, no. 6, 27 p.

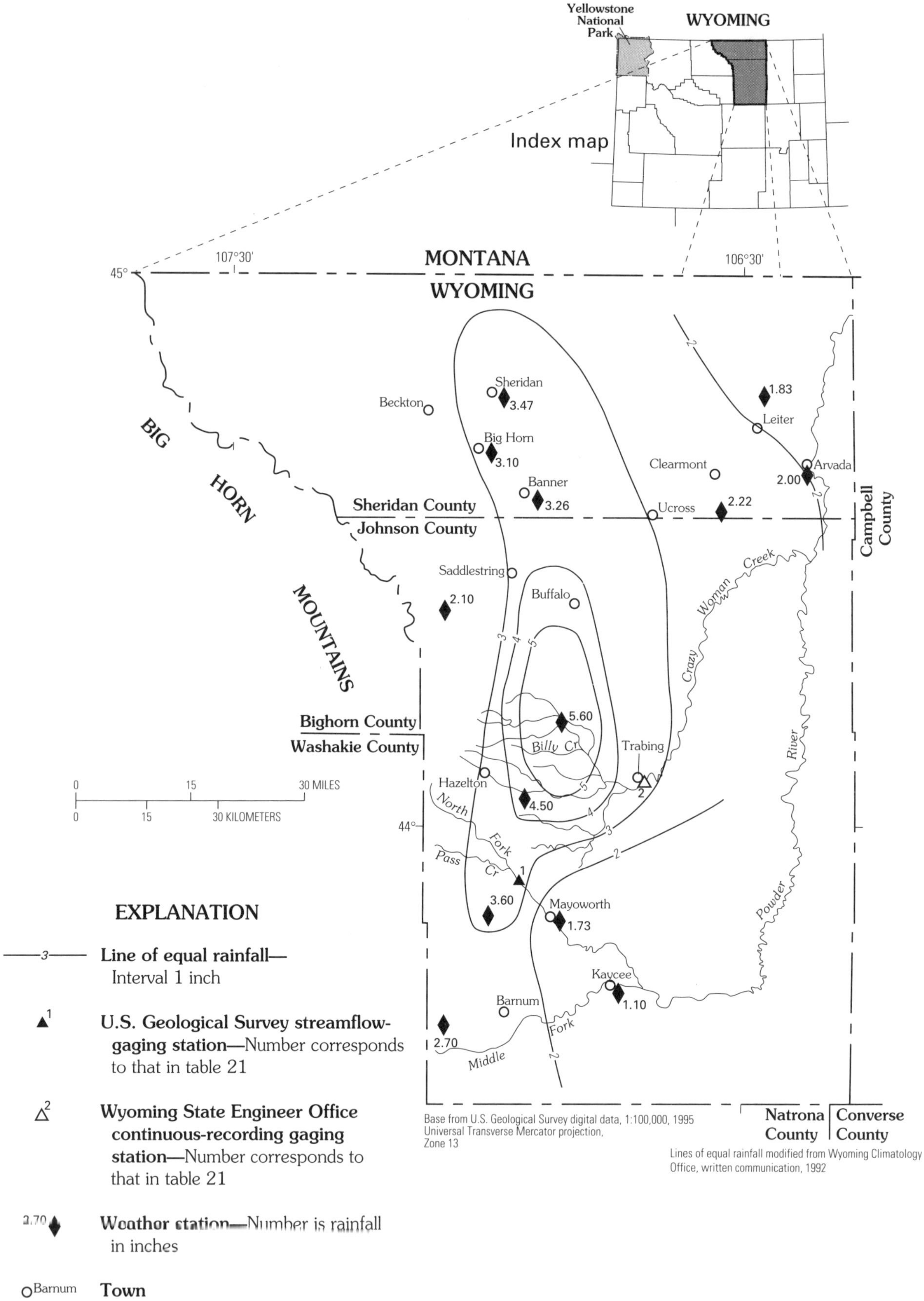

Figure 38. Lines of equal rainfall and location of selected streamflow-gaging stations in Johnson and Sheridan Counties, Wyoming, June 14–15, 1992.

Table 21. Maximum stages and discharges prior to and during June 14–15, 1992, North Fork Powder River below Pass Creek, near Mayoworth, Wyoming

[mi^2, square miles; ft, feet above an arbitrary datum; ft^3/s, cubic feet per second; --, not determined or not applicable. Source: Recurrence intervals calculated from U.S. Geological Survey data. Other data from U.S. Geological Survey reports or data bases]

				Maximum prior to June 14–15, 1992				Maximum during June 14–15, 1992			
Site no. (fig. 38)	Station no.	Stream and place of determination	Drainage area (mi^2)	Period	Year	Stage (ft)	Discharge (ft^3/s)	Day	Stage (ft)	Discharge (ft^3/s)	Discharge recur-rence interval (years)
1	06311400	North Fork Powder River below Pass Creek, near Mayoworth, WY	100	1973–92	1984	8.89	1,590	15	7.47	1,090	20
2	--	Crazy Woman Creek at Trabing Bridge, WY[1]	--	--	--	(+7-foot increase)	--	--	--	--	--

[1]Data on file with the Wyoming State Engineer's Office, Cheyenne.

June 18, 1992, in Central Kentucky

By Kevin J. Ruhl

High winds and locally excessive rains and flooding were experienced throughout the central part of Kentucky on the morning of June 18, 1992, as a result of a fast-moving frontal system. Widespread flooding occurred throughout the Beech Fork Basin in central Kentucky. Localized intense flooding occurred in the Cartwright Creek Basin in Washington County (fig. 39). Maximum discharge occurred on June 18 in Cartwright Creek near Fredericktown (site 4, fig. 39). However, on Beech Fork at Maud (site 3, fig. 39), the maximum discharges occurred on June 19.

A fast-moving frontal system also produced excessive rainfall throughout central Kentucky on June 18 and 19. Springfield received approximately 6 inches of rain in about 4 hours; 6.27 inches of rain fell from 0700 on June 18 to 0700 on June 19. Bardstown, Kentucky, received 3.13 inches of rain from 1800 on June 17 to 1800 on June 18. The area around Springfield received the largest rainfall amounts, as reflected by the lines of equal rainfall shown in figure 40. The lines were drawn on the basis of rainfall data from the National Oceanic and Atmospheric Administration (1992), which gives 24-hour precipitation values at selected locations.

The flooding on Cartwright Creek near Fredericktown was the most extreme in more than 200 years according to one elderly resident whose family has resided in the area for that long. Just upstream from Fredericktown, extensive walls of fieldstone erected in the flood plain during the late 1700's were toppled over by the force of the flowing water. The peak discharge for the June 18 flood was determined using indirect methods in this vicinity (site 4, table 22) immediately following the flood. Using techniques outlined by Choquette (1988), the recurrence interval of this flood magnitude was 100 years. The largest previously known flood on Cartwright Creek in recent times occurred in December 1978 and was approximately 2.5 feet lower in elevation than this flood at the location of the indirect measurement.

Flood-determination sites are shown in figure 39. The maximum-discharge information and date of occurrence are given in table 22. The maximum discharge for Rolling Fork near Lebanon (site 2) streamflow-gaging station had a 5-year recurrence interval but was not the maximum for the 1992 water year. The discharges at sites 1, 3, and 5 listed in table 22 are not extremely high, indicating that the excessive rainfall was localized. Recurrence-interval determinations were made from information by Choquette (1988) and Hannum (1976).

Residents in the Cartwright Creek Basin were adversely affected. About 100 people were evacuated from their homes in Springfield. Fredericktown, a small town located geographically midway between Bardstown and Springfield and at the mouth of Cartwright Creek, was especially hard hit. At least 12 families were evacuated, and several businesses were damaged. A state of emergency was declared in Washington County, and the National Guard was called in to help with the evacuations.

REFERENCES

Choquette, A.F., 1988, Regionalization of peak discharges for streams in Kentucky: U.S. Geological Survey Water-Resources Investigations Report 87–4209, 105 p.

Hannum, C.H., 1976, Technique for estimating magnitude and frequency of floods in Kentucky: U.S. Geological Survey Water-Resources Investigations 76–62, 81 p.

National Oceanic and Atmospheric Administration, 1992, Climatological data, Kentucky: Asheville, N.C., National Climatic Data Center, v. 87, no. 6, 22 p.

EXPLANATION

▲[2] **Flood-determination site**—Number corresponds to that in table 22

Figure 39. Location of flood-determination sites for flood of June 18, 1992, in central Kentucky.

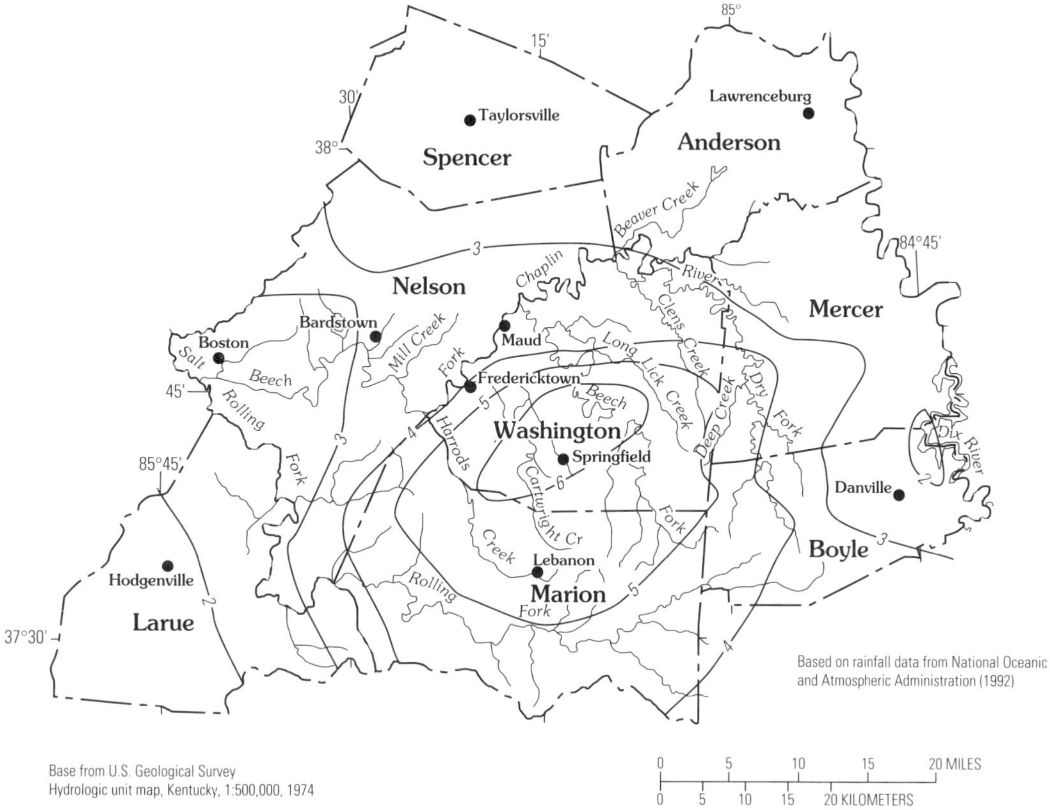

EXPLANATION

——5—— **Line of equal rainfall**—Interval 1 inch

Figure 40. Lines of equal rainfall for June 18 and 19, 1992, in central Kentucky.

Table 22. Maximum stages and discharges prior to and during June 18, 1992, in central Kentucky

[mi^2, square miles; ft, feet above an arbitrary datum; ft^3/s, cubic feet per second; <, less than; >, greater than; --, not determined or not applicable. Source: Recurrence interval calculated from U.S. Geological Survey data. Other data from U.S. Geological Survey reports or data bases]

Site no. (fig. 39)	Station no.	Stream and place of determination	Drainage area (mi^2)	Maximum prior to June 1992				Maximum during June 1992			
				Period	Year	Stage (ft)	Discharge (ft^3/s)	Day	Stage (ft)	Discharge (ft^3/s)	Discharge recurrence interval (years)
				Kentucky River Basin							
1	03285000	Dix River near Danville, KY	318	1943–92	1978	21.81	44,400	18	11.14	13,700	2
				Salt River Basin							
2	03299000	Rolling Fork near Lebanon, KY	239	1938–92	1970	24.51	54,800	19	19.21	24,200	5
3	03300400	Beech Fork at Maud, KY	436	1972–92	1978	26.16	33,300	19	23.64	22,200	<2
4	03300498	Cartwright Creek near Fredericktown, KY	82.2	--	--	--	--	18	--	31,300	>100
5	03301500	Rolling Fork near Boston, KY	1,299	1938–92	1989	52.62	66,500	20	41.14	24,700	<2

July 1992, in Eastern Kansas

By Seth E. Studley

Excessive rainfall during several days in July 1992 caused isolated flooding in parts of eastern Kansas (table 23). Intense isolated thunderstorms occurred on July 4–5 in Wabaunsee County, July 22 in the southern part of Riley County, July 24 in and near Bourbon County in southeast Kansas, and on July 25 near the Nebraska border in Marshall, Nemaha, and Washington Counties (fig. 41).

An isolated severe thunderstorm occurred on the afternoon of July 4, 1992, in Wabaunsee County, Kansas. Rainfall totals ranged from 6.74 inches at McFarland to 3.55 inches at Eskridge (fig. 41) (National Oceanic and Atmospheric Administration, 1992). Mill Creek near Paxico (site 7, fig. 41) had a maximum discharge of 38,500 cubic feet per second with a recurrence interval of between 25 and 50 years. Damage was limited to crops in low-lying areas.

Another isolated thunderstorm produced about 4 inches of rain over the 4.1-square-mile drainage basin of Kings Creek in Riley County (site 2, fig. 41) on July 22. The resulting maximum discharge of 5,800 cubic feet per second set a new maximum for the 12-year period of record at this site and was determined to have a 100-year recurrence interval. The flood wave lifted a small footbridge from near the streamflow-gaging station and carried it 0.25 mile downstream, destroying the bridge and damaging some discharge-measuring equipment.

Severe thunderstorms moved across the upper reaches of the Marmaton River in southeast Kansas on July 23 and 24, producing rainfall in excess of 4 inches for the 2-day period (fig. 41). Marmaton River near Marmaton (site 8, fig. 41) recorded a discharge of 48,500 cubic feet per second, which is considered a 50-year recurrence-interval flood.

Rainfall of 5.31inches fell at Marysville in Marshall County, and more than 2 inches fell at nearby sites (fig. 41) on the evening of July 25. Five streamflow-gaging stations in the area (sites 1 and 3–6, fig. 41) recorded large discharges although little flood damage occurred.

REFERENCE

National Oceanic and Atmospheric Administration, 1992, Climatologic data, Kansas, July 1992: Asheville, N.C., National Climatic Data Center, v. 106, no. 7, 31 p.

Table 23. Maximum stages and discharges prior to and during July 1992, in eastern Kansas

[mi^2, square miles; ft, feet above an arbitrary datum; ft^3/s, cubic feet per second; >, greater than; --, not determined or not applicable. Source: Recurrence intervals calculated from U.S. Geological Survey data. Other data from U.S. Geological Survey reports or data bases]

				Maximum prior to July 1992				Maximum during July 1992			
Site no. (fig. 41)	Station no.	Stream and place of determination	Drainage area (mi^2)	Period	Year	Stage (ft)	Discharge (ft^3/s)	Day	Stage (ft)	Discharge (ft^3/s)	Discharge recurrence interval (years)
1	06814000	Turkey Creek near Seneca, KS	276	1949–92	1973	24.77	21,400	25	24.47	19,600	50–100
2	06879650	Kings Creek near Manhattan, KS	4.1	1980–92	1982	11.28	4,530	22	12.05	5,800	>100
3	06882510	Big Blue River at Marysville, KS	4,780	1984–92	1986	38.90	39,700	25	37.80	35,800	--
4	06884200	Mill Creek at Washington, KS	344	1959–92	1983	29.13	12,300	25	21.90	5,520	25–50
5	06884400	Little Blue River near Barnes, KS	3,320	1958–92	1973	27.70	53,700	26	25.82	41,900	10–25
6	06885500	Black Vermillion River near Frankfort, KS	410	1948, 1951, 1953–92	1948 1973	30.20 30.06	-- 36,400	25	29.58	16,200	5–10
7	06888500	Mill Creek near Paxico, KS	316	1951, 1954–92	1951	34.70	77,200	5	31.03	38,500	25–50
8	06917380	Marmaton River near Marmaton, KS	292	1971–92	1986	42.87	106,000	24	38.35	48,400	50

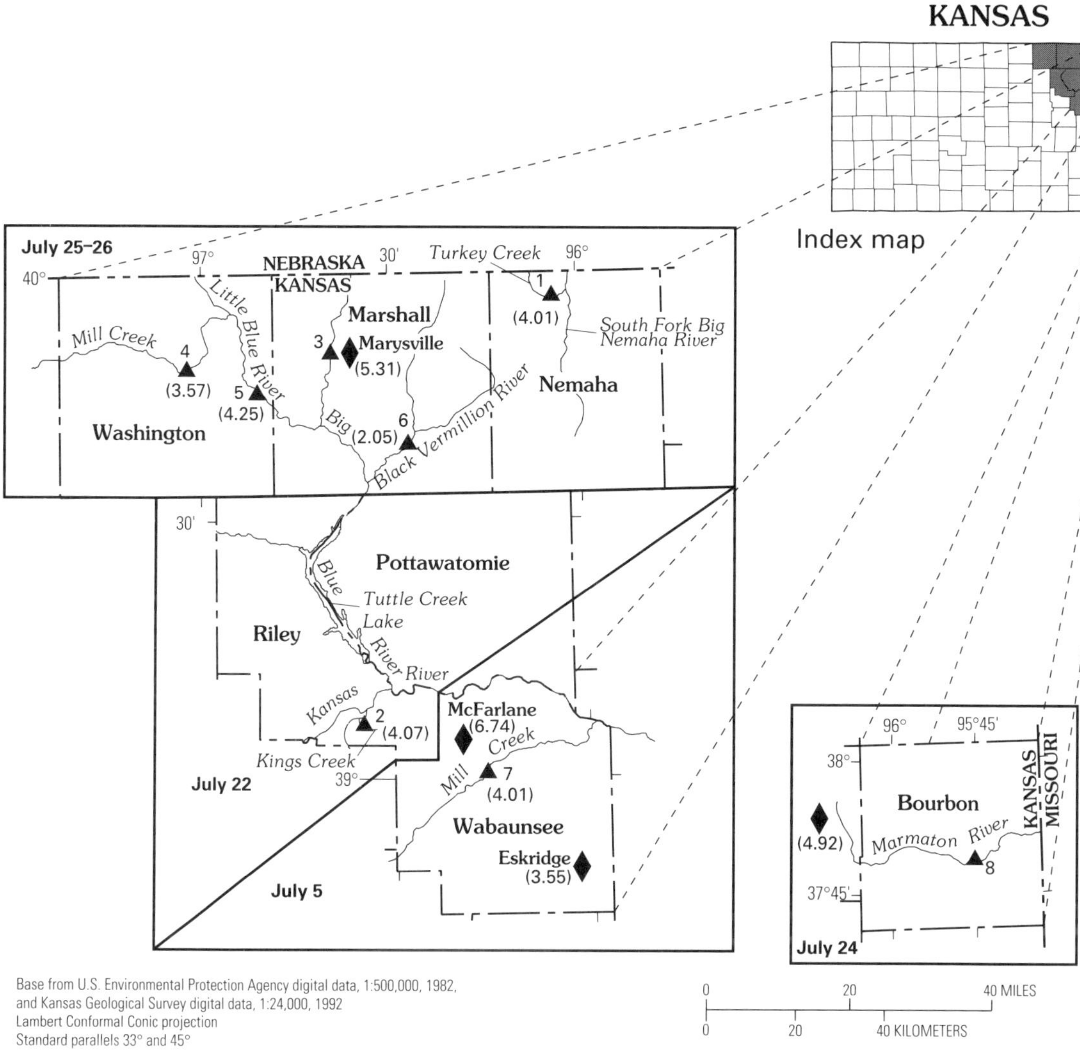

Base from U.S. Environmental Protection Agency digital data, 1:500,000, 1982, and Kansas Geological Survey digital data, 1:24,000, 1992
Lambert Conformal Conic projection
Standard parallels 33° and 45°

EXPLANATION

▲ 7 (4.01) **Flood-determination site**—Number corresponds to that in table 23. Number in parentheses is precipitation amount in inches

◆ (4.92) **National Weather Service precipitation station**—Number in parentheses is precipitation amount in inches

Figure 41. Location of flood-determination sites and precipitation amounts for floods of July 1992, in eastern Kansas.

July 1992, in Central and Western Ohio

By Harold Shindel

Severe thunderstorms and excessive rainfall started in mid-July 1992 following a period of drought over much of Ohio. On July 12–13, excessive rains in central and western Ohio caused moderate to severe small-stream and urban flooding. Rainfall amounts of more than 6 inches were reported in some areas (National Oceanic and Atmospheric Administration, 1992). Auglaize, Franklin, Mercer, and Shelby Counties (fig. 42), as well as parts of surrounding counties, were hardest hit.

Showers and thunderstorms were common statewide throughout the remainder of the month. Several other storms of note occurred. On July 16–17, excessive rains prompted flood warnings in central, western, and northwestern Ohio, as well as small-stream and urban flood warnings in many other areas of the State.

On the afternoon of July 26, an estimated 4 inches of rain fell within 2 hours in parts of south-central Ohio. The intense rains caused major flooding on Indian Creek, a tributary to the Scioto River that passes through the community of Massieville in Ross County. Floodwaters from Indian Creek rose so quickly that helicopters and boats were required to rescue some residents as they sat atop the roofs of houses and mobile homes.

Flood damage for the July 26 flood was extensive. Two people drowned, reportedly while attempting to aid others. Debris caused blockage of some stream crossings, which increased flood elevations and property damage (Federal Emergency Management Agency, 1992). In all, 35 mobile homes and 3 single-family homes were destroyed during the flood in the Massieville area. Another 40 mobile homes, 55 single-family homes, and 7 businesses sustained major damage (Baird, 1992). A number of residential propane tanks were dislodged from their moorings and were carried downstream by the flood, some of which lodged under bridges and near homes. Newspapers reported that several tanks exploded (Columbus Dispatch, 1992); however, there were no reports of explosion damage.

A maximum stage of 20.65 feet was determined by survey of high-water marks at the previously discontinued partial-record streamflow-gaging station on Indian Creek at Massieville (site 1, fig. 42). The previous maximum stage during the 1947–77 operation of the station, 17.98 feet, occurred on June 25, 1971 (table 24). Computations by the U.S. Geological Survey for the 1992 flood at this site estimated the maximum discharge at about 8,200 cubic feet per second, which was slightly less than the 100-year recurrence interval flood. On the same day, storms caused some flooding conditions in Belmont, Jefferson, and Perry Counties. Storms on July 29–30 in northeastern Ohio caused moderate flooding in Mahoning and Counties.

REFERENCES

Baird, D., 1992, Floods rout hundreds in south-central Ohio: Columbus (Ohio) Dispatch, July 27, 1992, p. 1A–2A.

Columbus (Ohio) Dispatch, 1992, Two die in Ross County flooding: July 28, 1992, p. 1A–2A.

Federal Emergency Management Agency, 1992, Interagency hazard mitigation team report for Ohio: FEMA–951–DR–OH, 19 p.

National Oceanic and Atmospheric Administration, 1992, Climatological data, Ohio: Asheville, N.C., National Climatic Data Center, v. 97, no. 7, 27 p.

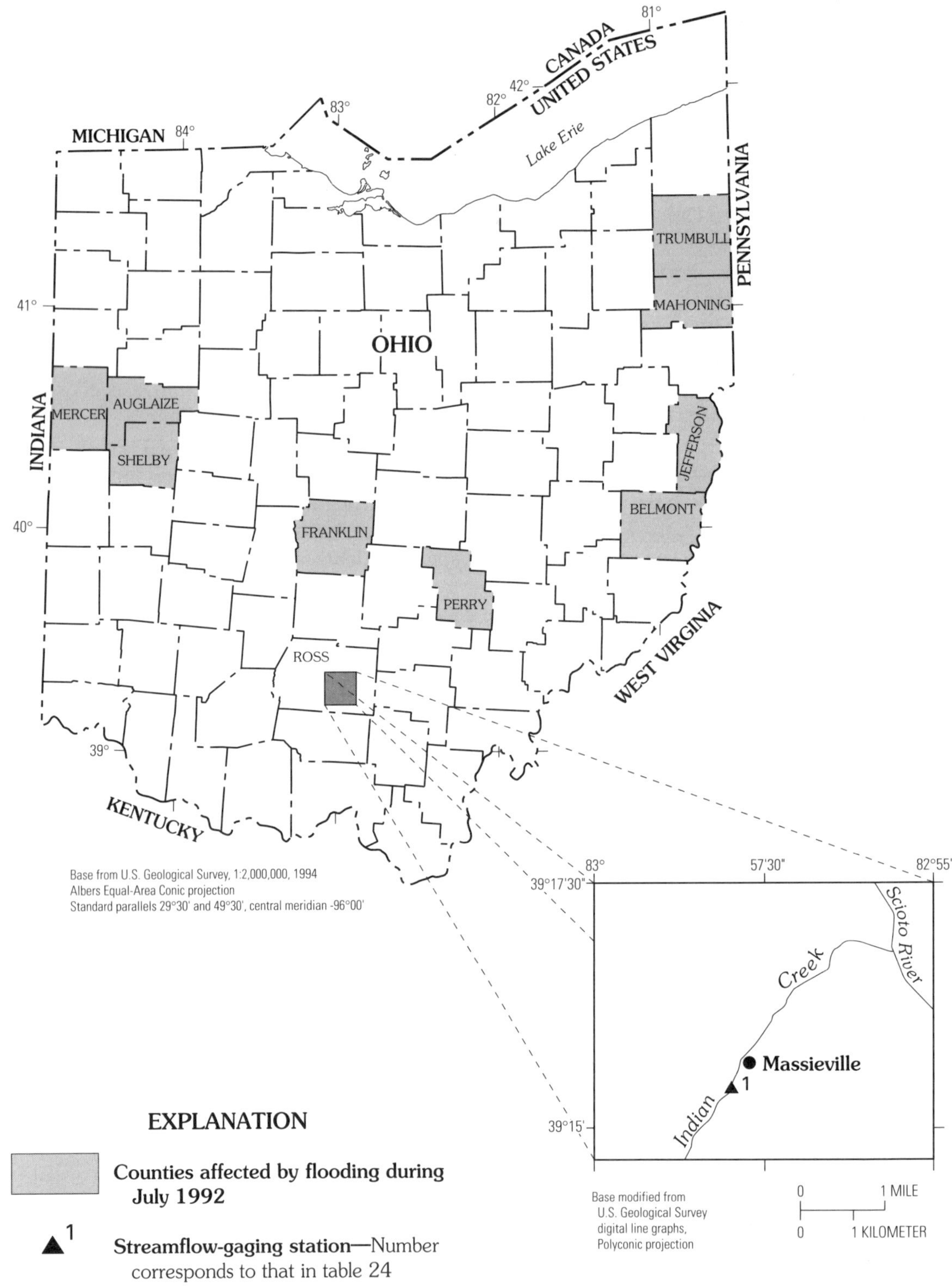

Figure 42. Ohio counties affected by flooding during July 1992 and location of flood-determination site for flood of July 26, 1992.

Table 24. Maximum stages and discharges prior to and during July 26, 1992, in Massieville, Ohio

[mi^2, square miles; ft, feet above an arbitrary datum; ft^3/s, cubic feet per second. Source: Recurrence intervals calculated from U.S. Geological Survey data. Other data from U.S. Geological Survey reports or data bases]

				Maximum prior to July 1992				Maximum during July 26, 1992			
Site no. (fig. 42)	Station no.	Stream and place of determination	Drainage area (mi^2)	Period	Year	Stage (ft)	Discharge (ft^3/s)	Day	Stage (ft)	Discharge (ft^3/s)	Discharge recurrence interval (years)
1	03234100	Indian Creek at Massieville, OH	9.6	1947–77	1953	16.95	5,640	26	20.65	8,200	100
					1971	17.98	4,500				

July 12–16, 1992, in Central Indiana

By Donald V. Arvin

As the month of July 1992 began in Indiana, the lack of precipitation began to show its effects. Low soil moisture resulted in pronounced crop stress. Water-supply managers started to initiate water conservation measures. However, thunderstorms and rain returned in July and resulted in total monthly precipitation ranging from 4 to more than 15 inches throughout the State (National Oceanic and Atmospheric Administration, 1992). These rains caused soils to become saturated and resulted in occasional lowland flooding. Intense rains on July 12–13 resulted in flooding along streams in Carroll, Howard, Grant, Blackford, Madison, Delaware, and Hancock Counties (fig. 43).

Soils of central Indiana became saturated with water from as much as 3 inches of rain that fell during the early morning hours of July 12. As a result of an additional 4 to 6 inches of rain that fell late July 12 and early July 13, many State and local roadways were closed.

This flood resulted in new maximum discharges for the period of record for at least two streamflow-gaging stations (table 25). The streamflow-gaging station on Pipe Creek at Frankton (site 5, fig. 43), which has a drainage area of 113 square miles and a period of record beginning in May 1968, had a record maximum discharge of 5,630 cubic feet per second on July 13, which had a 30-year recurrence interval. The streamflow-gaging station on Wildcat Creek near Jerome (site 3, fig. 43), which has a drainage area of 146 square miles and a period of record that dates back to July 1961, had a record maximum discharge of 7,120 cubic feet per second on July 14, 1992. The flood had a recurrence interval of 40 years.

REFERENCE

National Oceanic and Atmospheric Administration, 1992, Monthly report of river and flood conditions for July 1992: Indianapolis, Indiana, 3 p.

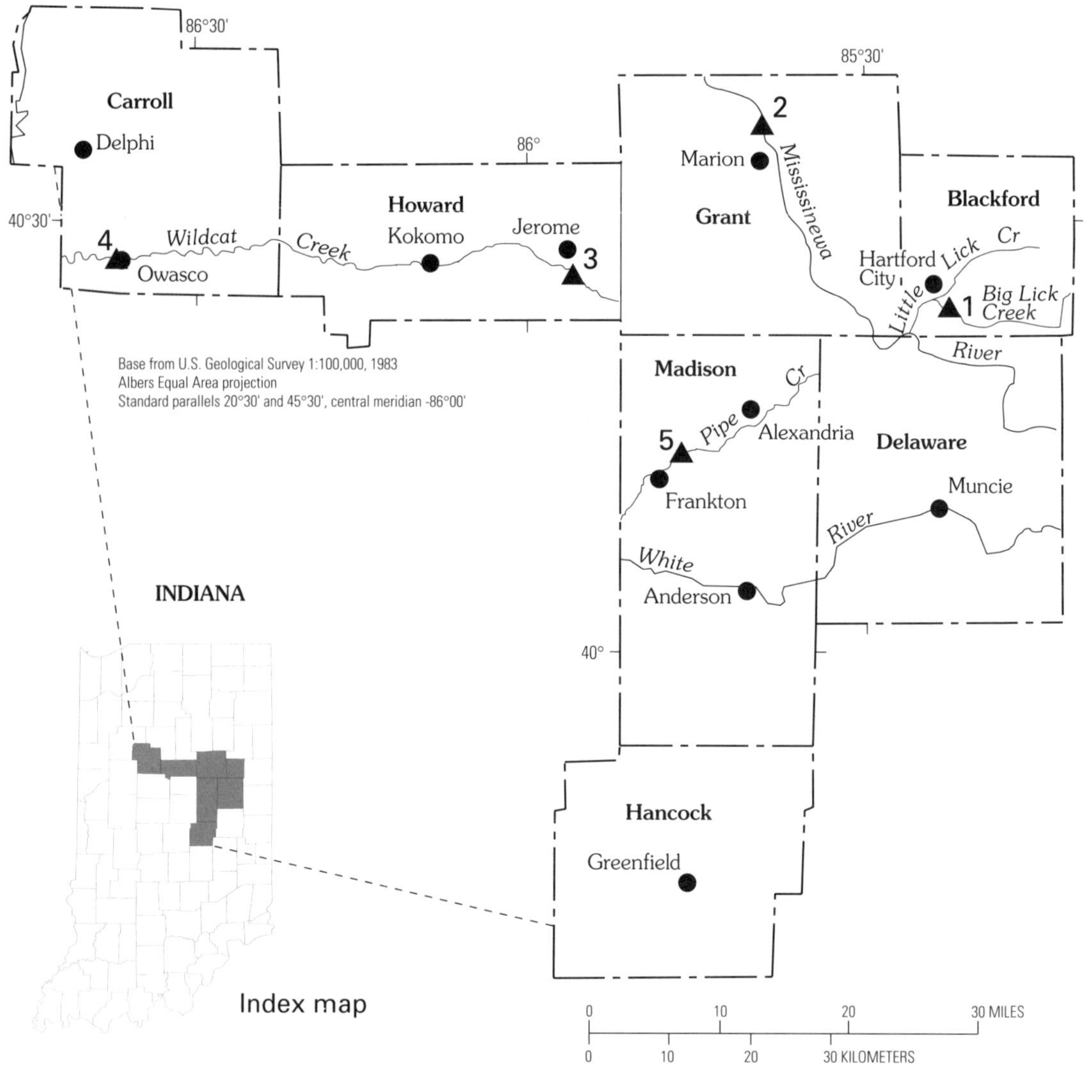

EXPLANATION

▲5 **Flood-determination site**—
Number corresponds to that in table 25

Figure 43. Location of flood-determination sites for flood of July 12–16, 1992, in central Indiana.

Table 25. Maximum stages and discharges prior to and during July 12–16, 1992, in central Indiana

[mi^2, square miles; ft, feet above an arbitrary datum; ft^3/s, cubic feet per second. Source: Recurrence intervals calculated from U.S. Geological Survey data. Other data from U.S. Geological Survey reports or data bases]

				Maximum prior to July 1992				Maximum during July 12–16, 1992			
Site no. (fig. 43)	Station no.	Stream and place of determination	Drainage area (mi^2)	Period	Year	Stage (ft)	Discharge (ft^3/s)	Day	Stage (ft)	Discharge (ft^3/s)	Discharge recurrence interval (years)
1	03326070	Big Lick Creek near Hartford City, IN	29.2	1971–92	1981	16.14	1,940	13	15.37	1,600	15–20
2	03326500	Mississinewa River at Marion, IN	682	1923–92	1927	17.40	25,000	15	15.16	19,600	10
3	03333450	Wildcat Creek near Jerome, IN	146	1961–92	1990	13.71	6,890	14	13.31	7,120	40
4	03334000	Wildcat Creek at Owasco, IN	396	1943–73, 1975–81, 1988–92	1950 1980	13.30 11.58	10,200 10,800	16	11.99	9,560	20
5	03348350	Pipe Creek at Frankton, IN	113	1968–92	1980 1989	14.78 14.71	4,340 4,920	13	15.00	5,630	30

July 25–August 2, 1992, in Southeastern Nebraska

By J.A. Boohar

Severe flooding in southeastern Nebraska occurred the latter part of July as a result of two separate storms. The first storm occurred July 25 and produced 7 inches of rain in much of the area, with unofficial reports of 11 inches in some places. The second storm occurred July 29 and produced 3 to 6 inches of rain on the already saturated ground (National Oceanic and Atmospheric Administration, 1992).

A list of the streamflow-gaging stations showing maximum discharges where major flooding was measured can be found in table 26. The location for these streamflow-gaging stations is shown in the figure 44.

Maximum discharge of the Little Blue River near Deweese (site 10, fig. 44) occurred on July 25 and was determined to have a 2-year recurrence interval. Other streamflow-gaging stations located downstream on the Little Blue River (fig. 44) near Alexandria (site 11) and near Fairbury, Nebraska (site 13), and at Hollenberg, Kansas (site 14), recorded new maximum discharges for the period of record as a result of the first storm (table 26). The streamflow-gaging station near Alexandria (site 11) recorded a maximum stage of 21.07 feet on July 25. Discharge was 32,600 cubic feet per second and was determined to be 1.1 times the 100-year recurrence-interval discharge. The stage and discharge exceeded the previous maximums by 3.77 feet and 7,000 cubic feet per second. The streamflow-gaging station near Fairbury (site 13, fig. 44) recorded a maximum stage of 24.33 feet on July 25. Discharge was 54,000 cubic feet per second and was determined to have a 100-year recurrence interval. The stage and discharge exceeded the previous maximums by 7.35 feet and 12,100 cubic feet per second. The streamflow-gaging station at Hollenberg, Kansas (site 14, fig. 44), recorded a maximum stage of 21.21 feet on July 26. The maximum discharge was 47,800 cubic feet per second and was determined to be 1.1 times the 100-year recurrence-interval discharge. The stage exceeded the previous maximum by 0.21 foot, but the discharge exceeded the previous maximum by 11,200 cubic feet per second. The streamflow-gaging station at Alexandria on Big Sandy Creek (site 12, fig. 44), a tributary of the Little Blue River, recorded a maximum discharge on July 25, which was determined to have a recurrence interval of 10 years.

July was a wet month in the area around Falls City. This area consists of relatively impermeable soils with rolling hills and stream valleys. Most of the streamflow in this area is a result of overland runoff. The average July rainfall for Falls City is 4.33 inches; the average annual rainfall is 34.72 inches (National Oceanic and Atmospheric Administration, 1992). The storms that were responsible for severe flooding along the Little Blue River contributed to the 1992 July total of 22.3 inches at Falls City. The Big Nemaha River at Falls City (site 4, fig. 44) recorded two peak stages. The first was recorded on July 26 with a stage of 25.85 feet and a discharge of 32,200 cubic feet per second. Four days later, a second peak was recorded as a result of the second storm; this maximum had a stage of 28.03 feet and a discharge of 40,900 cubic feet per second. This discharge was determined to have a 5-year recurrence interval. The streamflow-gaging station on the Little Nemaha River at Auburn (site 2, fig. 44) recorded two peak discharges, one on July 25 having a recurrence interval of 7 years and a second lesser discharge on July 30. The streamflow-gaging station on the North Fork Big Nemaha River at Humboldt (site 3, fig. 44) recorded a peak discharge on July 25 and a second, greater peak discharge on July 30 having a recurrence interval of 10 years.

These same thunderstorms also caused damaging floods along Salt Creek. The streamflow-gaging station on Salt Creek at Roca (site 1, fig. 44) recorded a maximum discharge on July 25 that was determined to have a recurrence interval of 20 years.

Streamflow-gaging stations along the Big Blue River and its tributaries also recorded maximums during this period. The streamflow-gaging station on the West Fork Big Blue River near Dorchester (site 5, fig. 44) and two streamflow-gaging stations along the Big Blue River, one near Crete (site 6) and the other at Beatrice (site 8), each recorded two maximum discharges. Meanwhile, the streamflow-gaging station on Turkey Creek near Wilber (site 7, fig. 44) and Big Blue River at Barneston (site 9) recorded one peak. The recurrence interval for these peaks were 5 years or less.

Table 26. Maximum stages and discharges prior to and during July 25–August 2, 1992, in southeastern Nebraska

[mi^2, square miles; ft, feet above an arbitrary datum; ft^3/s, cubic feet per second; --, not determined or not applicable; <, less than; >, greater than. Source: Recurrence intervals calculated from U.S. Geological Survey data. Other data from U.S. Geological Survey reports or data bases]

Site no. (fig. 44)	Station no.	Stream and place of determination	Drainage area (mi^2)	Maximum prior to July 1992				Maximum during July and August 1992			
				Period	Year	Stage (ft)	Discharge (ft^3/s)	Date (month/day)	Stage (ft)	Discharge (ft^3/s)	Discharge recurrence interval (years)
1	06803000	Salt Creek at Roca, NE	167	1951–92	1950	26.00	67,000	7/25	21.76	7,710	20
2	06811500	Little Nemaha River at Auburn, NE	793	1949–92	1950	27.65	164,000	7/25	25.02	40,300	7
								7/30	22.96	20,000	3
3	06814500	North Fork Big Nemaha River at Humboldt, NE	548	1952–92	1982	31.25	59,500	7/25	14.95	17,100	2
					1958	31.70	--	7/30	25.25	43,000	10
4	06815000	Big Nemaha River at Falls City, NE	1,340	1944–92	1974	31.40	71,600	7/26	25.85	32,200	4
								7/30	28.03	40,900	5
5	06880800	West Fork Big Blue River near Dorchester, NE	1,206	1958–92	1950	24.80	49,400	7/26	17.40	4,580	3
								7/30	13.50	2,330	<2
6	06881000	Big Blue River near Crete, NE	2,716	1945–92	1950	28.74	27,600	7/27	21.01	5,170	<2
					1986	29.86	--	7/31	21.91	5,880	<2
7	06881200	Turkey Creek near Wilber, NE	460	1959–92	1984	21.43	33,000	7/25	15.14	2,870	2
8	06881500	Big Blue River at Beatrice, NE	3,900	1974–92	1984	31.27	55,100	7/27	18.22	11,400	3
					1973	33.02	--	8/02	12.14	5,920	<2
9	06882000	Big Blue River at Barneston, NE	4,447	1932–92	1941	34.30	57,700	7/25	25.00	25,000	5
10	06883000	Little Blue River near Deweese, NE	979	1953–72, 1974–92	1969	18.57	25,100	7/25	9.17	4,550	2
11	06883570	Little Blue River near Alexandria, NE	1,557	1959–72, 1974–92	1960	17.30	25,600	7/25	21.07	[1]32,600	>100
12	06883940	Big Sandy Creek at Alexandria, NE	607	1979–92	1984	16.71	21,900	7/25	15.98	15,200	10
13	06884000	Little Blue River near Fairbury, NE	2,350	1908–15, 1928–92	1984	16.98	41,900	7/25	24.33	[1]54,000	100
14	06884025	Little Blue River at Hollenberg, KS	2,752	1974–92	1984	21.00	36,600	7/26	21.21	[1]47,800	>100
					1973	23.07	--				

[1]Maximum discharge for 1992 exceeds the previous maximum discharge.

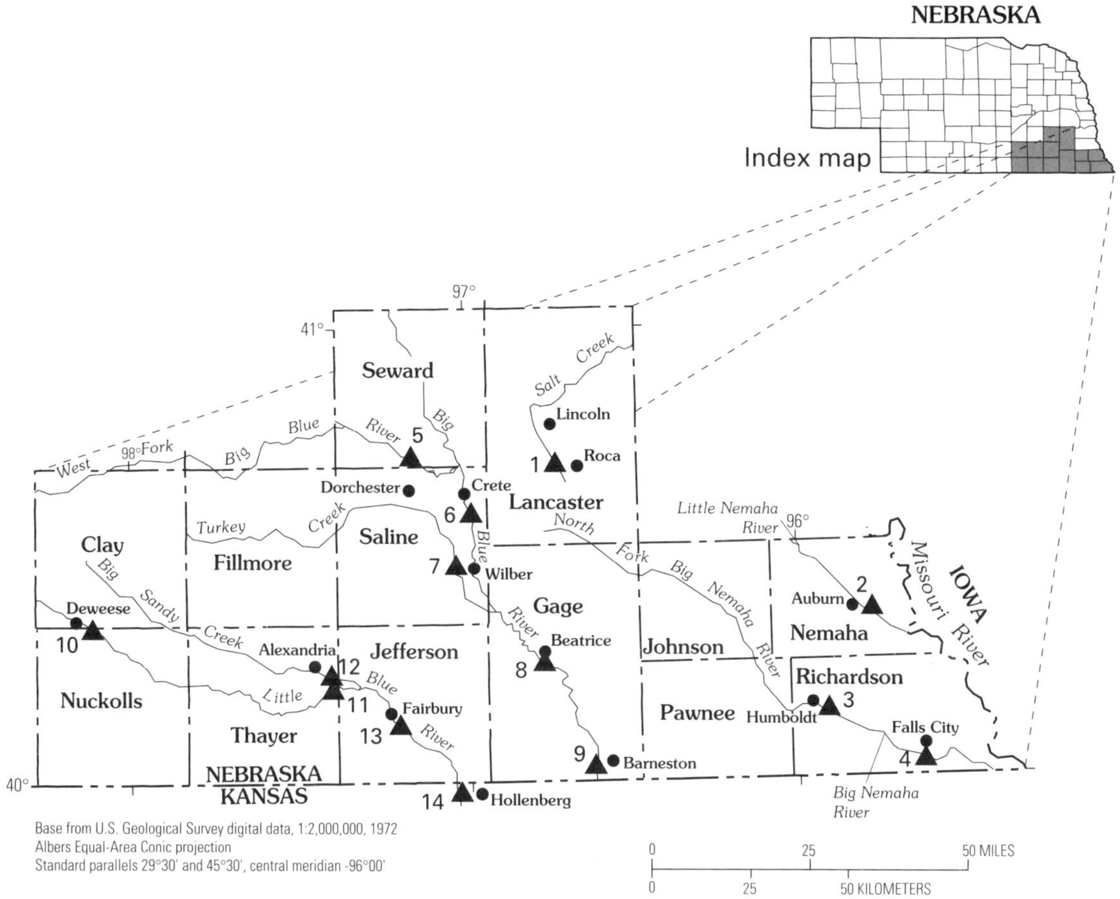

EXPLANATION

▲1 **Flood-determination site**—Number corresponds to that in table 26

Figure 44. Location of flood-determination sites for flood of July 25–August 2, 1992, in southeastern Nebraska.

Many homes were flooded, cattle were swept away, numerous roads were closed, and crops were damaged. No serious injuries were reported. Total losses in the State were estimated to be $6,683,000 (U.S. Army Corps of Engineers, 1994).

REFERENCES

National Oceanic and Atmospheric Administration, 1992, Climatological data, Nebraska, July 1992: Asheville, N.C., National Climatic Data Center, v. 97, no. 7, 17 p.

U.S. Army Corps of Engineers, 1994, Annual flood damage report to Congress for fiscal year 1993: Washington, D.C., 17 p.

August 8, 1992, in Southeastern and South-Central Indiana

By Donald V. Arvin

On the morning of August 8, 1992, torrential rains of 3 to 13 inches fell across southeastern and south-central Indiana, causing flash floods that forced people from their homes, damaged businesses, and claimed the life of one person, who drowned. Affected areas included Lawrence, Jackson, Jennings, Washington, Scott, Jefferson, and Clark Counties (fig. 45). Scott County was the hardest hit, receiving between 8 and 13 inches of rain (National Oceanic and Atmospheric Administration, 1992b).

Because of the flash flooding, numerous roads and bridges were destroyed. Nearly 65 miles of railroad tracks were washed out. A landslide, with boulders the size of automobiles, closed a local road near Hanover in Jefferson County. In Henryville in Clark County, people were forced to their rooftops by the rapidly rising water and were rescued by helicopter.

The torrential rain fell from thunderstorms that developed in warm, moist air during the early hours of August 8. The storm system remained nearly stationary from the time the intense rain began, just after 0300, until the storm dissipated in the early afternoon. The 100-year rainfall for a 6-hour period in this area of Indiana is about 5.3 inches (Glatfelter, 1984); the U.S. Soil Conservation Service received unofficial reports for this storm of more than 12.5 inches of rain during a 6-hour period in Scott County. A National Weather Service automated rain gage at Crothersville recorded 7 inches of rain during a 3-hour period (National Oceanic and Atmospheric Administration, 1992a).

The flash floods were most pronounced on the smaller tributaries and in the headwaters of the major streams. At the streamflow-gaging station, Muscatatuck River near Deputy (site 3, fig. 45), which has a drainage area of 293 square miles, the stream rose more than 29 feet in 12 hours. Maximum stages and maximum discharges at two other sites are listed in table 27. There were unofficial reports of local streams in Clark and Jefferson Counties rising 30 to 35 feet. Because flows in the larger receiving streams were relatively low prior to the storm, water that drained rapidly from the tributaries did not cause significant flooding along the larger streams and rivers.

REFERENCES

Glatfelter, D.R., 1984, Techniques for estimating magnitude and frequency of floods on streams in Indiana: U.S. Geological Survey Water-Resources Investigations Report 84–4134, 110 p.

National Oceanic and Atmospheric Administration, 1992a, Monthly report of river and flood conditions for August 1992: Indianapolis, Indiana, 3 p.

———1992b, Climatological data, Indiana, August 1992: Asheville, N.C., National Climatic Data Center, v. 97, no. 8, 28 p.

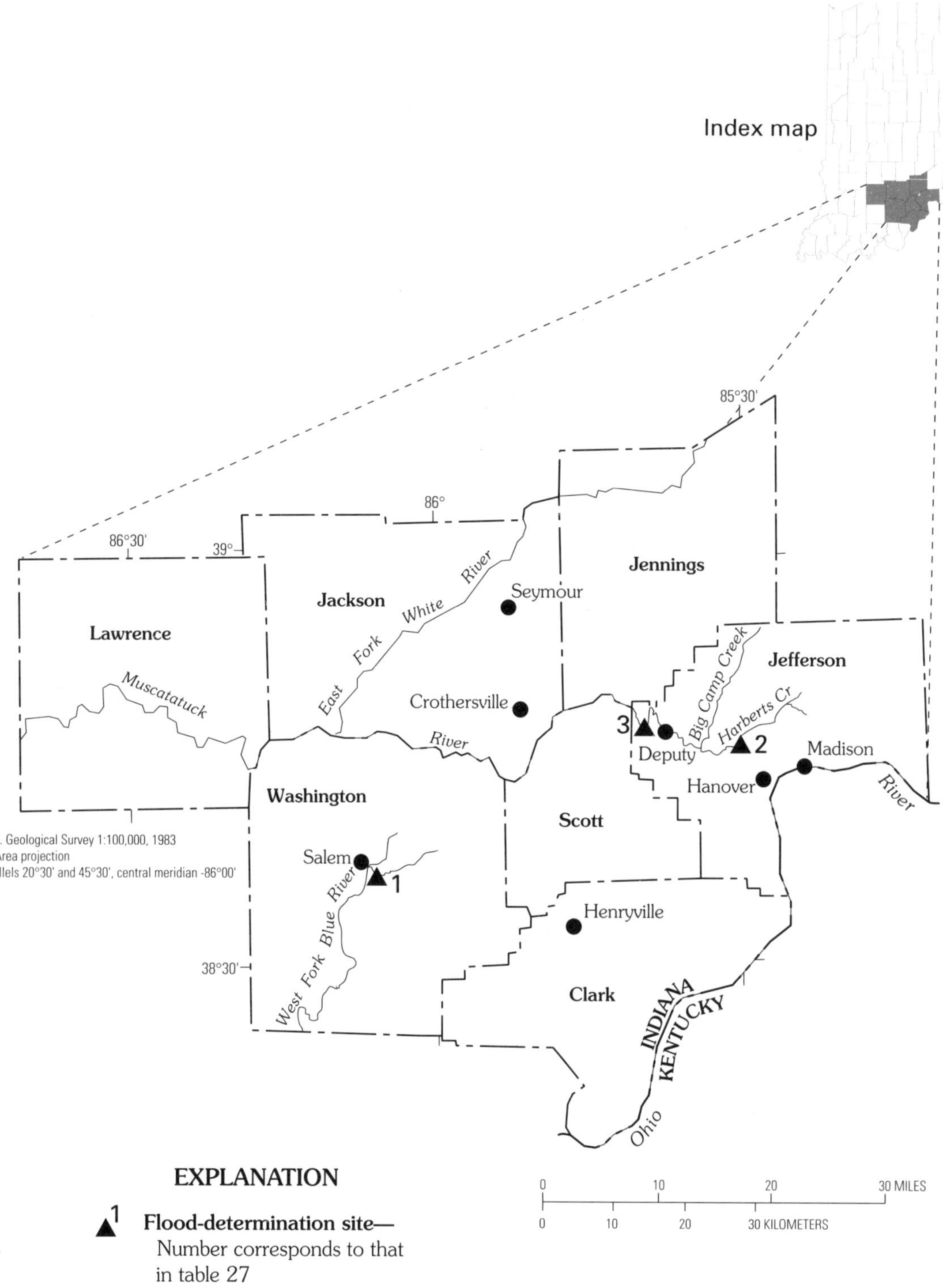

Base from U.S. Geological Survey 1:100,000, 1983
Albers Equal Area projection
Standard parallels 20°30' and 45°30', central meridian -86°00'

Figure 45. Location of flood-determination sites for flood of August 8, 1992, in southeastern and south-central Indiana.

Table 27. Maximum stages and discharges prior to and during August 8, 1992, in southeastern and south-central Indiana

[mi^2, square miles; ft, feet above an arbitrary datum; ft^3/s, cubic feet per second. Source: Recurrence intervals calculated from U.S. Geological Survey data. Other data from U.S. Geological Survey reports or data bases]

				Maximum prior to August 8, 1992				Maximum during August 8, 1992			
Site no. (fig. 45)	Station no.	Stream and place of determination	Drainage area (mi^2)	Period	Year	Stage (ft)	Discharge (ft^3/s)	Day	Stage (ft)	Discharge (ft^3/s)	Discharge recurrence interval (years)
1	03302680	West Fork Blue River at Salem, IN	19.0	1970–92	1990	15.58	9,240	8	11.28	3,530	10
2	03366200	Harberts Creek near Madison, IN	9.31	1968–92	1990	8.96	2,150	8	8.40	1,810	20
3	03366500	Muscatatuck River near Deputy, IN	293	1947–92	1959	34.30	52,200	8	31.70	27,000	15–20

September 1992, in South-Central Iowa

By Rodney Southard

Monthly rainfall totals varied greatly across Iowa during September 1992, from 1.22 inches in northern Iowa to 17.86 inches in Lucas County in south-central Iowa (National Oceanic and Atmospheric Administration, 1992). The south-central climatological division, which includes Lucas County, received an average of 10.82 inches or 266 percent of normal rainfall during September, making it the third wettest September in 103 years of record for south-central Iowa (National Oceanic and Atmospheric Administration, 1992).

Most of the rainfall in south-central Iowa fell on September 14 and 15. Intense thunderstorms developed over south-central Iowa about mid-morning on the 14th and redeveloped over the same areas until about 1200 on the 15th, resulting in 4 or more inches of rain over most of south-central Iowa (National Oceanic and Atmospheric Administration, 1992). The largest official rainfall totals were 12.06 inches at Derby and 11.05 inches near Promise City. The maximum 24-hour rainfall at Derby was 11.7 inches, which is the fifth largest official 24-hour rainfall total ever recorded in Iowa (National Oceanic and Atmospheric Administration, 1992). The largest unofficial rainfall total was 16 inches at Van Wert. The rainfall total at Van Wert was more than double the 100-year, 24-hour rainfall of 7.74 inches for that part of the State.

Runoff from the intense rainfall resulted in major flooding in the Thompson, Chariton, and South Fork Chariton River Basins (fig. 46, table 28). The maximum stage and discharge of the Thompson River at Davis City (site 7) were 24.29 feet and 57,000 cubic feet per second, respectively. The maximum discharge for this flood was substantially higher than the previous maximum discharge at Davis City of 30,000 cubic feet per second. The maximum stage for the Chariton River near Chariton (site 9, fig. 46) exceeded the previous maximum recorded by 6.18 feet and was more than double the previous maximum discharge. Excessive rainfall also occurred in the headwaters of the South Fork Chariton River. At South Fork Chariton River near Promise City (site 10, fig. 46), the maximum stage was 4.89 feet above the previous maximum of July 4, 1981, and the maximum discharge was more than 3.6 times the 100-year discharge. A listing of other streams considerably affected by the flooding is included in table 28.

REFERENCE

National Oceanic and Atmospheric Administration, 1992, Climatological data, Iowa, September 1992: Asheville, N.C., National Climatic Data Center, v. 103, no. 9, 36 p.

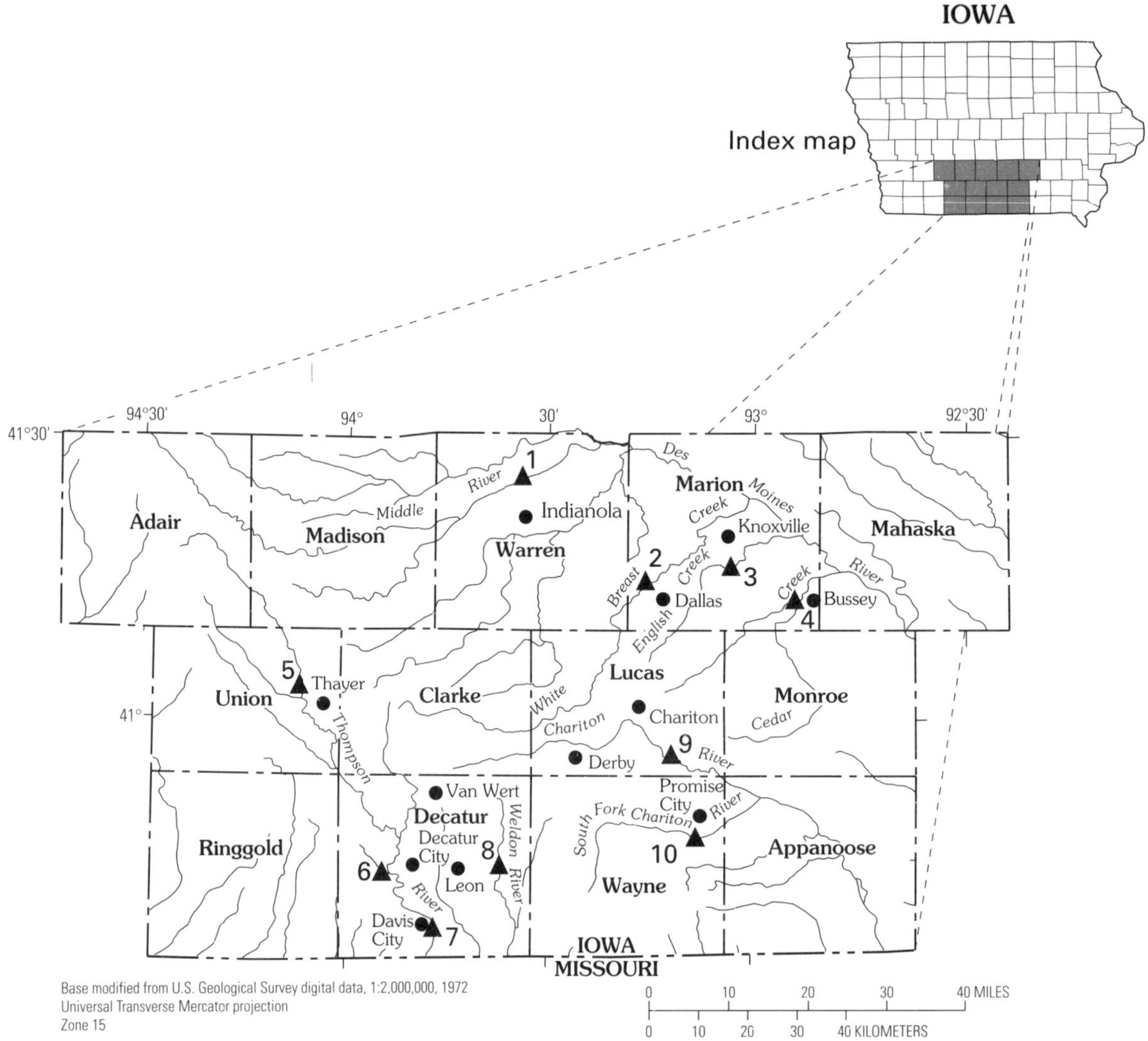

EXPLANATION

▲6 **Flood-determination site—**
Number corresponds to that in table 28

Figure 46. Location of flood-determination sites for flood of September 1992, in south-central Iowa.

Table 28. Maximum stages and discharges prior to and during September 1992, in south-central Iowa

[mi^2, square miles; ft, feet above an arbitrary datum; ft^3/s, cubic feet per second ; >, greater than; --, not determined or not available. Source: Recurrence intervals calculated from U.S. Geological Survey data. Other data from U.S. Geological Survey reports or data bases]

				Maximum prior to September 1992				Maximum during September 1992			
Site no. (fig. 46)	Station no.	Stream and place of determination	Drainage area (mi^2)	Period	Year	Stage (ft)	Discharge (ft^3/s)	Day	Stage (ft)	Discharge (ft^3/s)	Discharge recurrence interval (years)
1	05486490	Middle River near Indianaola, IA	503	1940–92	1947	28.27	34,000	16	23.23	13,500	10
2	05487980	White Breast Creek near Dallas, IA	342	1963–92	1982	33.45	37,300	16	31.57	30,300	>100
3	05488200	English Creek near Knoxville, IA	90.1	1985–92	1982	30.28	28,000	16	22.81	2,990	--
4	05489000	Cedar Creek near Bussey, IA	374	1948–92	1982	34.61	96,000	16	28.28	20,900	12
5	06897858	Seven Mile Creek near Thayer, IA	6.61	1991–92	1991	21.69	--	15	24.92	--	--
6	06897950	Elk Creek near Decatur City, IA	52.5	1968–92	1990	28.19	18,000	15	28.13	16,800	10
7	06898000	Thompson River at Davis City, IA	701	1918–25, 1941–92	1885	24.8	30,000	16	24.29	57,000	>100
8	06898400	Weldon River near Leon, IA (discontinued)	104	1959–91	1959	25.27	48,600	15	28.69	82,500	>100
9	06903400	Chariton River near Chariton, IA	182	1966–92	1981	23.14	16,600	15	29.32	37,700	>100
10	06903700	South Fork Chariton River near Promise City, IA	168	1968–92	1981	29.95	28,000	15	34.84	70,600	>100

September 5–6, 1992, on Cat Point Creek, Eastern Virginia

By Byron J. Prugh, Jr.

Runoff from locally excessive rains on September 5–6, 1992, in eastern Virginia caused the failure of Chandlers Millpond dam near Montross, Virginia (fig. 47). The resultant floodwaters generated a new maximum stage of record (1943–92) at the streamflow-gaging station on Cat Point Creek, 3.8 miles south of Montross and 4.5 miles downstream from the millpond. The maximum discharge of 5,240 cubic feet per second was the second highest discharge for the period of record. The highest discharge was 6,820 cubic feet per second recorded on August 20, 1969. The maximum stage of the September 5–6, 1992, flood (10.86 feet) was 0.41 foot higher than the maximum stage of August 1969 because of bridge and highway reconstruction in the 1970's that changed the geometry of the channel and road embankment. A natural flood of the magnitude recorded in September 1992 would have a recurrence interval of approximately 50 years. Figure 47 shows the location of the streamflow-gaging station, the precipitation gages, and total rainfall for September 5–6, 1992.

Varying amounts of rainfall were recorded in Virginia between September 3 and 7 as a frontal system spawned numerous thunderstorms as it moved across the State. The largest amounts of rainfall were recorded between Norfolk in the south and Fredericksburg in the north (fig. 47). More than 6 inches of rainfall were measured at some locations, with the largest amounts received late on September 5 and early on September 6. At Kilmarnock, 38 miles to the southeast of Montross, the 2-day total was 6.41 inches; at Colonial Beach, 15 miles to the northwest, the total was 2.09 inches; and at Warsaw, 10 miles to the southeast, the total was 3.49 inches (National Oceanic and Atmospheric Administration, 1992). Local reports indicated that 6 inches of precipitation fell in the Montross area between midnight and dawn on September 6.

A 150-to-200-foot wide breach in Chandlers Millpond dam also closed State Route 647, which crossed the crest of the dam. No deaths or serious injuries were reported.

REFERENCE

National Oceanic and Atmospheric Administration, 1992, Climatological data, Virginia, September 1992: Asheville, N.C., National Climatic Data Center, v. 102, no. 9, 23 p.

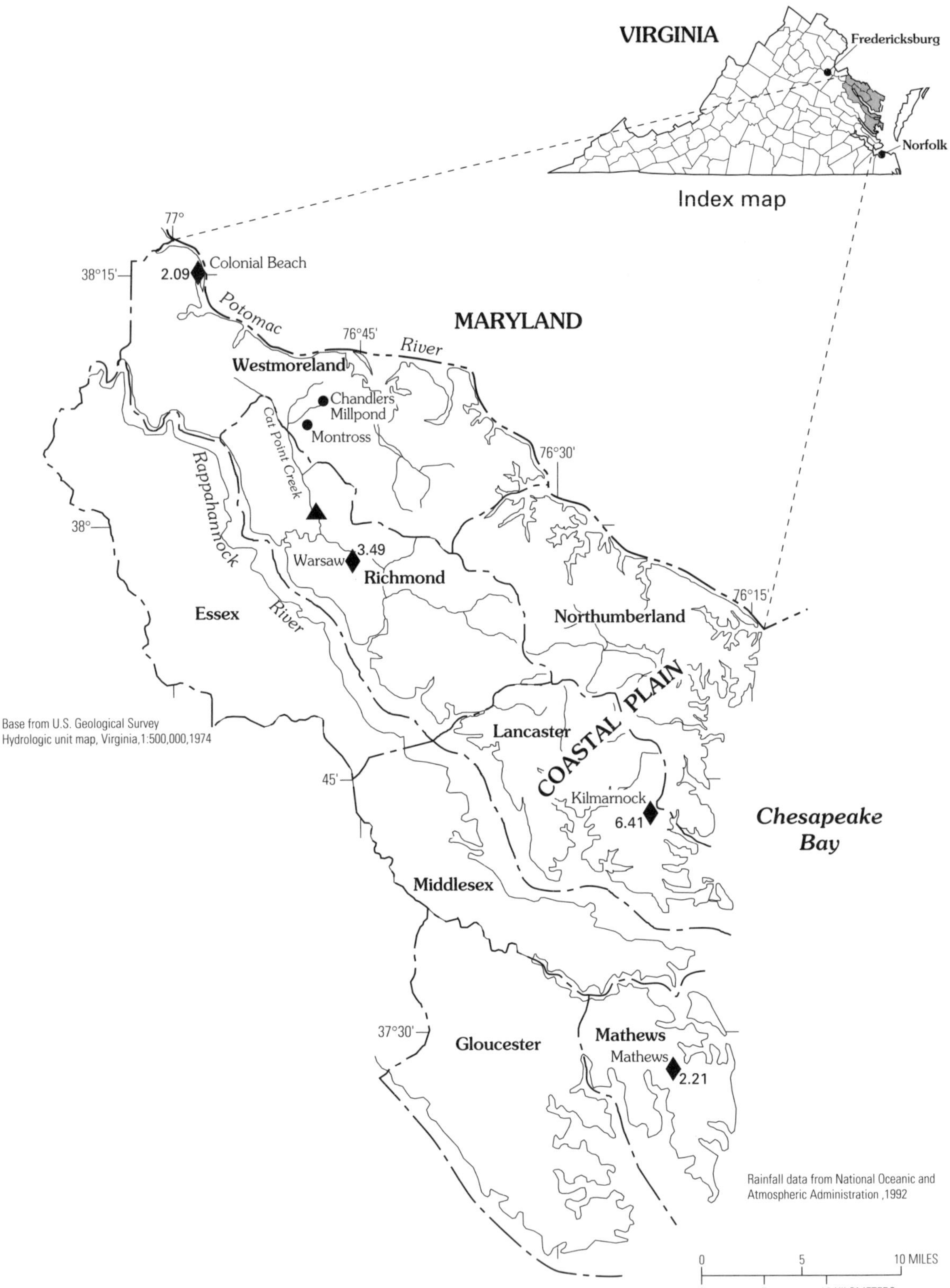

EXPLANATION

Warsaw ◆3.49 **Precipitation gage**—Number is total rainfall, in inches

▲ **Streamflow-gaging station**

Figure 47. Location of precipitation gages and streamflow-gaging station and total rainfall for September 5–6, 1992, in eastern Virginia.

September 10, 1992, Swain County, North Carolina

By W. Harold Eddins *and* Thomas J. Zembrzuski, Jr.

A severe flash flood swept along Raven Fork near Smokemont, North Carolina (fig. 48), between 1900 and 2200 on September 10, 1992, and caused serious damage to bridges, homes, campgrounds, commercial trout ponds, and maintenance facilities. Property damage was estimated in the millions of dollars, but due to timely warnings from the National Weather Service (NWS) and local emergency management personnel, there were no deaths or injuries. About 800 people were evacuated from their homes, and hundreds of tourists in campgrounds along the stream were warned to seek higher ground. Witnesses reported seeing a wall of water roaring down the stream at approximately 1930 and having only minutes to run for safety.

The flood was the result of about 3 hours of intense rainfall from a small but strong thunderstorm centered over the higher elevations of Mt. Guyot (elevation 6,621 feet above sea level) and Mt. Hardison (elevation 6,134 feet above sea level) at the headwaters of Raven Fork (fig. 48). The NWS determined from radar data that the storm began before 1900, and bucket surveys indicated approximately 5 to 8 inches of rain fell (R.C. Jones, National Weather Service, oral commun., March 1994). The excessive rainfall was limited to the upstream reaches of Raven Fork; little rain fell in the downstream, more densely populated part of the basin (below elevation 2,600 feet above sea level) where property damages were sustained from the floodwaters.

Raven Fork drains 74.3 square miles and is tributary to the Oconaluftee River. Maximum discharge was determined at a site near Smokemont, North Carolina (site 1, table 29). Flow was computed by the slope-area method (Dalrymple and Benson, 1976) to be 16,200 cubic feet per second. Maximum depths in the main channel ranged from 14 to 16 feet, and the average flow velocity ranged from 13 to 14 feet per second.

Data collected at the streamflow-gaging station on Oconaluftee River at Birdtown (site 2, fig. 48) provided further evidence of how localized the Raven Fork flood was. At the gage, the Oconaluftee River drains 184 square miles, which includes the entire Raven Fork drainage basin. Maximum discharge recorded at the gage was only 10,700 cubic feet per second, cresting at 2300, about 3.5 hours after the crest of Raven Fork.

The recurrence interval of the Raven Fork flood was approximately 500 years at site 1. By the time the crest had traveled downstream to the gage on Oconaluftee River, the discharge had declined to less than that of a 5-year recurrence interval.

REFERENCE

Dalrymple, Tate, and Benson, M.A., 1976, Measurement of peak discharge by the slope-area method: U.S. Geological Survey Techniques of Water-Resources Investigations, book 3, chap. A2, 12 p.

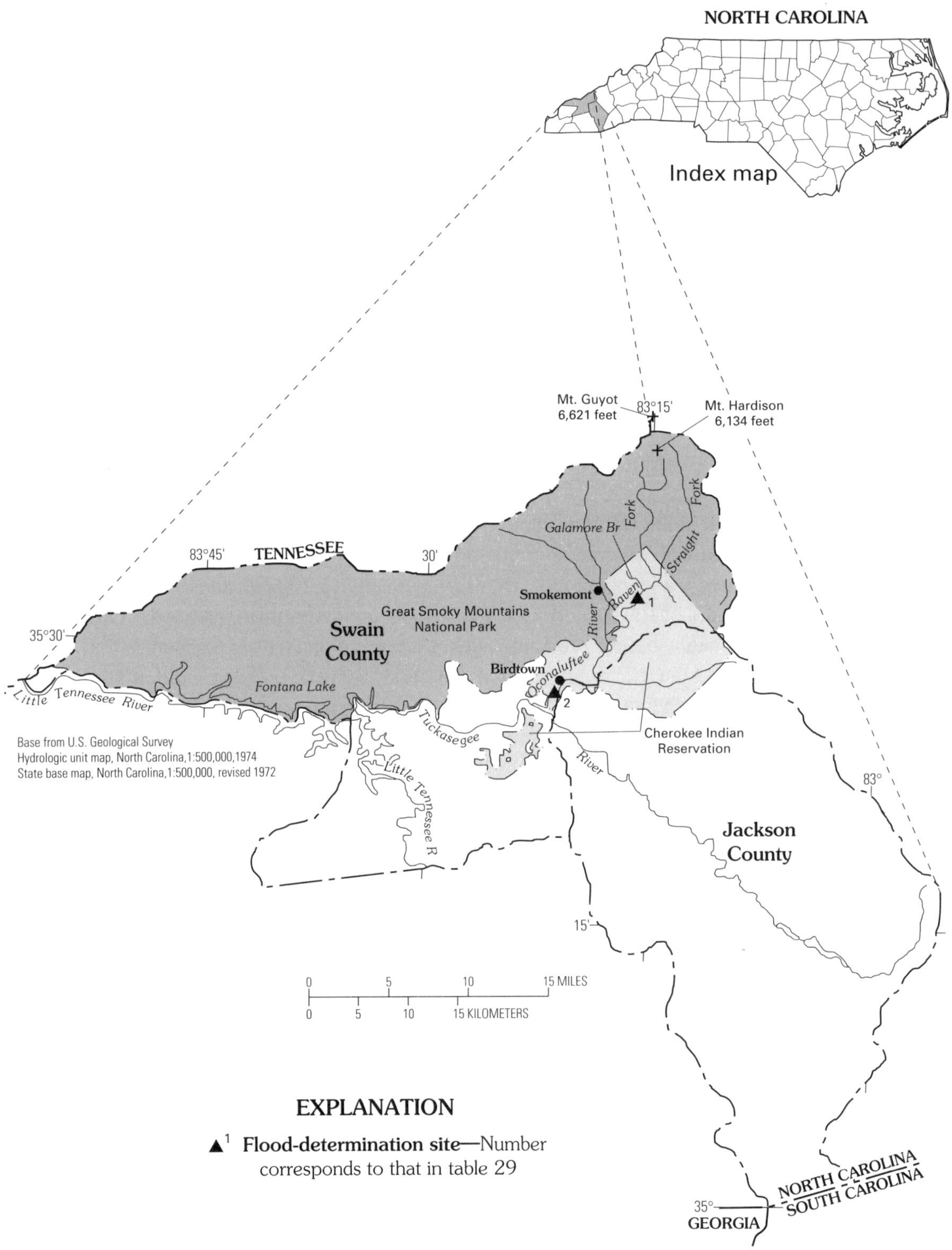

Figure 48. Location of flood-determination sites for flood of September 10, 1992, in Swain County, North Carolina.

Table 29. Maximum stages and discharges prior to and during September 10, 1992, Swain County, North Carolina

[mi^2, square miles; ft, feet above an arbitrary datum; ft^3/s, cubic feet per second; --, not determined or not available; <, less than. Source: Recurrence intervals calculated from U.S. Geological Survey data. Other data from U.S. Geological Survey reports or data bases]

Site no. (fig. 48)	Station no.	Stream and place of determination	Drainage area (mi^2)	Maximum prior to September 10, 1992: Period	Year	Stage (ft)	Discharge (ft^3/s)	Maximum on September 10, 1992: Day	Stage (ft)	Discharge (ft^3/s)	Discharge recurrence interval (years)
1	0351092910	Raven Fork above Galamore Branch near Smokemont, NC	64.9	--	1969	--	[1]9,000	10	--	16,200	500
2	03512000	Oconaluftee River at Birdtown, NC	184	1946, 1949–92	1969	12.46	15,900	10	9.44	10,700	<5

[1]Unpublished data from Tennessee Valley Authority.

October 9–10, 1992, in South Carolina

By Curtis L. Sanders, Jr.

Flooding occurred on October 9–10, 1992, as a result of excessive rainfall in Allendale, Hampton, and Colleton Counties in southern South Carolina. Total rainfall amounts at the towns of Allendale and Hampton, in Hampton County, were 7.44 and 5.02 inches, respectively, between 1700 on October 7 and 0800 on October 9, 1992 (National Oceanic and Atmospheric Administration, 1992).

Flood data for selected sites (fig. 49) in the Coosawhatchie, Salkehatchie, and Savannah River Basins are listed in table 30. Flooding on the Coosawhatchie River at State Secondary Road S–21 near Fairfax (site 6, fig. 49) was greatly increased by flow from a breached dam on a 60-acre pond on a tributary that joins the Coosawhatchie River just upstream from the streamflow-gaging station (site 6).

Maximum discharges at streamflow-gaging stations in the basins had recurrence intervals ranging from about 60 years to more than 100 years. Maximum discharges equaled or exceeded the 100-year recurrence interval at nine sites. Three persons lost their lives, and three were seriously injured in flood-related accidents.

REFERENCE

National Oceanic and Atmospheric Administration, 1992, Climatological data, South Carolina: Asheville, N.C., National Climate Data Center, v. 95, no. 10, 24 p.

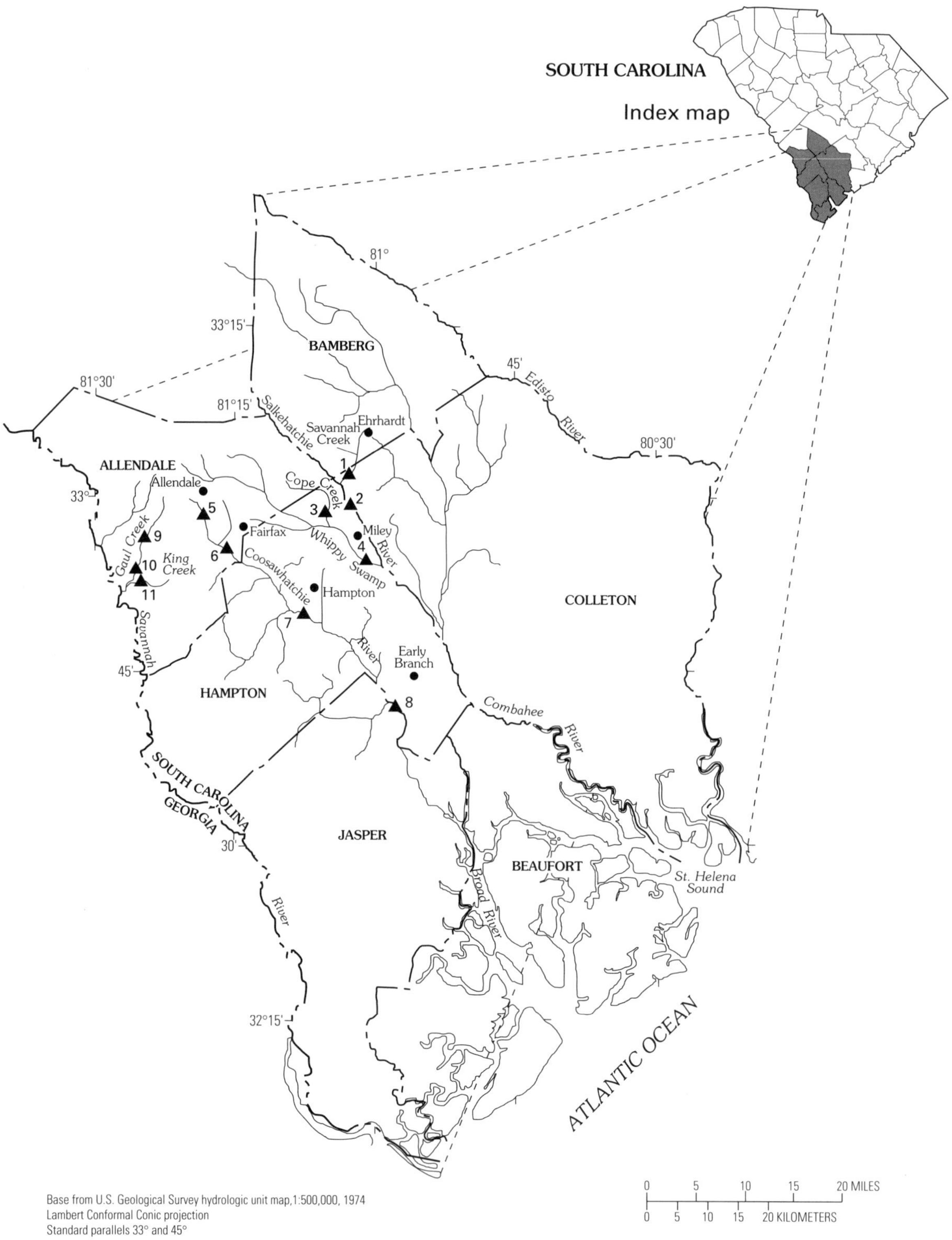

EXPLANATION

▲10 **Flood-determination site**—Number corresponds to that in table 30

Figure 49. Location of flood-determination sites for flood of October 9–10, 1992, in South Carolina.

Table 30. Maximum stages and discharges prior to and during October 9–10, 1992, in South Carolina

[mi^2, square miles; ft, feet above arbitrary datum; ft^3/s, cubic feet per second; >, greater than; --, not available. Source: Recurrence intervals calculated from U.S. Geological Survey data. Other data are from U. S. Geological Survey reports or data bases]

Site no. (fig. 49)	Station no.	Stream and place of determination	Drainage area (mi^2)	Maximum prior to October 1992				Maximum during October 9–10, 1992			
				Period	Date (month/day/year)	Stage (ft)	Discharge (ft^3/s)	Day	Stage (ft)	Discharge (ft^3/s)	Discharge recurrence interval (years)
					Salkehatchie River Basin						
1	02175450	Savannah Creek near Ehrhardt, SC	12.4	1964–74, 1975–93	03/13/80	8.30	895	9	9.33	1,200	>100
2	02175500	Salkehatchie River near Miley, SC	341	1951–93	03/13/80	5.44	3,300	9	5.79	4,360	100
3	--	Cope Creek at State Secondary Road 13 near Miley, SC	12.8	--	--	--	--	9	--	2,330	>100
4	--	Whippy Swamp at State Secondary Road 13 near Miley, SC	134	--	--	--	--	9	--	10,100	>100
					Coosawhatchie River Basin						
5	--	Coosawhatchie River at State Secondary Road 47 near Allendale, SC	10.5	--	--	--	--	9	--	1,100	60
6	--	Coosawhatchie River at State Secondary Road 21 near Fairfax, SC	48.1	--	--	--	--	9	--	[1]11,900	>100

Table 30. Maximum stages and discharges prior to and during October 9–10, 1992, in South Carolina—Continued

				Maximum prior to October 1992				Maximum during October 9–10, 1992			
Site no. (fig. 49)	Station no.	Stream and place of determination	Drainage area (mi^2)	Period	Date (month/ day/year)	Stage (ft)	Discharge (ft^3/s)	Day	Stage (ft)	Discharge (ft^3/s)	Discharge recurrence interval (years)
				Coosawhatchie River Basin—Continued							
7	02176500	Coosawhatchie River near Hampton, SC	203	1951–93	09/02/69	8.39	8,160	10	7.92	8,820	>100
8	--	Coosawhatchie River at State Secondary Road 36 near Early Branch, SC	382	--	--	--	--	10	--	14,100	>100
				Savannah River Basin							
9	--	Gaul Creek at State Secondary Road 107 near Allendale, SC	8.5	--	--	--	--	9	--	2,240	>100
10	--	Gaul Creek at State Highway 3 near Allendale, SC	17.9	--	--	--	--	9	--	4,320	>100
11	--	King Creek at State Highway 3 near Allendale, SC	17.9	--	--	--	--	9	--	1,560	70

[1]Flooding was greatly increased by flow from a breached dam on a 60-acre pond on a tributary just upstream from the streamflow-gaging station.

December 11–12, 1992, in New Jersey

By Thomas P. Suro

An intense, slow-moving "nor'easter" storm hit the eastern coast of New Jersey during December 11 and 12, 1992. This storm produced strong winds and record and near-record flooding along the entire Atlantic Coast of New Jersey from Bergen County to Cape May (fig. 50). Two deaths were attributed to the storm. The President of the United States declared Bergden, Essex, Hudson, Somerset, Union, Middlesex, Monmouth, Ocean, Salem, Atlantic, Cumberland, and Cape May Counties a disaster area. The State was granted $46 million in disaster relief funds for public damages (A.S. Mangeri, New Jersey State Police, Emergency Management Office, oral commun., 1994) and $265 million for insured damage (National Weather Service, 1994) that occurred as a result of this storm. (No information on uninsured private damages is available.) The hardest hit areas were near Raritan, Newark, Sandy Hook, and Upper New York Bay. The effects of this storm were documented by the U.S. Geological Survey (USGS) from records at tidal crest-stage gages and from streamflow-gaging stations, and from high-water marks located along the entire coast. The storm's effects also were recorded by several National Ocean Service (NOS) tide gages along the New Jersey coast and in adjacent States.

Data were collected from 11 tidal crest-stage gages, 1 streamflow-gaging station, 2 USGS tide gages, and 5 NOS tide gages to document the effects of the storm on the New Jersey coast. After the storm had passed, the USGS identified high-water marks to document tidal flooding within the State. The elevations of these high-water marks were determined by the New Jersey Department of Transportation and several counties.

The location of and coincident storm elevations at the USGS tide gages and crest-stage gages, as well as the NOS tide gages, are shown in figure 50. The highest tide elevations during the storm occurred in the northern part of the State near Raritan Bay (by crest-stage gages) at Perth Amboy (site 2; 10.4 feet) and Keyport (site 3; 10.13 feet above sea level). The NOS tide gages at Cape May (site 17), Atlantic City (site 13), and Sandy Hook (site 4) recorded maximum tide elevations of 6.83, 7.20, and 8.68 feet above sea level (the highest tide elevations recorded at these sites during their periods of record).

The northern parts of the State sustained more extensive flooding and higher observed tide elevations than other parts of the State as a result of the timing between the normal tide cycle and the storm surge. The storm surge occurred nearly at low tide at Cape May (see fig. 51) and nearly at high tide at Sandy Hook (fig. 52). High-water marks indicate that parts of Monmouth and Middlesex Counties experienced even higher tide elevations than those recorded by these gages. Record elevations were observed at Keyport (site 3), Manahawkin (site 6), Beach Haven (site 7), and Ocean City (site 15). Several of the crest-stage gages that did not record a new maximum for the period of record did record the highest maximum since the "nor'easter" of March 7, 1962. The adjacent Coastal States of New York and Delaware, as well as Maryland, also experienced record and near-record tide elevations as a result of this storm. The USGS tide gages and crest-stage gages and the NOS tide gages that recorded elevations during the December 1992 storm are listed in table 31.

This storm produced precipitation in the coastal areas that ranged from 0.6 inch at Cape May to more than 2.4 inches at Sandy Hook. Inland precipitation ranged from more than 1.70 inches at Somerville to more than 3.80 inches at Mooristown (National Oceanic and Atmospheric Administration,1993). Most of the flooding and damage were the result of increased tide elevations. Some nontidal streamflow-gaging stations recorded large discharges associated with the storm as the result of runoff from excessive rainfall. The most severe flooding and damage occurred in Monmouth and Middlesex Counties near Raritan Bay. Major highways and railroads were closed from a few hours to several days. At one point, the Garden State Parkway was closed in the area of Cheesequake. Parts of New Jersey Routes 30, 35, 36, 40, 47, and 72 also were closed during the storm. In addition, the New Jersey Transit's North Coast line was shut down due to flooding and debris. The PATH subway line also was flooded for many days in the Newark area because of this storm. Flooding was less severe in the southern part of New Jersey because the surge from the storm

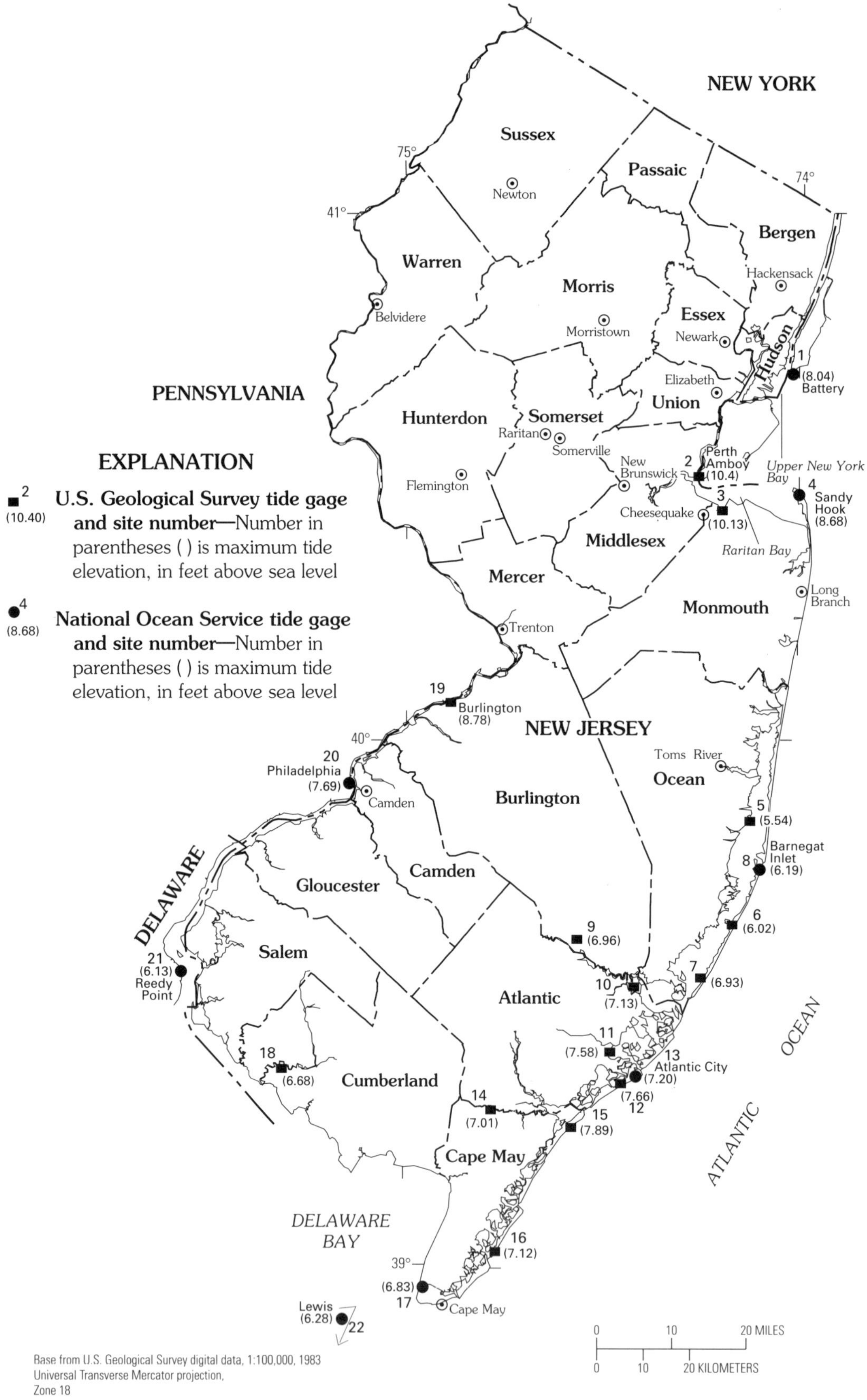

Figure 50. Location of U.S. Geological Survey and National Oceanic Service tide gages and maximum tide elevations recorded for storm of December 11–12, 1992, in New Jersey.

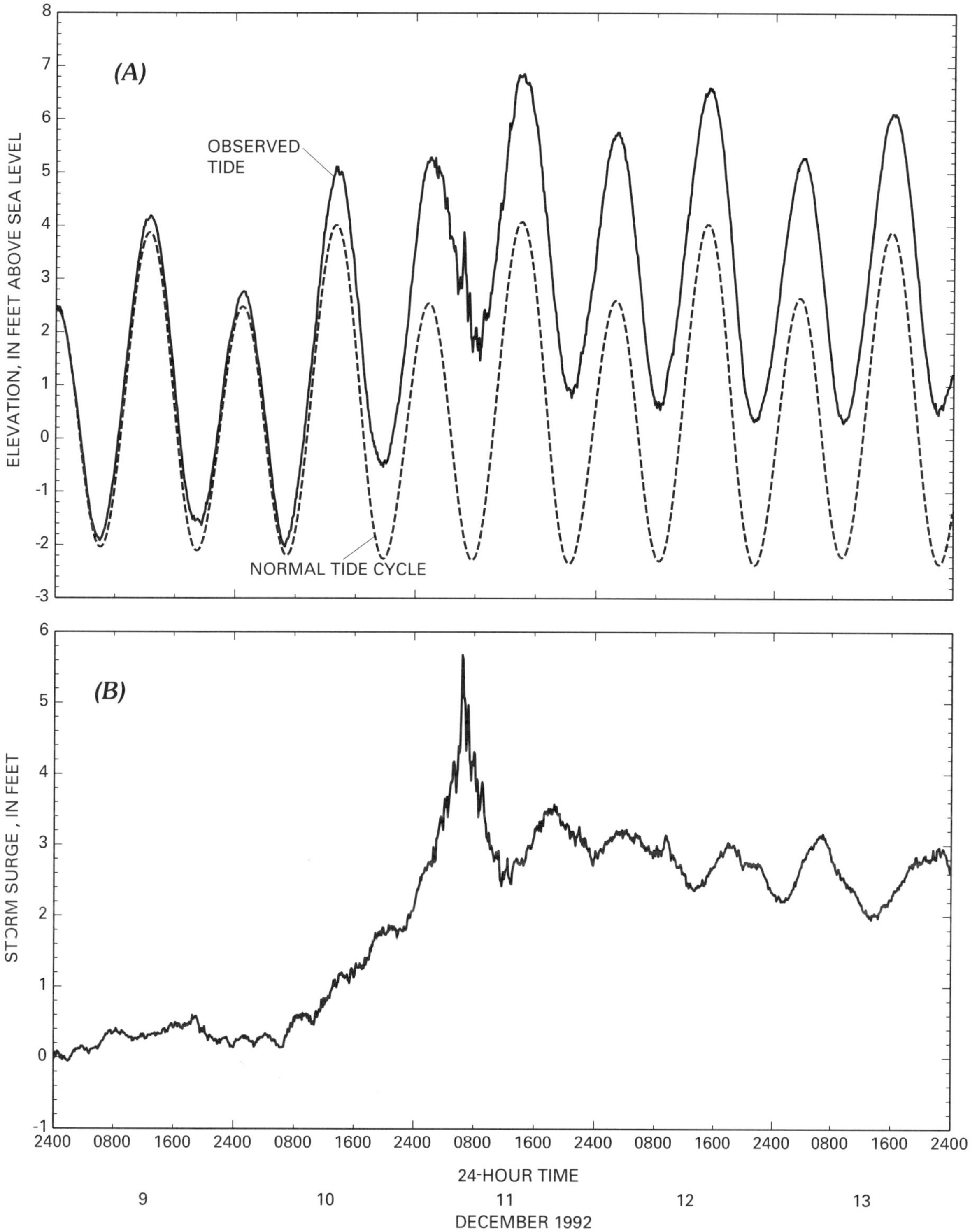

Figure 51. *(A)* Observed tide levels, normal tide cycle, and *(B)* storm surge at National Oceanic Service tide gage at Cape May, New Jersey, December 9–13, 1992. (Data from Len Hickman, National Ocean Service, National Oceanic and Atmospheric Administration, written commun., 1994.)

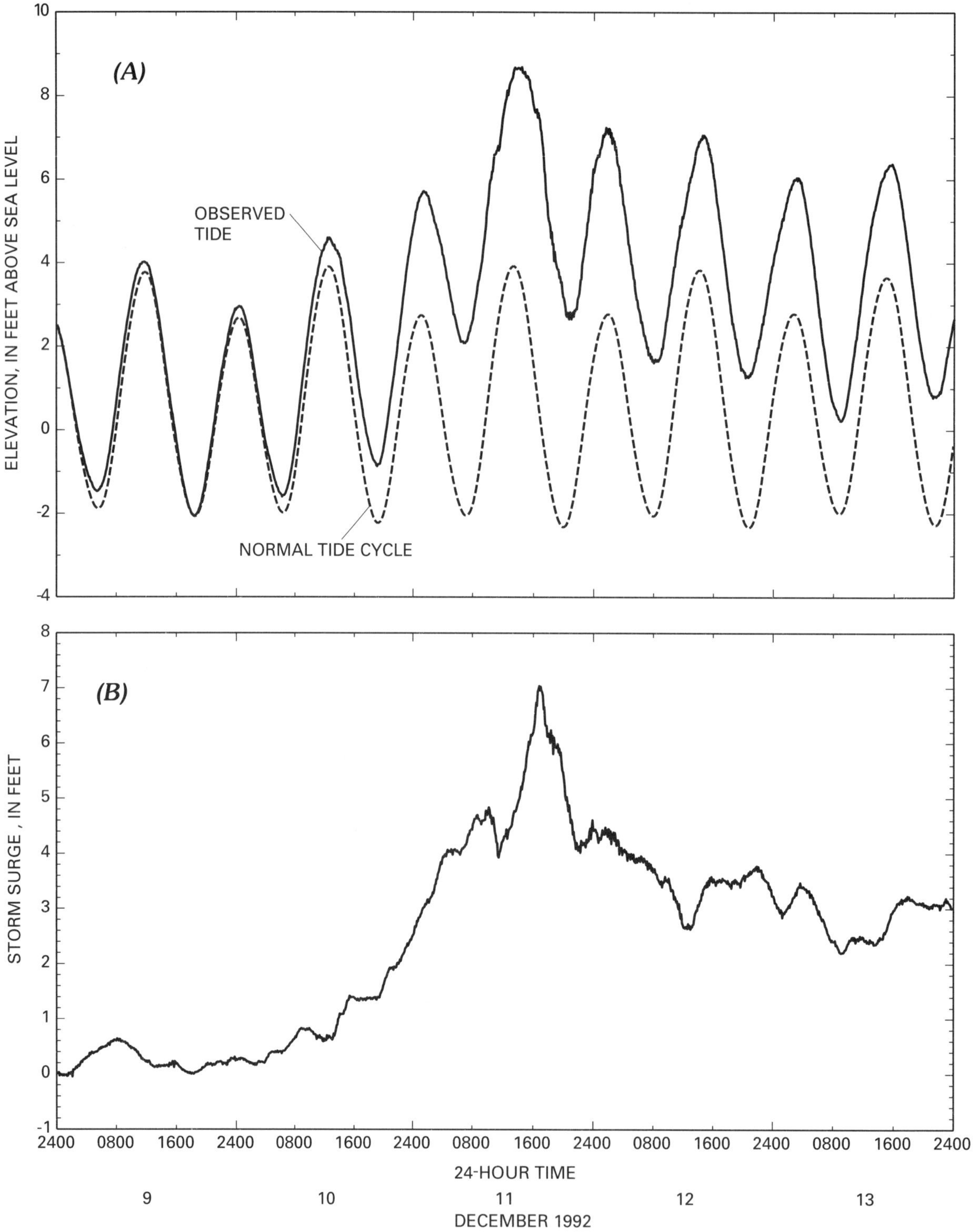

Figure 52. *(A)* Observed tide levels, normal tide cycle, and *(B)* storm surge at National Ocean Service tide gage at Sandy Hook, New Jersey, December 9–13, 1992. (Data from Len Hickman, National Ocean Service, National Oceanic and Atmospheric Administration, written commun., 1994.)

Table 31. Maximum tide elevations prior to and during December 11–12, 1992, in New Jersey

[ft, feet above sea level; --, not determined or not applicable. Sources: U.S. Geological Survey reports or data bases. National Ocean Service, National Oceanic and Atmospheric Administration]

Site no. (fig. 50)	Station no.	Site name and place of determination	Maximum prior to December 1992: Period (water years)	Maximum prior to December 1992: Date (month/day/year)	Maximum prior to December 1992: Tide elevation (ft)	Maximum during December 11–12, 1992: Tide elevation (ft)
1	[1]8518750	National Ocean Service tide gage at The Battery, NY	1920–92	9/12/60	[3]8.35	8.04
2	[2]01406700	Raritan River at Perth Amboy, NJ	[4]1938, 1944, 1950, 1953, 1955, 1960, 1966–69, 1979–92	9/12/60	10.0	10.4
3	[2]01407030	Luppatatong Creek at Keyport, NJ	1944, 1950, 1960, 1980–92	9/12/60	10.3	10.13
4	[1]8531680	National Ocean Service tide gage at Sandy Hook, NJ	1933–92	3/6/62	[3]8.56	8.68
5	[2]01409000	Cedar Creek at Lanoka Harbor, NJ	1936–37, 1944–45, 1950, 1955, 1979–85, 1992	2/18/36	6.45	5.54
6	[2]01409145	Manahawkin Bay near Manahawkin, NJ	1965–92	3/29/84	5.36	6.02
7	[2]01409285	Little Egg Harbor at Beach Haven, NJ	1979–92	3/29/84	6.19	6.93
8	[1]8533615	National Ocean Service tide gage at Barnegat Inlet, NJ	1965–80, 1984–92	8/9/76	6.16	6.19
9	[2]01409510	Batsto River at Pleasant Mills, NJ	1958–92	3/7/62	7.2	6.96
10	[2]01410100	Mullica River near Port Republic, NJ	1962, 1965–92	3/6/62	7.9	7.13
11	[2]01410500	Absecon Creek at Absecon, NJ	[4]1924–29, [4]1933–40, [4]1946–84, 1985–92	3/29/84	7.77	7.58
12	[2]01410570	Beach Thorofare at Atlantic City, NJ	1944, 1950, 1960, 1962, [4]1978, 1969–92	3/6/62	8.3	7.66
13	[1]8534720	National Ocean Service tide gage at Atlantic City, NJ	1911–92	9/14/44	[3]7.56	7.20
14	[2]01411300	Tuckahoe River at Head of River, NJ	[4]1979–92	3/29/84	7.00	7.01
15	[2]01411320	Great Egg Harbor Bay at Ocean City, NJ	1965–92	3/29/84	7.53	7.89

Table 31. Maximum tide elevations prior to and during December 11–12, 1992, in New Jersey—Continued

Site no. (fig. 50)	Station no.	Site name and place of determination	Maximum prior to December 1992			Maximum during December 11–12, 1992
			Period (water years)	Date (month/day/year)	Tide elevation (ft)	Tide elevation (ft)
16	[2]01411360	Great Channel at Stone Harbor, NJ	1965–92	3/29/84	7.33	7.12
17	[1]8536110	National Ocean Service tide gage at Cape May, NJ	1965–92	9/27/85	[3]7.09	6.83
18	[2]01413038	Cohansey River at Greenwich, NJ	1950, 1979–92	11/25/50	8.8	6.68
19	[2]01464598	Delaware River at Burlington, NJ	[4,5]1921–26, [4,5]1931–40, [4,5] 1951–55, [4,5] 1957–64, [4]1965–92	8/20/55	10.8	8.78
20	[1]8545530	National Ocean Service tide gage at Philadelphia, PA	1900–92	11/25/50	[3]8.50	7.69
21	[1]8551910	National Ocean Service tide gage at Reedy Point, DE	1919–92	10/25/80	[3]6.68	6.13
22	[1]8557380	National Ocean Service tide gage at Lewes, DE	1956–92	3/7/62	[3]7.76	6.28

[1]Data from Deitemyer (1993).
[2]Data from Bauersfeld and others (1994).
[3]National Ocean Service, National Oceanic and Atmospheric Administration, written commun., 1994.
[4]Operated as a continuous-record streamflow-gaging station.
[5]Operated by U.S. Army Corps of Engineers, Philadelphia District.

hit the coast closer to low tide there (fig. 51*B*). By the time the storm reached Middlesex and Monmouth Counties, the surge associated with the storm was nearly in phase with the high-tide cycle (fig. 52), which resulted in much more intense flooding than in parts of southern New Jersey. As a result of the full moon on December 9 and an unusually long storm duration that affected several tide cycles (Deitemyer, 1993), the resulting tide caused the most severe flooding along the New Jersey coast in nearly 30 years.

REFERENCES

Bauersfeld, W.R., Moshinsky, E.W., and Gurney, C.E., 1994, Water-resource data for New Jersey—water year 1993, v. 1, Surface-water data: U.S. Geological Survey Water-Data Report NJ–93–1, 523 p.

Deitemyer, D.H., 1993, Effects of December 1992 northeaster on water levels, data report: Silver Spring, Md., National Oceanic and Atmospheric Administration, National Ocean Service, NOAA Technical Memorandum NOS OES 006, 51 p., 2 app.

National Oceanic and Atmospheric Administration, 1993, Climatological data—New Jersey, December 1992: Asheville, N.C., National Climatic Data Center, v. 97, no. 12, 11 p.

National Weather Service,1994, Disaster survey report, The Great Nor'easter of December 1992: National Weather Service Eastern Region, 59 p., 1 app.

December 30, 1992–January 15, 1993, in Northern Indiana

By Donald V. Arvin

Rain falling on snow-covered, frozen ground created flood conditions in northern Indiana during the last week of December 1992 and the first two weeks of January 1993. Flooding occurred on many rivers including the Elkhart, Tippecanoe, Yellow, and Kankakee Rivers (fig. 53). Flooding also occurred on numerous lakes in northeast Indiana and in both St. Joseph River Basins.

On Christmas Eve temperatures plunged, causing the ground to freeze throughout northern Indiana. Snow cover was 3 to 7 inches in the area that later experienced flooding (National Oceanic and Atmospheric Administration, 1992). Rains of 2 to 3 inches fell December 30–31, melting the snow and causing rapid surface runoff. Discharge on December 31 at the streamflow-gaging station, Fish Creek at Hamilton (site 6, fig. 53), reached a maximum of 757 cubic feet per second, exceeding the 100-year recurrence interval (table 32). Just as floodwaters were beginning to recede, an additional 1.5 to 2.5 inches of rain fell across the entire State on January 3–4 (National Oceanic and Atmospheric Administration, 1993). On January 5, discharge at Fish Creek at Hamilton reached a maximum of 729 cubic feet per second, again exceeding the 100-year recurrence interval. Substantial flooding during this storm also occurred in other nearby basins. Discharges corresponded to about the 50-year recurrence interval for Pigeon Creek near Angola (site 1, fig. 53), and about the 25-year recurrence interval for North Branch Elkhart River at Cosperville (site 3, fig. 53) and for the Yellow River at Plymouth (site 10, fig. 53). On January 5, the discharge of 13,400 cubic feet per second was the maximum for the period of record at St. Joseph River near Fort Wayne (site 8, fig. 53).

Because of this winter flood, many local and State roads were closed, and many homes and businesses were threatened. The Tippecanoe River flooded nearly 200 homes in Carroll County. Nearly 150 homes at a resort in Newton County were saved from flood damage when sandbagging efforts by the National Guard prevented failure of a weakened dike along the Kankakee River. Sandbagging operations by local officials prevented flood damage in the business districts of Plymouth and South Bend. In total, at least 1,000 homes were evacuated during the flood.

REFERENCES

National Oceanic and Atmospheric Administration, 1992, Climatological data, Indiana, December 1992: Asheville, N.C., National Climatic Data Center, v. 97, no. 12, 32 p.

———1993, Climatological data, Indiana, January 1993: Asheville, N.C., National Climatic Data Center, v. 98, no. 1, 32 p.

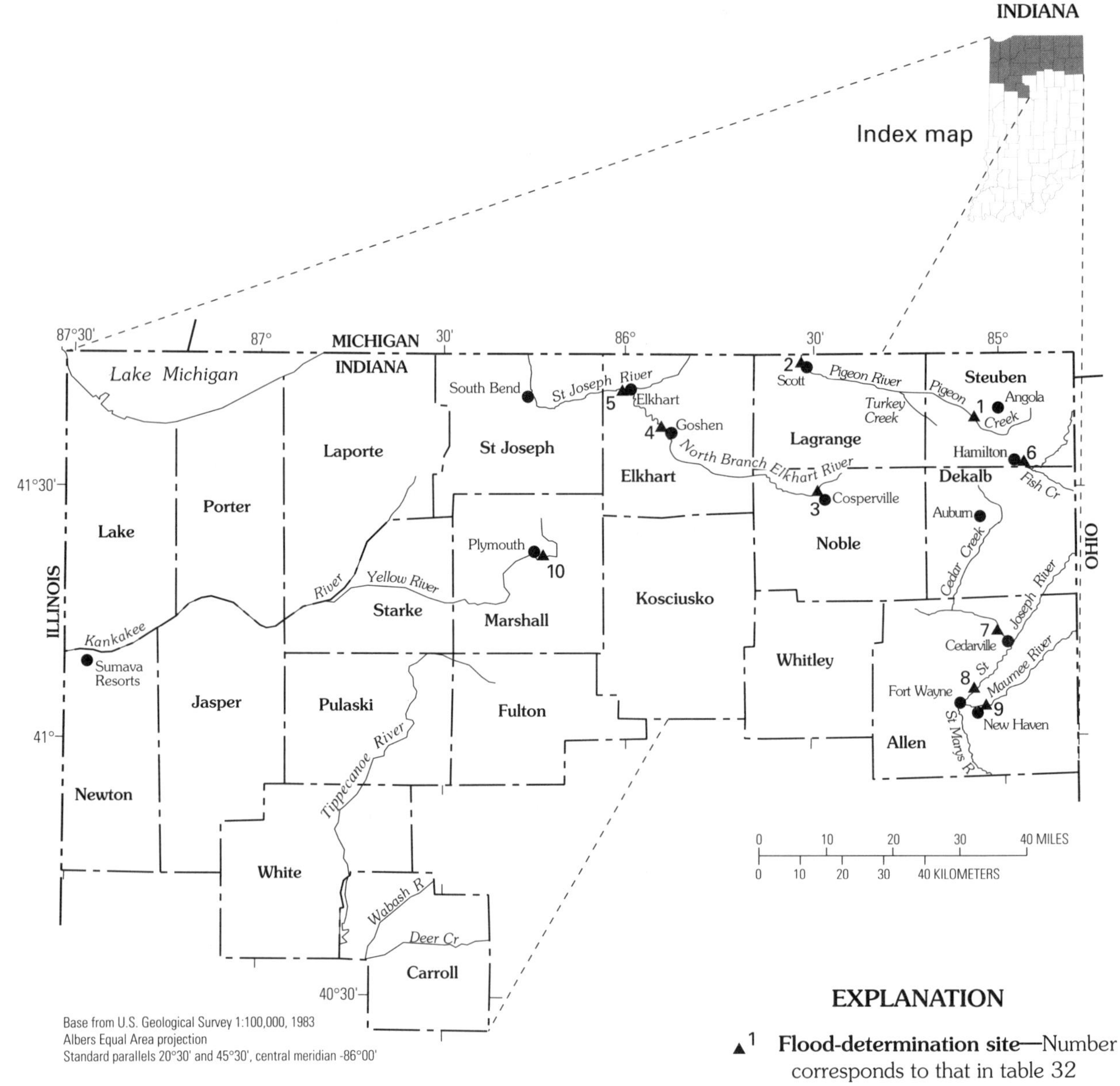

Figure 53. Location of flood-determination sites for flood of December 30, 1992–January 15, 1993, in northern Indiana.

Table 32. Maximum stages and discharges prior to and during December 30, 1992–January 15, 1993, in northern Indiana

[mi^2, square miles; ft, feet above an arbitrary datum; ft^3/s, cubic feet per second; >, greater than. Source: Recurrence intervals calculated from U.S. Geological Survey data. Other data from U.S. Geological Survey reports or data bases]

				Maximum prior to December 1992				Maximum during December 1992–January 1993			
Site no. (fig. 53)	Station no.	Stream and place of determination	Drainage area (mi^2)	Period	Year	Stage (ft)	Discharge (ft^3/s)	Date (month/day)	Stage (ft)	Discharge (ft^3/s)	Discharge recurrence interval (years)
1	04099510	Pigeon Creek near Angola, IN	106	1945–93	1982	13.90	795	1/7	10.21	784	50
2	04099750	Pigeon River near Scott, IN	361	1968–93	1982	7.85	2,370	1/6	7.16	1,930	10
3	04100222	North Branch Elkhart River at Cosperville, IN	142	1971–93	1982	8.12	919	1/7	7.60	839	25
4	04100500	Elkhart River at Goshen, IN	594	1931–93	1985	11.87	6,360	1/5	9.69	4,600	10
					1982	11.94	6,180				
5	04101000	St. Joseph River at Elkhart, IN	3,370	1947–93	1985	27.05	18,800	1/5	25.75	14,700	10
					1982	27.91	18,600				
6	04177720	Fish Creek at Hamilton, IN	37.5	1969–93	1990	9.75	683	12/31	10.28	757	>100
					1985	11.95	654				
7	04180000	Cedar Creek near Cedarville, IN	270	1946–93	1990	13.38	5,580	1/5	12.11	4,840	10
8	04180500	St. Joseph River near Fort Wayne, IN	1,060	1983–93	1985	17.79	13,200	1/5	18.40	13,400	25
					1989	17.86	13,200				
9	04183000	Maumee River at New Haven, IN	1,967	1946–93	1982	25.49	26,600	1/6	22.03	19,700	10
10	05516500	Yellow River at Plymouth, IN	294	1948–93	1954	17.13	5,390	1/6	14.25	3,200	25

SUMMARY OF FLOODS OF 1993

January and February 1993, in Southern California

By James C. Bowers

THE STORMS AND RESULTING FLOODS

From January 6 to February 28, 1993, a series of storms produced 20 to 40 inches of rain over much of the southern California coastal and mountain areas and more than 52 inches at some precipitation gages in the San Bernardino Mountains (National Oceanic and Atmospheric Administration, 1993). These storms, which coincided with a reappearance of weak warming of sea-surface temperatures in the tropical regions of the Pacific Ocean, were driven by a regional atmospheric low-pressure system off the coast of northern California and Oregon (National Oceanic and Atmospheric Administration, 1994). In southern California, precipitation intensified because a high-pressure area that extended over Alaska, the Gulf of Alaska, and the western States concentrated this low-pressure system farther south than usual and held it in place just offshore, sending a series of storms into southwestern part of the United States. Tropical moisture was pulled up from the south by the jetstream, which crossed the west coast from the southwest at about the latitude of San Diego. Arriving storms from the jetstream helped generate intense rain and snow. Rainfall in southern California was further intensified by the orographic effect of the east-west-trending mountains, which helped to enhance precipitation in the San Gabriel and San Bernardino Mountains. Precipitation also was excessive in the Laguna Mountains of San Diego County (fig. 54).

Substantial peak streamflows occurred January 6–7 as a result of excessive rainfall on a fairly substantial snowpack that had accumulated in December 1992. The recurrence intervals of discharges in the Mojave and Santa Ana River Basins were about 25 to 50 years (table 33). In the Victorville area, the Mojave River overtopped a levee, causing damage to a housing tract. This reach of the river channel had become extensively overgrown with vegetation as a result of the drought conditions of the past 6 years and the lack of channel-clearing high flows.

Rain continued throughout the first 2 weeks of January, and a second major runoff occurred late on January 16 as the low-pressure system moved slightly south before moving to the east on January 18. The most severe flooding was in the Santa Margarita and San Luis Rey River Basins (fig. 55) in northern San Diego and southern Riverside Counties. In the 24-hour period beginning at 0800 on January 16, 6.80 inches of rain were recorded at the Santa Rosa Plateau precipitation gage (shown in fig. 55) in the Santa Margarita River Basin, and similar rainfall totals were reported throughout the area (National Oceanic and Atmospheric Administration, 1993).

A nearly identical storm pattern developed in early February as a stationary atmospheric low-pressure system centered off the Oregon coast. Major storms and resultant runoff occurred February 8 and from February 18–19. Although the maximum discharges did not exceed the 50-year recurrence-interval magnitudes, substantial local flooding occurred because of the saturated conditions in the basins due to the January storms. Major streambank failures occurred along the Mojave River in the Silver Lakes area (fig. 54), about 10 miles north of Victorville, as a result of sustained high flow in the normally dry channel.

Rainfall totals decreased substantially as the storms came onshore and moved from the southwest to the northeast (fig. 54). During this 2-month storm series, more than 200 percent of the average annual rainfall fell on the mountains nearest the coast; farther inland, the north side of the San Bernardino Mountains received 127 percent of the average annual rainfall (National Oceanic and Atmospheric Administration, 1993).

The highest maximum discharges in the Mojave and Santa Ana Rivers (see stage hydrographs, fig. 56) occurred January 6–7 as a result of rapid melting of the accumulated snowpack. The three subsequent maximums were slightly lower, but nearly of the same magnitude. By comparison, the highest stage in De Luz Creek in the Santa Margarita River Basin (fig. 56) occurred January 16; its discharge was nearly an order of magnitude higher than the other three major maximum discharges. Maximum discharges on January 16 exceeded a 100-year recurrence-interval flood at several streamflow-gaging stations in the basin.

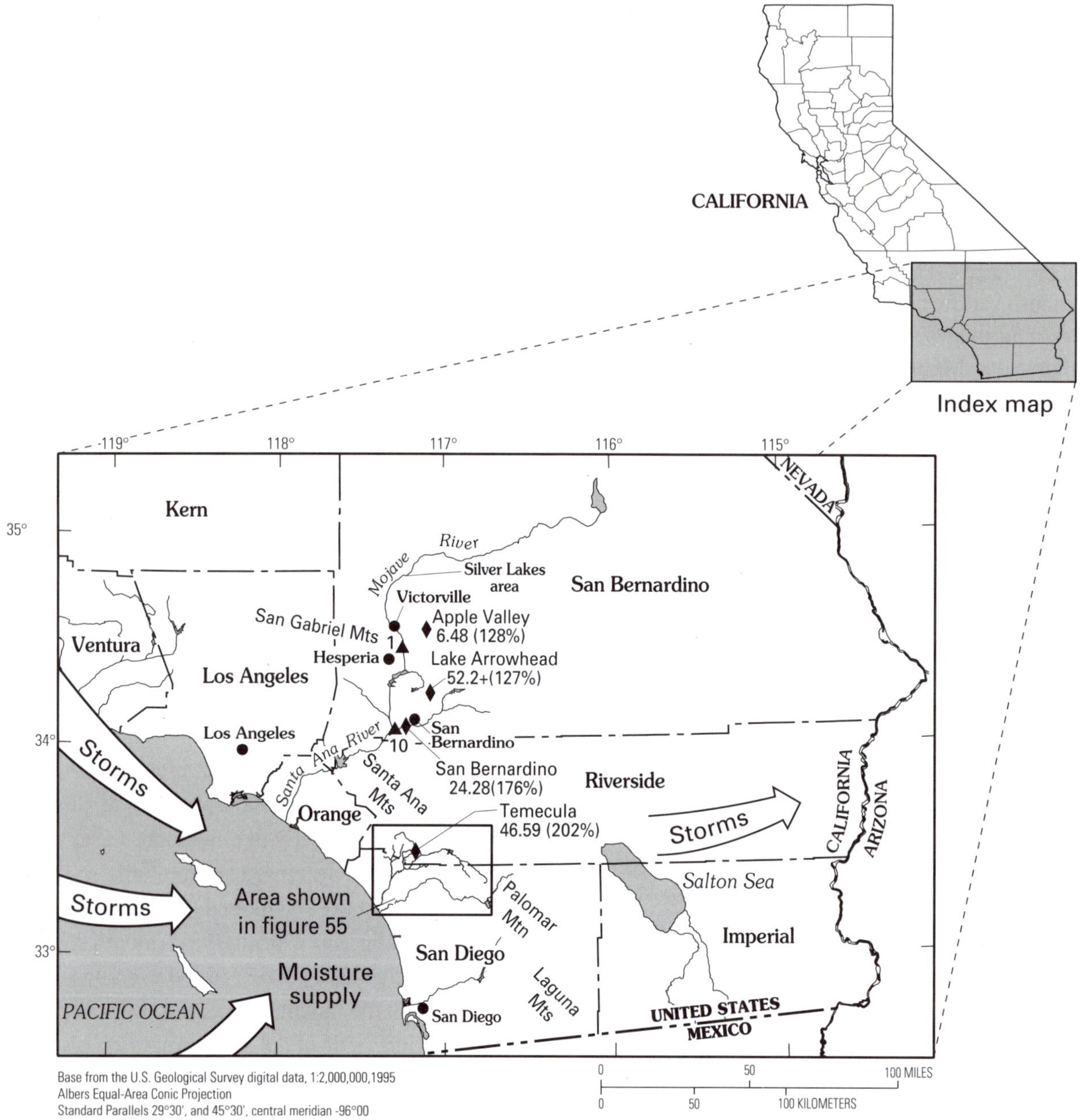

Base from the U.S. Geological Survey digital data, 1:2,000,000,1995
Albers Equal-Area Conic Projection
Standard Parallels 29°30', and 45°30', central meridian -96°00
U.S. Gological Survey digital data, 1:100,000,1995
Universal Transverse Mercator projection,
Zone 11

EXPLANATION

♦ Apple Valley 6.48 (128%) **Precipitation gage**—Number is total January–February 1993 rainfall, in inches. Number in parentheses ( ) is percentage of average annual rainfall

▲1 **Streamflow-gaging station**—Number corresponds to that in table 33

Figure 54. January–February 1993 rainfall at selected precipitation gages in southern California (precipitation data from National Oceanic and Atmospheric Administration, 1993).

Table 33. Maximum stages and discharges prior to and during January–February 1993, in southern California

[mi², square miles; ft, feet above an arbitrary datum; ft³/s, cubic feet per second; --, not determined or not applicable. Source: Recurrence intervals calculated from U.S. Geological Survey data. Other data from U.S. Geological Survey reports or data bases]

Site no. (figs. 54 or 55)	Station no.	Stream and place of determination	Drainage area (mi²)	Maximum prior to January 1993: Period	Year	Stage (ft)	Discharge (ft³/s)	Maximum during January 1993: Day	Stage (ft)	Discharge (ft³/s)	Discharge recurrence interval (years)
1	10261100	Mojave River below Forks Reservoir, near Hesperia, CA	209	1972–74, 1981–93	1983	--	11,700	--	7.61	21,300	25–50
2	11039800	San Luis Rey River near Pala, CA	[1]166 364	1988–93	1991	5.56	1,700	16	--	14,000	--
3	11042400	Temecula Creek near Aguanga, CA	131	1958–93	1980	12.00	4,200	16	--	8,100	--
4	11043000	Murrieta Creek near Temecula, CA	[2]170 588	1931–93	1980	13.70	21,800	16	17.24	25,000	25–50
5	11044000	Santa Margarita River near Temecula, CA	[3]268 740	1925–93	1927	18.00	25,000	16	22.50	31,000	25
6	11044250	Rainbow Creek near Fallbrook, CA	10.8	1925–93	1927	--	33,100	16	--	8,000	--
7	11044350	Sandia Creek near Fallbrook, CA	21.4	1990–93	1991	8.74	2,100	--	17.60	5,100	--
8	11044800	De Luz Creek near De Luz, CA	33.0	1993	--	--	--	16	15.13	9,700	--
9	11046000	Santa Margarita River at Ysidora, CA	[2]351 740	1924–28, 1931–52, 1954–93	1927	18.00	33,600	16	20.47	[4]44,000	25–50
10	11059300	Santa Ana River near San Bernardino, CA	359	--	1969	11.90	28,000	17	6.86	15,300	25–50

[1]Excludes drainage area upstream from Lake Henshaw that did not overflow.
[2]Excludes drainage area upstream from Skinner Reservoir that did not overflow.
[3]Excludes drainage area upstream from Vail Lake Reservoir that did not overflow.
[4]Preliminary estimate based on discharge-drainage area relations at other streamflow-gaging stations in the Santa Margarita Basin.

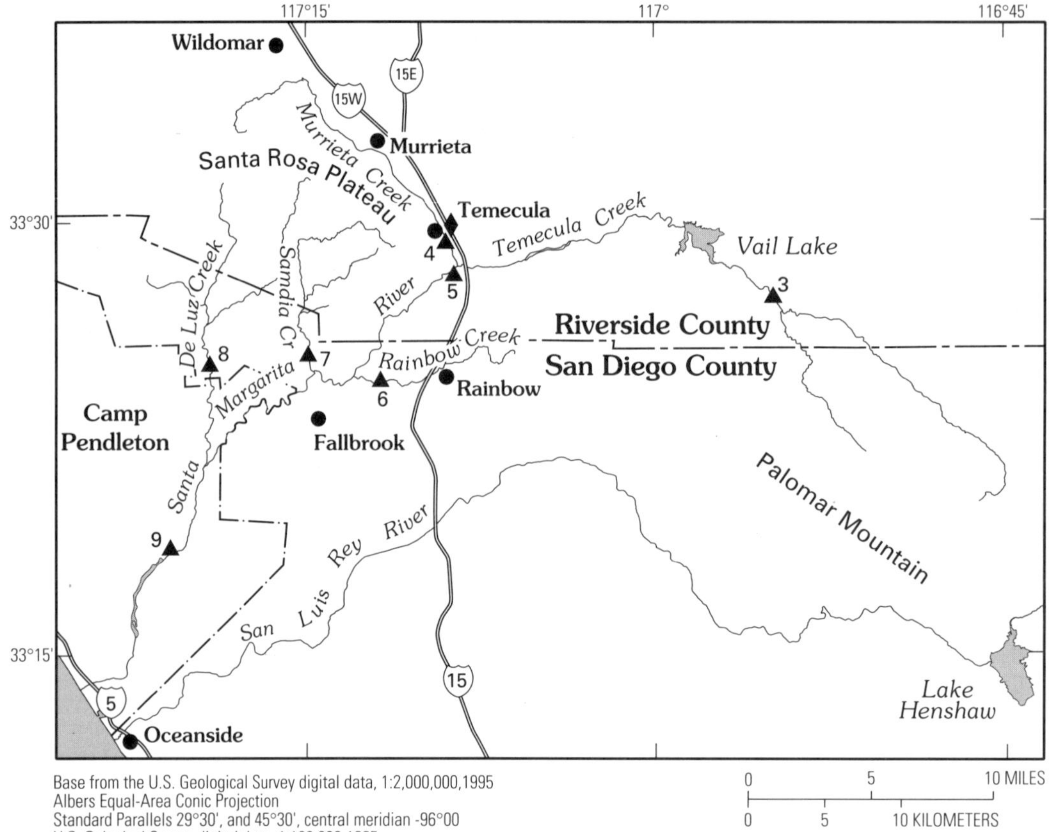

Base from the U.S. Geological Survey digital data, 1:2,000,000,1995
Albers Equal-Area Conic Projection
Standard Parallels 29°30', and 45°30', central meridian -96°00
U.S. Gologicai Survey digital data, 1:100,000,1995
Universal Transverse Mercator projection,
Zone 11

EXPLANATION

▲[3] Streamflow-gaging station—Number corresponds to that in table 33

♦ Precipitation gage

Figure 55. Location of streamflow-gaging stations and precipitation gages in the San Luis Rey and Santa Margarita River Basins in southern California.

THE SANTA MARGARITA RIVER BASIN FLOOD—JANUARY 16, 1993

The January 16, 1993, flooding in the Santa Margarita River Basin resulted from excessive rainfall that generally was localized over the upper reaches of the basin—principally in the Santa Rosa Plateau and Temecula areas (fig. 55), where the 6-hour rainfall was 126 and 114 percent, respectively, of the 100-year, 6-hour precipitation values (Miller and others, 1973). This intense precipitation was on a basin still saturated from the January 6–7 storm.

The most severe flooding during the January–February 1993 storms (see table 33) happened on January 16 in the Murrieta Creek flood plain at Temecula (fig. 55). At the Murrieta Creek streamflow-gaging station near Temecula (site 4), where flow overtopped the gage shelter, the stage was the maximum for the 68 years of record and exceeded the previous (February 21, 1980) record by more than 3 feet.

Maximums of record also were recorded on the Santa Margarita River near Temecula (site 5) and on other, smaller streams in the basin. Extensive flooding occurred along the Santa Margarita River as it passed through Camp Pendleton, the U.S. Marine Corps base near the mouth of the river. The floodwaters spread over the broad, flat flood plain on the base and deposited large quantities of sediment and debris. The Santa Margarita River streamflow-gaging station at Ysidora (site 9) was damaged as the debris-laden river washed out the bridge. The estimated discharge (see table 33) of 44,000 cubic feet per second exceeded the maximum discharge for the 68 years of record (33,600 cubic feet per second on February 16, 1927) by 34 percent.

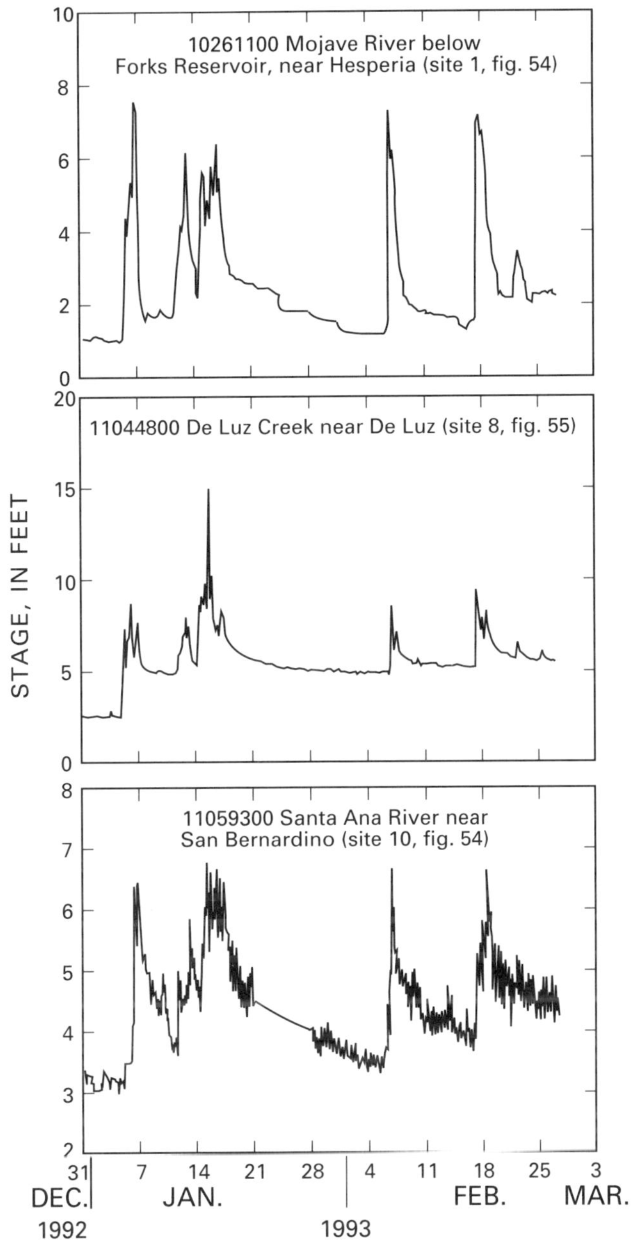

Figure 56. Stages at selected streamflow-gaging stations in southern California, December 31, 1992–February 28, 1993.

REFERENCES

Miller, J.F., Frederick, R.H., and Tracey, R.J., 1973 [1974], Precipitation-frequency atlas of the Western United States, volume 11—California: U.S. Department of Commerce, National Oceanic and Atmospheric Administration Atlas 2, 71 p.

National Oceanic and Atmospheric Administration, 1993, Climatological data, California: Asheville, N.C., National Climatic Data Center, v. 97, no. 1–2.

———1994, Fifth annual climate assessment, 1993: Camp Springs, Maryland, U.S. Department of Commerce, National Weather Service, Climate Analysis Center, 111 p.

January and February 1993, in Arizona

By C.F. Smith, K.M. Sherman, G.L. Pope, *and* P.D. Rigas

An unusual series of storms from the Pacific Ocean starting on January 6, 1993, and continuing through February 28, 1993, caused excessive and prolonged precipitation across Arizona. These excessive rains resulted in the most widespread and severe flooding in Arizona since the turn of the century (Mac Nish and others, 1993). Winter storms normally originate in the Gulf of Alaska and dissipate most of their moisture before reaching Arizona. Because the position of the jetstream during January was farther south than was normal, subtropical moisture from the Pacific Ocean west of the coast of Baja California was directed toward Arizona, creating excessive rains that resulted in widespread flooding and loss of property. Eight deaths and 112 injuries were attributed to the floods. Estimated damages to public and private property exceeded $400 million (U.S. Army Corps of Engineers, 1994).

The most intense rains fell in central Arizona north and east of Phoenix. This area includes most of the Verde and Salt River Basins. Precipitation was 520 percent of normal for January and 400 percent of normal for February in these areas (U.S. Army Corps of Engineers, 1994).

The prolonged rainfall during January and February caused record or near-record flood peaks and flood volumes on nearly every major river basin in the State. Figure 57 shows the network of selected streamflow-gaging stations operated by the U.S. Geological Survey in Arizona. Record maximum discharges occurred at 28 of the streamflow-gaging stations during the floods of 1993 (table 34). The following sections describe flooding in the unregulated portions of the Little Colorado River, upper Gila River, Santa Cruz River, Salt River, and Verde River Basins.

FLOODS IN THE LITTLE COLORADO RIVER BASIN

Beginning January 6 and again on February 19, 1993, major subtropical storm systems produced floods in the Little Colorado River Basin. Excessive rainfall, combined with snowmelt, caused severe and widespread flooding. Rainfall from a succession of smaller storms prior to the major storms resulted in near-saturated soil conditions. Precipitation from the two major storms was not distributed uniformly across the basin. Consequently, maximum-discharge flood frequency for streams in the Little Colorado River Basin varied from less than 2-year to greater than 25-year recurrence intervals.

Flooding in the Little Colorado River Basin generally occurred in tributaries draining the south-central mountains of the basin. Maximum discharges of record occurred at the gaging sites near Holbrook (site 5) and Winslow (site 7). Other streamflow-gaging stations within the basin recorded maximum discharges with recurrence intervals ranging from 5 to 25 years (table 34). Maximum discharges in the Little Colorado River attenuated between Winslow and Grand Falls (site 8). Consequently, streamflow-gaging stations downstream from Grand Falls recorded less frequent recurrence-interval floods; however, the duration of the floods were longer. The maximum discharge of the Little Colorado River near Cameron (site 10) on January 13 was 18,200 cubic feet per second, a 15-year recurrence interval. By contrast, the total flood volume for 15 consecutive days (January 10–24) was greater than the 100-year total volume recurrence interval.

Maximum discharges on the Little Colorado River during February generally were less than those during January. The streamflow-gaging station at Little Colorado River near Cameron (site 10) recorded a 10-year recurrence-interval maximum discharge on February 24; the discharge for January 13 was determined to have an approximate 15-year recurrence interval. Several tributaries that drain the south-central mountains within the basin sustained maximum discharges nearly equal to those recorded for the January flood.

FLOODS IN THE UPPER GILA RIVER BASIN

The abnormal weather pattern that caused excessive rainfall and flooding throughout Arizona persisted for approximately 2 months and sent a series of storms through Arizona that resulted in multiple maximum

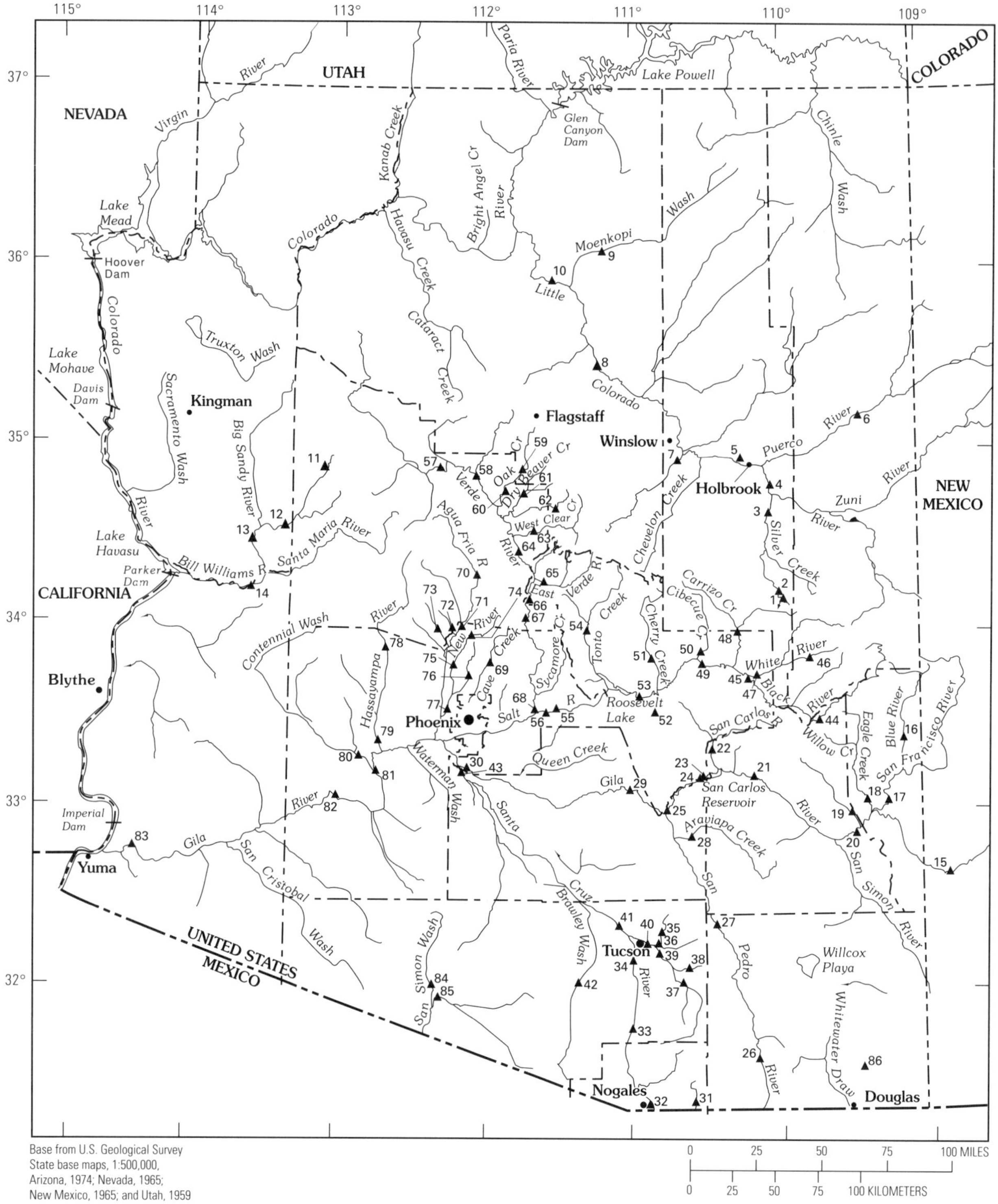

EXPLANATION

▲[1] **Streamflow-gaging station—** Number corresponds to that in table 34

Figure 57. Location of selected U.S. Geological Survey streamflow-gaging stations in Arizona used to determine maximum stages and discharges for January and February 1993.

Table 34. Maximum stages and discharges prior to and during January and February 1993, in Arizona

[mi^2, square miles; ft, feet above an arbitrary datum; ft^3/s, cubic feet per second; <, less than; --, not determined or not applicable; >, greater than. Source: Recurrence intervals calculated from U.S. Geological Survey data. Other data from U.S. Geological Survey reports or data bases]

				Maximum prior to January 1993				Maximum during January and February 1993			
Site no. (fig. 57)	Station no.	Stream and place of determination	Drainage area (mi^2)	Period	Year	Stage (ft)	Discharge (ft^3/s)	Date (month/day)	Stage (ft)	Discharge (ft^3/s)	Discharge recurrence interval (years)
		Little Colorado River Basin									
1	09390500	Show Low Creek near Lakeside, AZ	68.6	1953–93	1978	9.16	5,550	1/8	8.25	4,370	<25
2	09392000	Show Low Creek below Jaques Dam, AZ	73	1942–93	1984		4,800	1/8		1,840	--
3	09393500	Silver Creek near Snowflake, AZ	925	1951–93	1952	18.0	10,100	2/20	13.17	5,800	5
4	09394500	Little Colorado River at Woodruff, AZ	8,072	1905–93	1919	22.9	25,000	1/8	22.53	8,960	<10
5	0937100	LeRoux Wash near Holbrook, AZ	--	1980–93	1984	11.74	6,600	1/8	14.36	[1]8,460	10–20
6	09396100	Puerco River near Chambers, AZ	2,156	1971–93	1971	9.65	17,800	1/19	4.53	2,920	--
7	09397500	Chevelon Fork below Wildcat Canyon near Winslow, AZ	400	1948–70, 1978, 1982–93	1978	18.25	19,900	1/8	20.78	[1]24,700	25–50
8	09401000	Little Colorado River at Grand Falls, AZ	21,068	1926–93	1923	47.0	120,000	1/11	19.25	16,600	<5
9	09401260	Moenkopi Wash at Moenkopi, AZ	1,629	1976–93	1983	15.10	10,100	2/20	8.01	3,150	<2
10	09402000	Little Colorado River near Cameron, AZ	26,459	1947–93	1923		120,000	1/13	17.83	18,200	15
		Bill Williams River Basin									
11	09424432	Francis Creek near Bagdad, AZ	134	1985–93	1988	12.57	6,830	2/8	13.68	[1]12,500	--
12	09424447	Burro Creek at Old U.S. 93 Bridge, AZ	611	1980–93	1980	15.6	47,400	2/8	16.30	[1]55,300	--
13	09424450	Big Sandy River near Wikieup, AZ	2,742	1966–93	1980	12.51	38,500	2/9	16.00	[1]68,700	>50
14	09426000	Bill Williams River below Alamo Dam, AZ	4,633	1940–93	1951	30.80	65,100	[8]3/16	--	6,980	--

Table 34. Maximum stages and discharges prior to and during January and February 1993, in Arizona—Continued

Site no. (fig. 57)	Station no.	Stream and place of determination	Drainage area (mi^2)	Maximum prior to January 1993				Maximum during January and February 1993			
				Period	Year	Stage (ft)	Discharge (ft^3/s)	Date (month/day)	Stage (ft)	Discharge (ft^3/s)	Discharge recurrence interval (years)
				Upper Gila River Basin							
15	09432000	Gila River below Blue Creek, near Virden, NM	3,203	1891–1993	1978	29.0	58,700	1/11	20.97	30,000	50
16	09444200	Blue River near Clifton, AZ	506	1967–93	1972	22.56	30,000	1/19	18.43	17,000	>10
17	09444500	San Francisco River at Clifton, AZ	2,766	1870–1993	1983	19.72	90,900	1/18	25.27	42,900	20
18	09447000	Eagle Creek above pumping plant near Morenci, AZ	622	1944–93	1983	13.04	36,400	1/18	--	[1]36,800	>100
19	09447800	Bonita Creek near Morenci, AZ	302	1981–93	1983	17.3	19,400	1/18	16.50	[1]19,500	--
20	09448500	Gila River at head of Safford Valley, near Solomon, AZ	7,896	1904–93	1983	20.8	132,000	1/19	18.56	86,200	40
21	09466500	Gila River at Calva, AZ	11,470	1929–93	1983	23.1	150,000	1/20	19.68	109,000	60
22	09468500	San Carlos River near Peridot, AZ	1,026	1910–93	1941	11.4	40,600	1/8	12.12	[1]54,800	>100
23	09469000	San Carlos Reservoir at Coolidge Dam, AZ	12,886	1928–93	1980	--	(contents—[2]1,090,000 acre-feet)	1/20	--	(contents—[2]1,060,000 acre-feet)	--
24	09469500	Gila River below Coolidge Dam, AZ	12,886	1899–1993	1916	--	130,000	1/21	--	29,300	--
25	09470000	Gila River at Winkelman, AZ	13,268	1917–93	1944	[4]18.4	55,000	1/20	22.21	[3]37,200	--
26	09471000	San Pedro River at Charleston, AZ	1,234	1904–93	1926	21.9	98,000	1/19	8.99	11,500	<5
27	09472000	San Pedro River near Redington, AZ	2,927	1943–93	1926	29.0	90,000	1/19	18.22	19,100	8
28	09473000	Aravaipa Creek near Mammoth, AZ	537	1931–93	1983	16.76	70,800	1/11	9.09	13,000	15
29	09474000	Gila River at Kelvin, AZ	18,011	1911–93	1916	19.5	132,000	1/19	31.55	[3]74,900	--
30	09479500	Gila River near Laveen, AZ	20,615	1940–93	1983	12.08	[5]35,000	1/20	12.41	[1,3]41,600	--
				Santa Cruz River Basin							
31	09480000	Santa Cruz River near Lochiel, AZ	82.2	1942–93	1977	10.21	12,000	1/18	7.56	7,320	30

Table 34. Maximum stages and discharges prior to and during January and February 1993, in Arizona—Continued

Site no. (fig. 57)	Station no.	Stream and place of determination	Drainage area (mi²)	Maximum prior to January 1993: Period	Year	Stage (ft)	Discharge (ft³/s)	Maximum during January and February 1993: Date (month/day)	Stage (ft)	Discharge (ft³/s)	Discharge recurrence interval (years)
		Santa Cruz River Basin—Continued									
32	09480500	Santa Cruz River near Nogales, AZ	533	1907–93	1977	15.5	31,000	1/18	9.83	8,800	7
33	09482000	Santa Cruz River at Continental, AZ	1,682	1940–93	1983	16.34	45,000	1/19	14.75	32,400	75
34	09482500	Santa Cruz River at Tucson, AZ	2,222	1905–93	1983	22.2	52,700	1/19	11.67	37,400	>100
35	09484000	Sabino Creek near Tucson, AZ	35.5	1904–93	1970	10.21	7,730	1/8	7.60	[1]12,900	>100
36	09484500	Tanque Verde Creek at Tucson, AZ	219	1940–93	1978	7.33	12,700	1/8	11.85	[1]24,500	70
37	09484600	Pantano Wash near Vail, AZ	457	1959–93	1958	24.00	38,000	[8]7/11	8.10	1,840	<2
38	09485000	Rincon Creek near Tucson, AZ	44.8	1952–93	1971	10.50	9,660	1/8	7.62	3,720	8
39	09485450	Pantano Wash at Broadway Blvd., at Tucson, AZ	599	1958, 1979–93	1958	--	20,000	1/18	4.49	4,340	10
40	09485700	Rillito Creek at Dodge Blvd., at Tucson, AZ	871	1988–93	1990	9.20	10,600	1/8	14.84	[1]24,100	>50
41	09486500	Santa Cruz River at Cortaro, AZ	3,503	1939–93	1983	16.57	[5]65,000	1/19	9.91	--	--
42	09487000	Brawley Wash near Three Points, AZ	776	1966–93	1983	12.07	19,100	[8]8/27	14.95	10,800	15
43	09489000	Santa Cruz River near Laveen, AZ	8,581	1940–93	1983	19.74	[5]33,000	1/21	18.01	11,000	>25
		Salt River Basin									
44	09489500	Black River below pumping plant, near Point of Pines, AZ	560	1953–93	1972	18.0	17,900	1/8	11.37	8,940	10
45	09490500	Black River near Fort Apache, AZ	1,232	1912–93	1916	--	[6]50,000	1/8	28.10	[1]54,700	25
46	09492400	East Fork White River near Fort Apache, AZ	38.8	1957–93	1983	5.40	2,700	1/8	3.15	563	7
47	09494000	White River near Fort Apache, AZ	632	1957–93	1978	15.71	14,600	1/8	14.57	12,600	35
48	09496500	Carrizo Creek near Show Low, AZ	439	1951–93	1965	13.0	23,000	1/8	14.25	19,300	25
49	09497500	Salt River near Chrysotile, AZ	2,849	1906–93	1916	18.0	74,000	1/8	18.33	[1]76,600	40
50	09497800	Cibecue Creek near Chrysotile, AZ	295	1959–93	1977	17.3	22,200	1/8	10.96	10,600	10
51	09497980	Cherry Creek near Globe, AZ	200	1965–93	1979	--	15,700	1/8	12.42	10,100	10

Table 34. Maximum stages and discharges prior to and during January and February 1993, in Arizona—Continued

Site no. (fig. 57)	Station no.	Stream and place of determination	Drainage area (mi²)	Maximum prior to January 1993: Period	Year	Stage (ft)	Discharge (ft³/s)	Maximum during January and February 1993: Date (month/day)	Stage (ft)	Discharge (ft³/s)	Discharge recurrence interval (years)
		Salt River Basin—Continued									
52	09498400	Pinal Creek at Inspiration Dam, near Globe, AZ	195	1980–93	1981	5.05	2,920	1/11	8.50	[1]5,700	--
53	09498500	Salt River near Roosevelt, AZ	4,306	1906–93	1941	24.4	117,000	1/8	30.09	[1]143,000	40
54	09499000	Tonto Creek above Gun Creek, near Roosevelt, AZ	675	1941–93	1980	17.00	61,400	1/8	17.95	[1]72,500	40
55	09501000	Reservoir System on Salt River, at and below Roosevelt Dam, AZ	6,211	1910–93	1941	--	[2]1,764,000	1/20	--	[2]1,708,000	--
56	09502000	Salt River below Stewart Mountain Dam, AZ	6,232	1930–93	1980	25.0	75,200	1/20	19.92	34,500	--
		Verde River Basin									
57	09503700	Verde River near Paulden, AZ	2,507	1963–93	1980	12.72	15,700	2/20	14.25	[1]23,200	>50
58	09504000	Verde River near Clarkdale, AZ	3,503	1915–93	1920	19.1	50,600	2/20	26.39	[1]53,200	50
59	09504430	Oak Creek at Sedona, AZ	233	1982–93	1982	15.64	20,700	2/19	20.33	[1]23,200	--
60	09504500	Oak Creek near Cornville, AZ	355	1940–93	1980	16.30	26,400	2/20	19.15	26,000	25
61	09505350	Dry Beaver Creek near Rimrock, AZ	142	1961–93	1970	14.35	26,600	1/8	10.01	11,600	7
62	09505200	Wet Beaver Creek near Rimrock, AZ	111	1961–93	1980	13.96	10,900	1/8	17.21	16,000	100
63	09505800	West Clear Creek near Camp Verde, AZ	241	1965–93	1978	11.6	22,400	1/8	13.22	[1]24,800	50
64	09506000	Verde River near Camp Verde, AZ	5,009	1934–93	1938	26.1	97,000	2/20	28.36	[1]119,000	>50
65	09507980	East Verde River near Childs, AZ	331	1961–93	1970	22.5	23,500	1/8	19.80	20,100	>10
66	09508300	Wet Bottom Creek near Childs, AZ	36.4	1967–93	1980	16.0	6,830	1/8	18.36	[1]7,380	10
67	09508500	Verde River below Tangle Creek, AZ	5,858	1945–93	1891	--	150,000	1/8	23.40	145,000	>50
68	09511300	Verde River near Scottsdale, AZ	6,615	1961–93	1980	--	98,000	1/8	25.37	[1]127,000	--
		Lower Gila River Basin									
69	09512280	Cave Creek below Cottonwood Creek, near Cave Creek, AZ	82.7	1981–93	1991	7.24	3,900	1/8	15.24	[1]9,200	--
70	09512500	Agua Fria River near Mayer, AZ	585	1940–93	1980	15.76	33,100	1/8	14.58	27,400	>50

Table 34. Maximum stages and discharges prior to and during January and February 1993, in Arizona—Continued

Site no. (fig. 57)	Station no.	Stream and place of determination	Drainage area (mi²)	Maximum prior to January 1993: Period	Year	Stage (ft)	Discharge (ft³/s)	Maximum during January and February 1993: Date (month/day)	Stage (ft)	Discharge (ft³/s)	Discharge recurrence interval (years)
		Lower Gila River Basin—Continued									
71	09512800	Agua Fria River near Rock Springs, AZ	1,111	1970–93	1980	21.08	59,500	1/8	25.28	52,500	<25
72	09512830	Boulder Creek near Rock Springs, AZ	37.8	1983–93	1991	8.26	3,230	1/8	10.79	[1]10,000	--
73	09512860	Humbug Creek near Castle Hot Springs, AZ	59.9	1983–93	1991	9.00	3,220	1/8	10.81	[1]6,000	--
74	09513780	New River near Rock Springs, AZ	68.3	1962–93	1970	13.5	18,600	1/8	10.80	12,600	10
75	09513835	New River at Bell Road, AZ	185	1963–93	1967	13.5	14,600	1/8	5.04	2,760	<5
76	09513860	Skunk Creek near Phoenix, AZ at I-17	64.9	1960–93	1964	10.48	11,500	1/8	4.46	4,990	7
77	09513910	New River near Glendale, AZ	324	1961–93	1967	10.40	19,800	1/11	2.46	5,450	<5
78	09516500	Hassayampa River near Morristown, AZ	796	1939–93	1970	19.0	47,500	1/8	15.91	26,300	35
79	09517000	Hassayampa River near Arlington, AZ	1,471	1961–93	1970	8.40	39,000	1/8	12.50	11,400	8
80	09517490	Centennial Wash at SPRR Bridge, near Arlington, AZ	1,817	1980–93	1984	11.34	15,600	1/11	8.93	9,210	--
81	09519500	Gila River below Gillespie Dam, AZ	49,650	1891, [7]1921–93	1891 1980	-- 18.81	250,000 178,000	1/9	--	[3]130,000	--
82	09519800	Gila River below Painted Rock Dam, AZ	50,910	1959–93	1983	9.36	9,190	2/26	16.79	[1]32,000	--
83	09520500	Gila River near Dome, AZ	57,850	1903–93	1916	--	200,000	[8]3/3	26.81	28,900	--
		Other basins									
84	09535100	San Simon Wash near Pisinimo, AZ	569	1972–93	1976	10.82	12,500	1/8	7.30	1,050	<2
85	09535300	Vamori Wash at Kom Vo, AZ	1,250	1972–93	1983	10.54	10,400	[8]8/28	9.36	1,400	<5
86	09537200	Leslie Creek near McNeal, AZ	79.1	1969–77, 1982–93	1984	8.54	4,600	1/19	6.66	2,410	10

[1]Maximum of record.
[2]Contents in acre-feet.
[3]Maximum daily discharge.
[4]Site and datum then in use.
[5]Estimated on basis of flood routing.
[6]Estimated on basis of records for Salt River near Chrysotile.
[7]Period of regulated discharge.
[8]Maximum discharge for 1993 did not occur during January and February.

discharges in many of the rivers within the Gila River Basin in Arizona and New Mexico. Flooding that occurred in the upper Gila River Basin in New Mexico is discussed in a following article.

Most of the rainfall in the upper Gila River Basin was generated by four storms that occurred during January 6–8, 10–11, 14–15, and 17–18. Precipitation varied from about 4 inches near the Arizona-New Mexico border to approximately 7 inches at San Carlos Reservoir (National Oceanic and Atmospheric Administration, 1993). Two factors contributed to the severity of the flooding—first, the duration and intensity of the rain, which saturated and exceeded the infiltration capacity of the soil, maximizing runoff volumes; second, much of the rain fell over existing water-laden snowpack, accelerating snowmelt that combined with direct runoff from the storm resulting in flooding that set records for total runoff volume as well as maximum discharge. Recurrence intervals determined for maximum discharges for streams within the Gila River Basin ranged from less than 10 years to greater than 100 years. Although maximum-of-record discharges occurred at several stations, a notable aspect of this flooding was the total runoff volumes produced. The contents in San Carlos Reservoir increased to near-maximum storage levels, resulting in maximum gate releases as well as uncontrolled spillway discharges.

Flooding within the Gila River Basin was widespread and generated maximum-of-record discharges on several streams that drain into the Gila River from the north. Maximum discharge values at or greater than the 100-year recurrence interval occurred at Eagle Creek above pumping plant near Morenci (site 18), Bonita Creek near Morenci (site 19), and San Carlos River near Peridot (site 22). The maximum discharges occurred on January 8, 1993, on the San Carlos River near Peridot and on January 18, 1993, at Eagle Creek above pumping plant near Morenci and Bonita Creek near Morenci. The peak discharge, which attenuated considerably along the main stem of the Gila River, reached the Gila River at Calva streamflow-gaging station (site 21) on January 20, with a maximum discharge of 109,000 cubic feet per second and a recurrence interval of 60 years. San Carlos Reservoir reached maximum capacity from this flood of 1,060,000 acre-feet on January 20, 1993, and resulted in a maximum reservoir outflow of 29,300 cubic feet per second, which is the highest outflow since 1940.

FLOODS IN THE SANTA CRUZ RIVER BASIN

The precipitation for January 5–19 for the Santa Cruz River Basin ranged from 2 inches in the western part of the watershed to 6 inches in the southern part. Most of the precipitation occurred on January 17 and 18 and caused severe flooding on the Santa Cruz River. The streamflow-gaging station on the Santa Cruz River near Lochiel (site 31) had a maximum discharge of 7,320 cubic feet per second with a recurrence interval of 30 years. The Santa Cruz River near Nogales streamflow-gaging station (site 32) had a maximum discharge of 8,800 cubic feet per second with a recurrence interval of 7 years. Runoff from about 6 inches of rain just north of Nogales produced a maximum discharge of 32,400 cubic feet per second with a recurrence interval of 75 years at the Santa Cruz River at Continental streamflow-gaging station (site 33). This was the second highest maximum discharge since 1892. The streamflow-gaging station on the Santa Cruz River at Tucson (site 34) recorded a maximum discharge of 37,400 cubic feet per second with a recurrence interval of more than 100 years. This was the second highest maximum discharge since 1892. The Santa Cruz River near Laveen streamflow-gaging station (site 43) recorded a maximum discharge of 11,000 cubic feet per second with a recurrence interval of more than 25 years.

The precipitation totals for January 5–19 for the Rillito Creek Basin near Tuscon ranged from 4 inches in the lower elevations to 6 inches in the higher elevations. The majority of the precipitation occurred on January 5 through 7, which resulted in record flooding on January 8th for Sabino Creek (site 35), Tanque Verde Creek (site 36), and Rillito Creek (site 40). The streamflow-gaging station Sabino Creek near Tucson (site 35) recorded a maximum discharge of 12,900 cubic feet per second with a recurrence interval of more than 100 years. The maximum discharge recorded at the Tanque Verde Creek at Tucson streamflow-gaging station (site 36) was 24,500 cubic feet per second with a recurrence interval of 70 years. The Rillito Creek at Dodge Blvd. streamflow-gaging station (site 40) recorded a maximum discharge of 24,100 cubic feet per second with a recurrence interval of more than 50 years. This was the highest maximum discharge since October 1983. The maximum discharge at a downstream site on Rillito Creek having 21 additional square miles of drainage area was 29,700 cubic feet per second recorded in October 1983.

FLOODS IN THE SALT RIVER BASIN

Precipitation totals for January 5–19, 1993, were highest in the western and central areas of the Salt River Basin. The precipitation total for these areas was approximately 12 inches. Tributaries that drain the north and eastern areas of the basin had precipitation totals that ranged from 12 inches in the downstream part of the basin to 6 inches in the upstream part. This produced less-severe flooding in the upstream part of the basin compared to record-setting flooding that occurred on the main stem of the Salt River.

The gaged tributaries in which the precipitation totals ranged from 6 to 12 inches were Black River, White River, Carrizo Creek, Cibecue Creek and Cherry Creek. Black River below pumping plant, near Point of Pines (site 44), had a maximum discharge of 8,940 cubic feet per second with a recurrence interval of 10 years. Black River near Fort Apache (site 45) had a maximum discharge of 54,700 cubic feet per second with a recurrence interval of 25 years. This was the largest maximum discharge since 1906. White River near Fort Apache (site 47) had a maximum discharge of 12,600 cubic feet per second with a recurrence interval of 35 years. Carrizo Creek near Show Low (site 48) had a maximum discharge of 19,300 cubic feet per second with a recurrence interval 25 years.

The precipitation total for the Tonto Creek Basin was 12 inches for January 5–19. The streamflow-gaging station at Tonto Creek above Gun Creek (site 54) recorded a maximum discharge of 72,500 cubic feet per second with a recurrence interval of 40 years. This was the highest maximum discharge since 1941.

Salt River near Chrysotile (site 49) and the Salt River near Roosevelt (site 53) both had maximum discharges for the period of record. Maximum discharge recorded at Salt River at Chrysotile was 76,600 cubic feet per second with a recurrence interval of 40 years. This was the highest maximum discharge since 1916. Maximum discharge at Salt River near Roosevelt was 143,000 cubic feet per second with a recurrence interval of 40 years. This was the highest maximum discharge since 1941.

FLOODS IN THE VERDE RIVER BASIN

Flooding in the upper Verde River Basin was more severe during February than during January, with maximum discharges occurring on February 20. The streamflow-gaging stations Verde River near Clarkdale (site 58) and at Verde River near Paulden (site 57) recorded maximum discharges with 50-year recurrence intervals or greater. Oak Creek near Cornville streamflow-gaging station (site 60) reached a maximum discharge nearly equal to the 1980 flood, which had a recurrence interval of 25 years. As floodwaters progressed downstream, a discharge of 119,000 cubic feet per second was reached at the Verde River near Camp Verde streamflow-gaging station (site 64), the highest discharge on record. Although some of these maximum discharges were high, the most notable aspect of the floods was the total volume of runoff produced. For example, flood volumes for a consecutive 15-day period during January for the Verde River below Tangle Creek (site 67) and Little Colorado River near Cameron (site 10) were determined to have a 70-year and a greater than 100-year recurrence interval, respectively.

In the Verde River Basin, the most severe flooding occurred in tributaries in the center of the basin and on the downstream main stem of the Verde on January 8. Streamflow-gaging stations at Wet Beaver Creek near Rimrock (site 62), West Clear Creek near Camp Verde (site 63), and Wet Bottom Creek near Childs (site 66) all recorded maximum discharges with recurrence intervals ranging from 10 to 100 years. Other tributaries to the Verde River contributed significant discharges (7- to 25-year recurrence intervals), which combined to produce a maximum discharge at Verde River below Tangle Creek (site 67) of 145,000 cubic feet per second (highest discharge since 1891), with a recurrence interval of greater than 50 years.

REFERENCES

Mac Nish, R.D., Smith, C.F., and Goddard, K.E., 1993, Floods in Arizona, January 1993: U.S. Geological Survey Open-File Report 93–54, 2 p.

National Oceanic and Atmospheric Administration, 1993, Climatological data, Arizona, January 1993: Asheville, N.C., National Climatic Data Center, v. 97, no. 1.

U.S. Army Corps of Engineers, 1994, Flood damage report, State of Arizona, floods of 1993: U.S. Army Corps of Engineers, Los Angeles District, 107 p.

January–March, July, and August 1993, in New Mexico

By Scott D. Waltemeyer

Large discharges and high stages occurred on several streams in New Mexico during January–March, July, and August 1993. Winter floods occurred at sites 1, 5, 6, 7, 8, and 9 (fig. 58) as a result of excessive precipitation that occurred in western New Mexico. The summer floods occurred in isolated areas (sites 2, 3, and 4) as a result of summer monsoon thunderstorms during July and August. Six streamflow-gaging stations experienced maximum discharges with recurrence intervals of 25 to 50 years (table 35). The greatest recurrence-interval flood occurred at streamflow-gaging station Los Esteros Creek above Santa Rosa Lake (site 3); the maximum discharge was the largest since data collection began in 1973.

No property damage or loss of life was reported to the Emergency Management Planning and Coordination office of the Department of Public Safety for New Mexico.

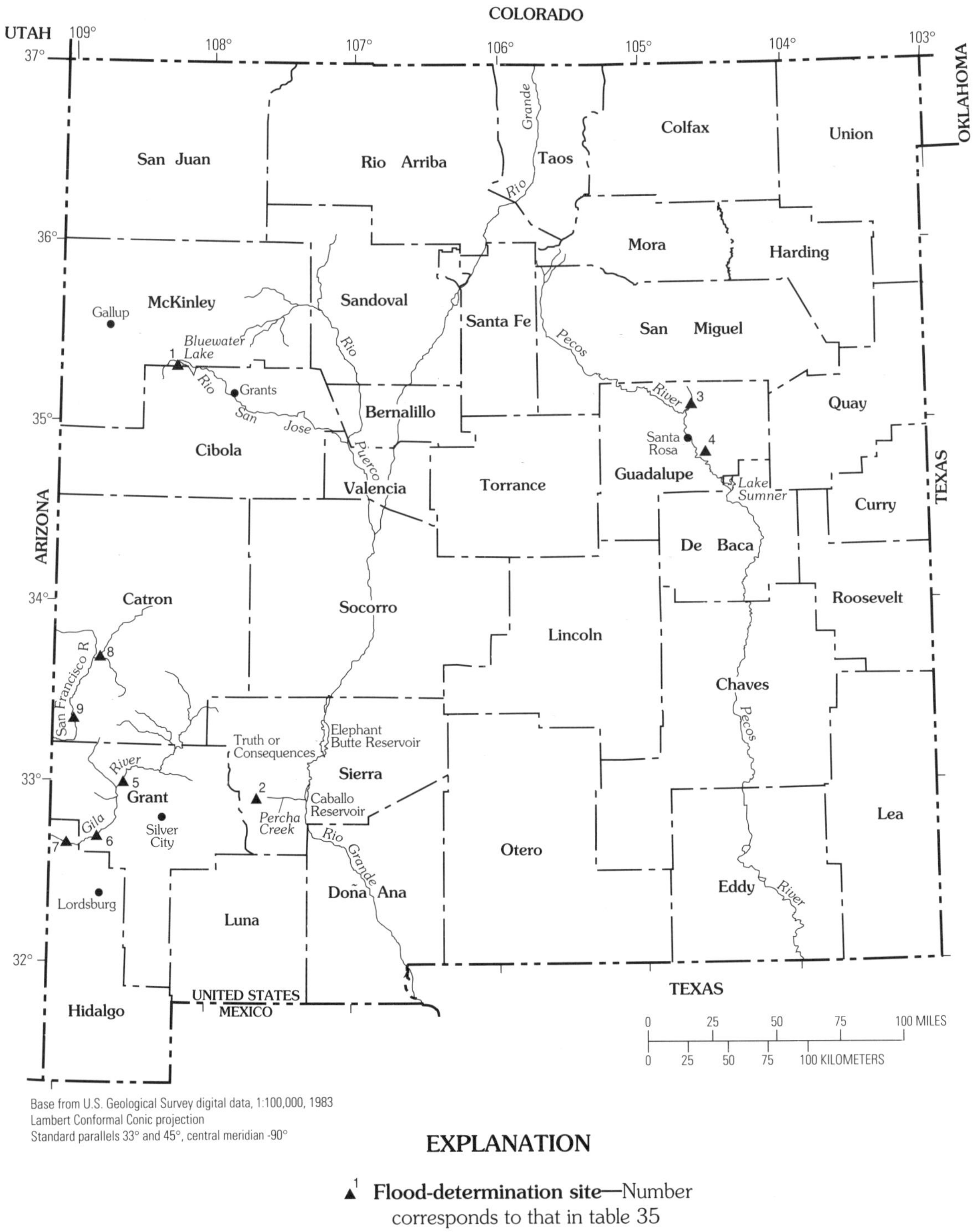

Figure 58. Location of selected U.S. Geological Survey streamflow-gaging stations in New Mexico used to determine maximum stages and discharges for January–March, July, and August 1993.

Table 35. Maximum stages and discharges prior to and during January–March, July, and August 1993, in New Mexico

[mi^2, square miles; ft, feet above an arbitrary datum; ft^3/s, cubic feet per second; --, not determined or not applicable. Source: Recurrence intervals calculated from U.S. Geological Survey data. Other data from U.S. Geological Survey reports or data bases]

				Maximum prior to January 1993				Maximum during January–March, July, and August 1993			
Site no. (fig. 58)	Station no.	Stream and place of determination	Drainage area (mi^2)	Period	Year	Stage (ft)	Discharge (ft^3/s)	Date (month/day)	Stage (ft)	Discharge (ft^3/s)	Discharge, recurrence interval (years)
1	08341300	Bluewater Creek above Bluewater Dam, near Bluewater, NM	75	1953–78, 1990–93	1953	8.99	3,570	3/19	4.05	1,270	25
2	08361700	Percha Creek near Hillsboro, NM	35.4	1957–78, 1980–93	1972	11.70	12,200	8/16	8.51	5,900	30
3	08382730	Los Esteros Creek above Santa Rosa Lake, NM	65.6	1973–93	1976	9.30	3,900	8/3	13.22	9,200	45
4	08383370	Pecos River tributary near Puerto de Luna, NM	.37	1961–93	1987	15.89	2,000	7/14	14.64	1,350	40
5	09430500	Gila River near Gila, NM	1,864	1928–93	1984	13.00	35,200	2/20	9.46	14,200	15
6	09431500	Gila River near Red Rock, NM	2,829	1905, 1911, 1929–93	1978	29.8	48,000	1/19	20.20	25,500	25
7	09432000	Gila River below Blue Creek near Virden, NM	3,203	1927–93	1978	29.00	58,700	1/11	20.97	30,000	50
8	09442680	San Francisco River near Reserve, NM	350	1959–93	1983	11.71	9,830	2/20	4.38	1,920	--
9	09444000	San Francisco River near Glenwood, NM	1,653	1928–93	1983	18.15	37,100	1/19	11.53	12,700	--

January 8–21, 1993, in Southeastern Georgia

By Timothy C. Stamey

Major flooding occurred throughout much of southeastern Georgia as a result of excessive rain on January 8–13, 1993. Selected streamflow-gaging stations located in the flood-affected area of the State are shown in figure 59. Total rainfall amounts in southeastern Georgia for this period were as much as 6.5 to 7 inches (National Oceanic and Atmospheric Administration, 1993). According to rainfall data furnished by the National Oceanic and Atmospheric Administration, areas in southeastern Georgia received about 3 inches of rain on January 8 and 9 and another 3 to 3.5 inches on January 12 and 13.

High water, resulting from the excessive rain falling on already saturated ground in southeastern Georgia, flooded agricultural and timberlands in the low-land areas adjacent to creeks and rivers. However, the high water did not produce severe or extensive flood damage to property or livestock. The most extensive flooding occurred in parts of the Altamaha, Ogeechee, Satilla, and Little River Basins (fig. 59).

Maximum discharges at selected active and discontinued streamflow-gaging stations had recurrence intervals ranging from 2 to 100 years. Recurrence intervals for the maximum discharges produced by the storms at the selected stations are shown in table 36. The range in the maximum discharge recurrence intervals for this flood was similar to maximum discharges that occurred during the March 1984, February 1986, and January 1991 floods, and which occurred in the same general area of the State.

The most severe flooding occurred at sites 4, 5, 6, 10, 12, 15, and 17–19 (fig. 59, table 36). Maximum discharges at these sites had recurrence intervals ranging from 25 to 100 years. Also, new maximum stages and discharges for the period of known floods (based on U.S. Geological Survey records) were recorded at sites 10, 17, and 18 (table 36). The maximum discharge of 23,100 cubic feet per second at site 12 was the fourth highest in 89 years of record, and the maximum discharge of 12,000 cubic feet per second at site 15 was the third highest since 1928.

REFERENCE

National Oceanic and Atmospheric Administration, 1993, Climatological data, Georgia, January 1993: Asheville, N.C., U.S. Department of Commerce, National Climatic Data Center, v. 97, no. 1, 22 p.

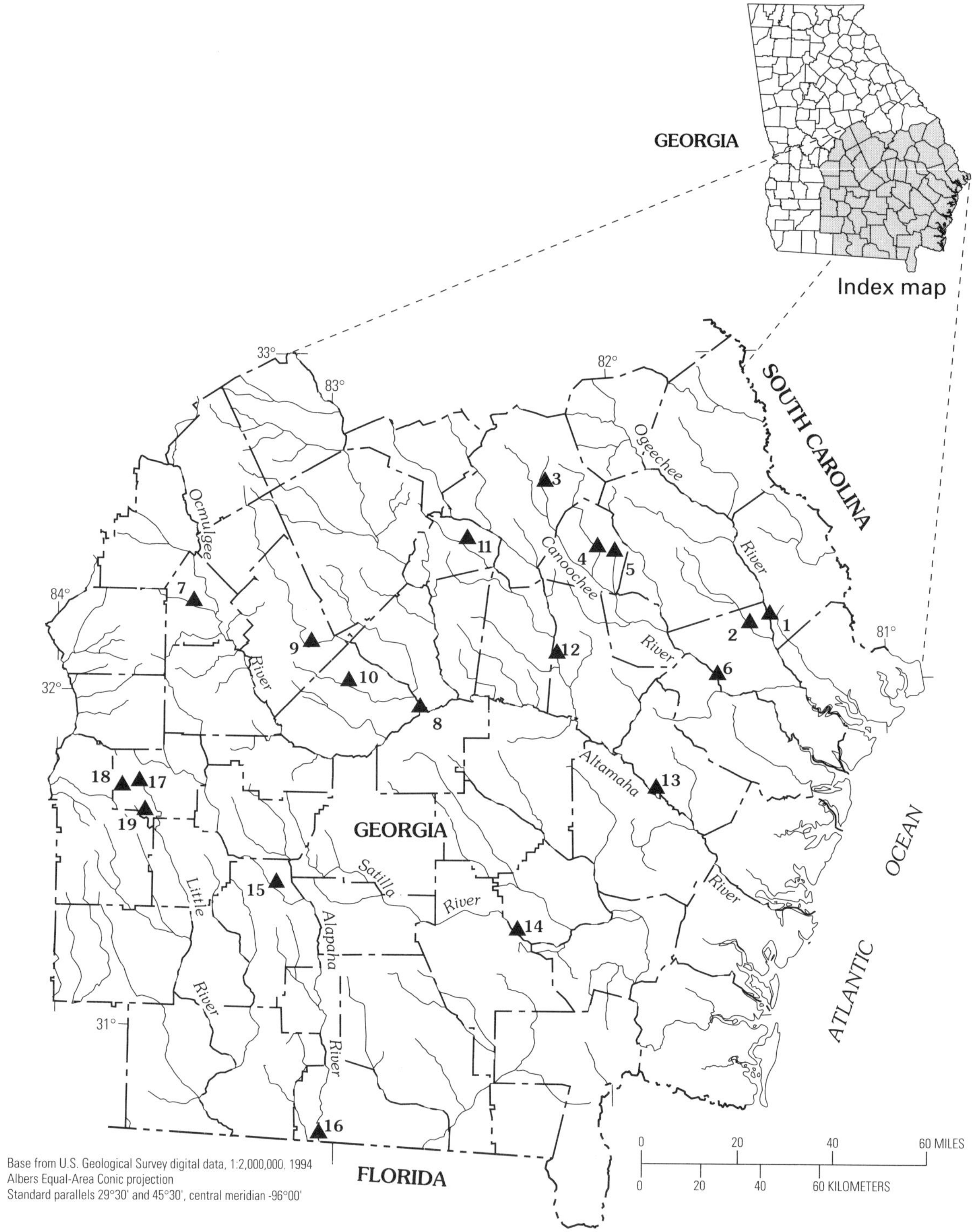

EXPLANATION

▲10 **Flood-determination site**—Number corresponds to that in table 36

Figure 59. Extent of flooding and location of selected streamflow-gaging stations for flood of January 8–21, 1993, in southeastern Georgia.

Table 36. Maximum stages and discharges prior to and during January 8–21, 1993, in southeastern Georgia

[mi^2, square miles; ft, feet above an arbitrary datum; ft^3/s, cubic feet per second. Source: Recurrence intervals calculated from U.S. Geological Survey data. Other data from U.S. Geological Survey reports or data bases]

				Maximum prior to January 1993				Maximum during January 8–21, 1993			
Site no. (fig. 59)	Station no.	Stream and place of determination	Drainage area (mi^2)	Period	Year	Stage (ft)	Discharge (ft^3/s)	Day	Stage (ft)	Discharge (ft^3/s)	Discharge recurrence interval (years)
				Ogeechee River Basin							
1	02202500	Ogeechee River near Eden, GA	2,650	1898–1993	1929	20.00	78,000	15	14.64	28,000	15
2	02202600	Black Creek near Blitchton, GA	232	1980–93	1990	12.91	5,620	14	12.88	5,560	15
3	02202820	Reedy Creek near Twin City, GA	8.99	1929, 1965–74, 1980, 1991, 1993	1929	7.70	1,830	12	4.26	550	6
4	02202900	Fifteen Mile Creek near Metter, GA	147	1962–83, 1991, 1993	1966	8.96	6,400	12	8.27	4,900	25
5	02202910	Ten Mile Creek tributary at Pulaski, GA	1.14	1965–87, 1991, 1993	1966	7.67	599	12	5.56	300	25
6	02203000	Canoochee River near Claxton, GA	555	1925–93	1925	17.80	20,500	14	16.60	13,500	40
				Altamaha River Basin							
7	02215100	Tucsawhatchee Creek near Hawkinsville, GA	163	1984–93	1991	14.13	4,740	9	13.19	3,760	10
8	02215500	Ocmulgee River at Lumber City, GA	5,180	1891–1993	1925	26.30	98,400	14	17.58	39,000	4
9	02215800	Gum Swamp Creek near Chauncey, GA	221	1984–93	1991	9.91	4,940	12	7.68	1,920	2
10	02216180	Turnpike Creek near McRae, GA	49.2	1983–93	1985	10.56	2,220	12	[1]11.04	[1]2,740	25
11	02225350	Reedy Creek tributary near Soperton, GA	1.68	1965–88, 1991, 1993	1973, 1984	3.47	290	12	3.30	261	10
12	02225500	Ohoopee River near Reidsville, GA	1,110	1904–93	1925	28.40	47,000	14	22.44	23,100	40
13	02226000	Altamaha River at Doctortown, GA	13,600	1925–93	1925	18.60	300,000	17	14.21	91,700	5
				Satilla River Basin							
14	02226500	Satilla River near Waycross, GA	1,200	1928, 1937–93	1948	22.40	39,000	14	17.05	9,320	4

Table 36. Maximum stages and discharges prior to and during January 8–21, 1993, in southeastern Georgia—Continued

Site no. (fig. 59)	Station no.	Stream and place of determination	Drainage area (mi^2)	Maximum prior to January 1993: Period	Year	Stage (ft)	Discharge (ft^3/s)	Maximum during January 8–21, 1993: Day	Stage (ft)	Discharge (ft^3/s)	Discharge recurrence interval (years)
				Suwannee River Basin							
15	02316000	Alapaha River near Alapaha, GA	663	1928–93	1928	19.00	16,000	18	16.80	12,000	25
16	02317500	Alapaha River at Statenville, GA	1,400	1928–93	1948	29.80	27,300	21	28.78	16,300	20
17	02317770	Newell Branch near Ashburn, GA	6.48	1965–75, 1991, 1993	1975	4.95	412	12	[1]6.54	[1]900	100
18	02317775	Daniels Creek near Ashburn, GA	1.11	1965–87, 1991, 1993	1983	3.60	231	12	[1]4.86	[1]366	100
19	02317780	Lime Sink Creek near Sycamore, GA	.68	1965–84, 1991, 1993	1982	6.08	252	12	5.95	244	30

[1]New maximum for period of record.

January 20–23, 1993, in Southern Mississippi

By W. Trent Baldwin

During a 2-day period from the morning of January 19 through the morning of January 21, 1993, a storm system swept through southern Mississippi producing rainfall totals in excess of 8 inches (fig. 60). The town of White Sand received 10.8 inches of rainfall in a 24-hour period (National Oceanic and Atmospheric Administration, 1993). The 50-year, 24-hour rainfall at White Sand is about 10.5 inches (Hershfield, 1961).

Maximum stages and discharges were documented at 25 flood-determination sites (fig. 60) that experienced a flood with a 2-year recurrence interval or greater. These floods are summarized in table 37. The most severe flooding occurred on West Hobolochitto Creek (site 24) where the flood was the largest on record and was estimated to have a 200-year recurrence interval.

REFERENCES

Hershfield, D.M., 1961, Rainfall frequency atlas of the United States: U.S. Department of Commerce, Weather Bureau, Technical Paper No. 40, 115 p.

National Oceanic and Atmospheric Administration, 1993, Climatological data, Mississippi, January 1993: Asheville, N.C., National Climatic Data Center, v. 98, no. 1, 27 p.

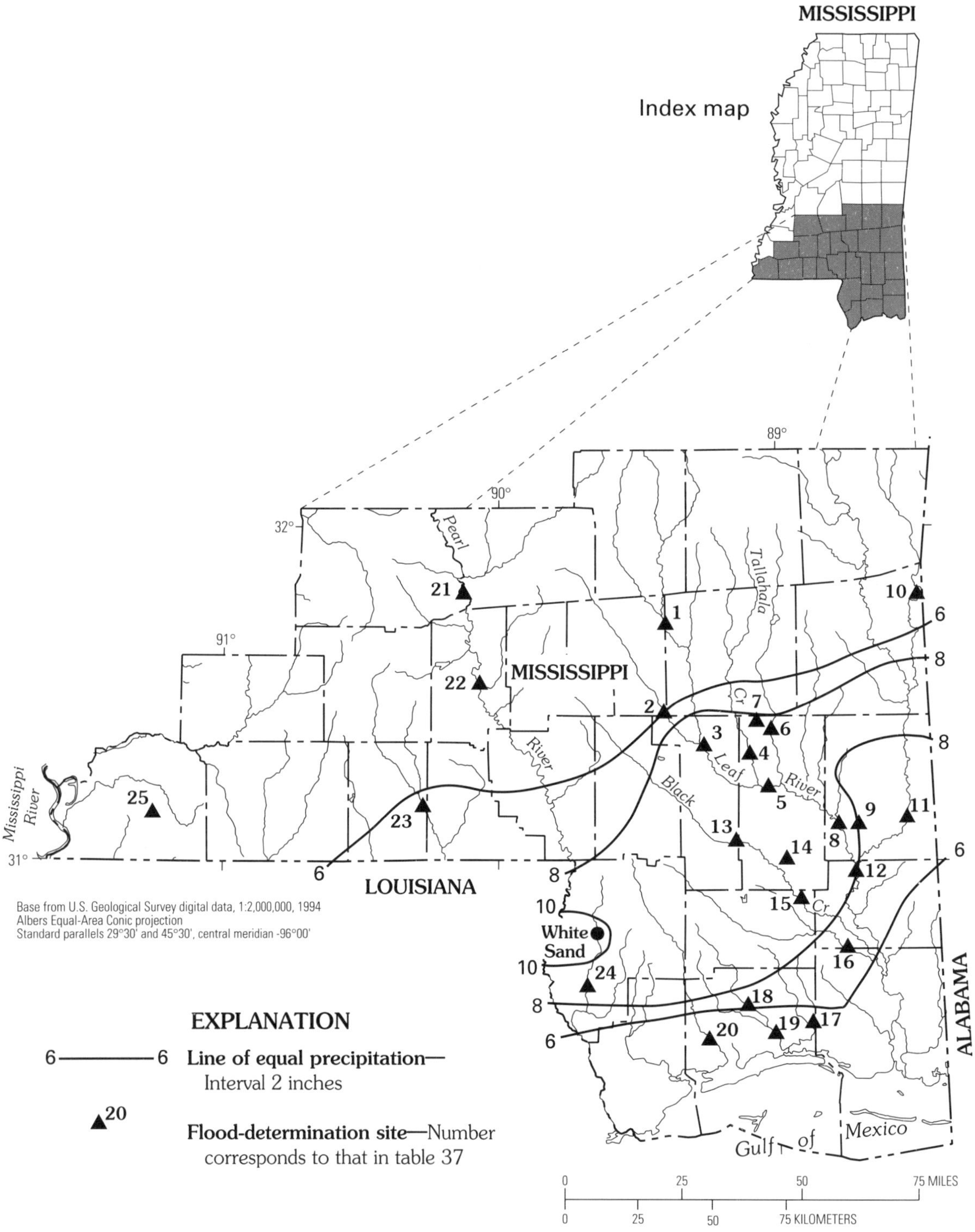

Figure 60. Location of flood-determination sites for flood of January 20–23, 1993, and lines of equal precipitation for storm January 19–21, 1993, in southern Mississippi (precipitation data from National Oceanic and Atmospheric Administration, 1993).

Table 37. Maximum stages and discharges prior to and during January 20–23, 1993, in southern Mississippi

[mi^2, square miles; ft, feet above an arbitrary datum; ft^3/s, cubic feet per second; --, not determined or not applicable. Source: Recurrence intervals calculated from U.S. Geological Survey data. Other data from U.S. Geological Survey reports or data bases]

Site no. (fig. 60)	Station no.	Stream and place of determination	Drainage area (mi^2)	Maximum prior to January 1993: Period	Year	Stage (ft)	Discharge (ft^3/s)	Maximum during January 20–23, 1993: Day	Stage (ft)	Discharge (ft^3/s)	Discharge recurrence interval (years)
				Pascagoula River Basin							
1	02472000	Leaf River near Collins, MS	743	1856–1993	1856	33.00	56,000	23	21.19	14,800	2
					1974	32.60	54,200				
2	02472500	Bouie Creek near Hattiesburg, MS	304	1900–93	1900	[1]33.50	[1]50,000	21	18.14	7,620	3
					1974	28.18	45,500				
3	02473000	Leaf River at Hattiesburg, MS	1,750	1900–93	1974	34.03	121,000	21	21.17	29,000	3
4	02474500	Tallahala Creek near Runnelstown, MS	612	1885–1993	1900	[2]30.50	38,000	21	21.59	11,000	3
					1961	[2]24.84	32,800				
5	02474560	Leaf River near New Augusta, MS	2,540	1900–93	1900	36.00	120,000	22	26.28	42,400	5
6	02474600	Bogue Homo near Richton, MS	344	1971–93	1973	27.63	21,900	21	20.08	9,440	3
7	02474650	Buck Creek near Runnelstown, MS	20.8	1951–93	1961	94.89	3,900	21	12.58	2,370	2
					1979	[2]91.41	5,700				
8	02475000	Leaf River near McLain, MS	3,500	1900–93	1900	31.80	131,000	22	25.34	61,000	5
					1961	31.64	128,000				
9	02475050	Waterfall Branch near McLain, MS	.65	1955–93	1959	11.71	764	20	7.13	300	2
10	02477990	Buckatunna Creek near Denham, MS	492	1972–93	1979	34.90	12,200	21	25.59	6,490	2
11	02478500	Chickasawhay River at Leakesville, MS	2,690	1900–93	1900	38.00	125,000	23	28.10	29,600	3
					1961	33.52	73,600				
12	02479000	Pascagoula River at Merrill, MS	6,590	1852–1993	1900	32.50	230,000	23	25.46	91,700	4
					1916	31.00	187,000				
13	02479130	Black Creek near Brooklyn, MS	355	1961–93	1983	29.96	42,500	21	25.29	20,300	10
14	02479155	Cypress Creek near Janice, MS	52.6	1959–93	1959	32.06	[1]16,700	21	27.56	8,980	15
15	02479160	Black Creek near Wiggins, MS	701	1916–93	1916	30.50	[1]60,000	22	26.79	32,300	15

Table 37. Maximum stages and discharges prior to and during January 20–23, 1993, in southern Mississippi—Continued

Site no. (fig. 60)	Station no.	Stream and place of determination	Drainage area (mi²)	Maximum prior to January 1993: Period	Year	Stage (ft)	Discharge (ft³/s)	Maximum during January 20–23, 1993: Day	Stage (ft)	Discharge (ft³/s)	Discharge recurrence interval (years)
		Pascagoula River Basin—Continued									
16	02479300	Red Creek at Vestry, MS	441	1959–93	1987	21.48	28,000	22	19.93	21,200	10
		Tchoutacabouffa River Basin									
17	02480500	Tuxachanie Creek near Biloxi, MS	92.4	1906–93	1906	23.20	21,000	21	16.82	4,890	2
					1987	23.63	19,600				
		Biloxi River Basin									
18	02481000	Biloxi River at Wortham, MS	96.2	1916–93	1983	25.30	10,300	21	22.96	7,000	5
19	02481130	Biloxi River near Lyman, MS	251	1957–93	1957	[1]21.50	[1]35,000	21	19.55	16,900	5
		Wolf River Basin									
20	02481510	Wolf River near Landon, MS	308	1971–93	1983	21.53	18,400	22	27.12	16,600	10
		Pearl River Basin									
21	02488000	Pearl River at Rockport, MS	4,560	1874–1993	1979	42.83	[1]123,000	21	27.97	33,900	2
22	02488500	Pearl River near Monticello, MS	4,990	1874–1993	1874	34.50	--	22	24.97	40,100	3
					1979	34.08	122,000				
23	02490500	Bogue Chitto near Tylertown, MS	492	1936–93	1936	34.70	49,000	21	20.76	13,800	2
					1983	34.62	64,200				
24	02492360	West Hobolochitto Creek near McNeill, MS	175	1966–93	1983	23.25	18,000	21	23.43	27,800	200
		Thompson Creek Basin									
25	07373550	Moores Branch near Woodville, MS	.21	1955–93	1973	9.90	455	20	7.06	285	4

[1]Estimated.
[2]Site and datum then in use.

January 21–23, 1993, in Southeastern Louisiana

By Brian E. McCallum

During January 1993, greater-than-normal rainfall occurred in southeastern Louisiana due to a stationary front over the Mississippi River Valley. The excessive rain kept the soil saturated and rivers at flood stage. A slow-moving cold front interacting with warm, moist air from the Gulf of Mexico caused excessive rainfall in the Amite River Basin on January 20. Rainfall on January 19–21 ranged from 3.30 inches at Liberty, Mississippi, to 11.23 inches just north of Baton Rouge, Louisiana (fig. 61). The average basin rainfall was 8.54 inches (U.S. Department of Commerce, 1993).

Maximum stages and discharges for the period of record at six streamflow-gaging stations within the Amite River Basin are given in table 38. These stations are located in figure 61. The discharge on the Comite River near Comite (site 4, fig. 61) was the greatest since records began in 1944. The flood crested at the Denham Springs site (site 5, fig. 61) on January 22; the stage was 38.15 feet above sea level. The maximum discharge had a recurrence interval of 10 to 25 years.

Damages to more than 1,500 dwellings were reported for Ascension, East Baton Rouge, East Feliciana, and Livingston Parishes. According to guidelines established by the American Red Cross, the flooding destroyed 16 dwellings within the basin, caused major damage to 680 single-family dwellings, apartments, and mobile homes, and minor damage to 807 dwellings. The American Red Cross offered relief to nearly 1,300 people following the flood at a cost of more than $1 million (Margaret McGarity, American Red Cross, written commun., 1994). No loss of life was reported due to the flooding in the basin despite the rapid rise of the rivers and the number of dwellings damaged.

Total property damage from the flood was estimated at more than $2 million by the Louisiana Office of Emergency Preparedness (Shawn Fontenot, Louisiana Office of Emergency Preparedness, oral commun., 1994). This amount does not include damage costs claimed by individuals with flood insurance.

REFERENCE

U.S. Department of Commerce, National Weather Service, March 1993, Monthly report of river and flood conditions: New Orleans, Louisiana, Weather Service Forecast Center.

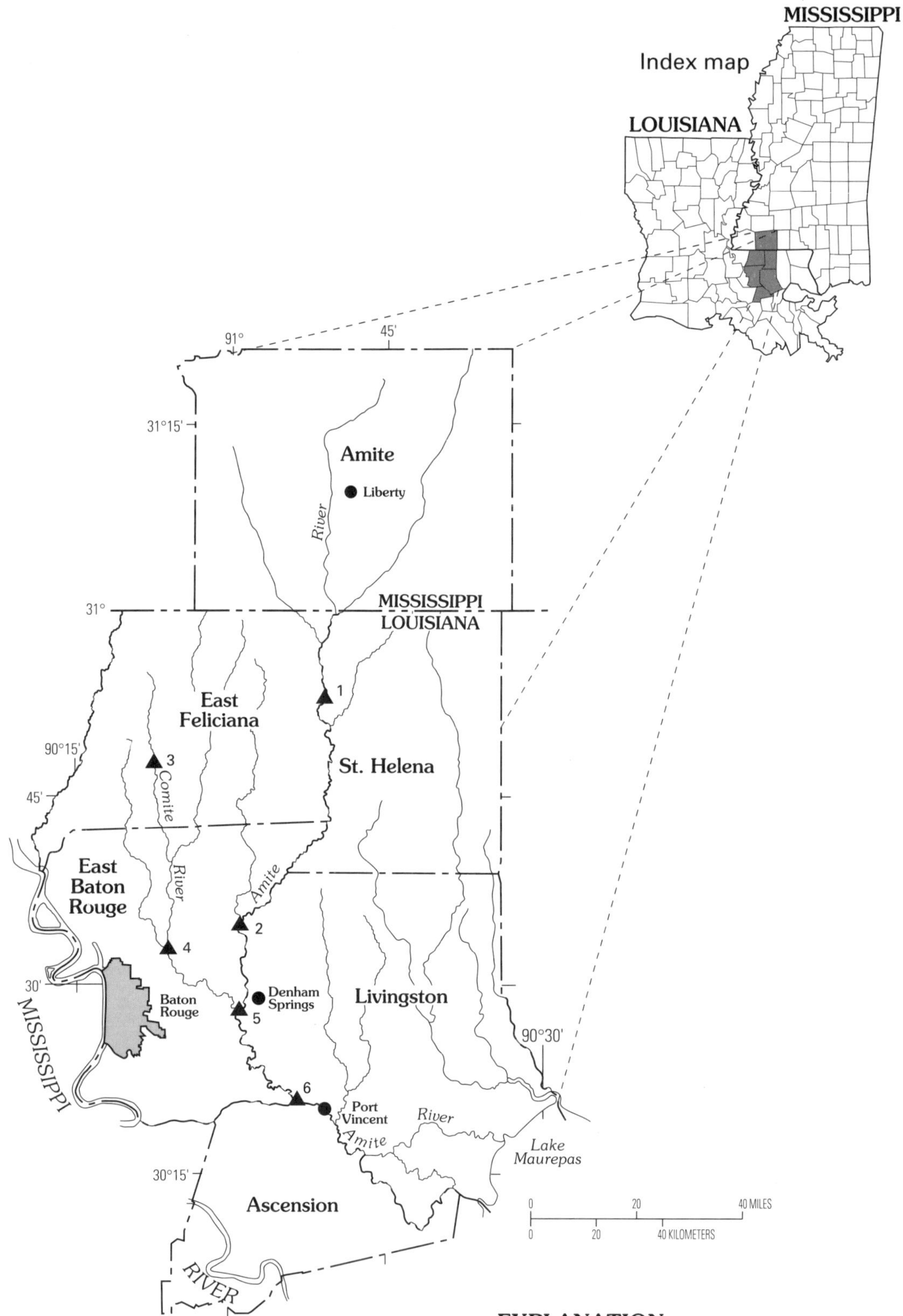

Base from U.S. Geological Survey hydrologic unit map, 1:500,000, 1974
Lambert Conformal Conic projection
Standard parallels 33° and 45°

EXPLANATION

▲6 Flood-determination site—Number corresponds to that in table 38

Figure 61. Location of streamflow-gaging stations in Amite River Basin, southeastern Louisiana.

Table 38. Maximum stages and discharges prior to and during January 21–23, 1993, in Amite River Basin, in southeastern Louisiana

[mi^2, square miles; ft, feet above an arbitrary datum; ft^3/s, cubic feet per second; --, not determined or not applicable. Source: Recurrence intervals calculated from U.S. Geological Survey data. Other data from U.S. Geological Survey reports or data bases]

				Maximum prior to January 21–23, 1993				Maximum during January 21–23, 1993			
Site no. (fig. 61)	Station no.	Stream and place of determination	Drainage area (mi^2)	Period	Year	Stage (ft)	Discharge (ft^3/s)	Day	Stage (ft)	Discharge (ft^3/s)	Discharge recurrence interval (years)
1	07377000	Amite River near Darlington, LA	580	1949–93	1990	22.05	104,000	21	16.08	19,400	5–10
2	07377300	Amite River at Magnolia, LA	884	1949–83, 1992–93	1977	51.91	85,100	21	49.08	59,600	10–25
3	07377500	Comite River near Olive Branch, LA	145	1942–93	1961	23.37	19,900	21	14.42	13,600	5–10
4	07378000	Comite River near Comite, LA	284	1944–93	1953	30.64	20,500	21	27.58	30,400	25–50
5	07378500	Amite River near Denham Springs, LA	1,280	1921–93	1983	41.50	112,000	22	38.15	81,900	10–25
6	07380120	Amite River at Port Vincent, LA	1,596	1946–93	1983	14.65	--	23	11.87	--	--

March 1993, in Virginia

By Byron J. Prugh, Jr.

MARCH 4–5, 1993

On March 4 and 5, 1993, a low-pressure center moved from South Carolina across southeastern Virginia. Moisture-laden air associated with the center was driven by southeasterly winds that were forced up the eastern slope of the Blue Ridge Mountains and produced excessive rains that were concentrated over the central part of Virginia. Precipitation varied from 2 inches along the coastal plain, to between 3 and 4 inches over much of the Piedmont and Shenandoah River Valley, to 1.5 to 2 inches in the southwestern corner of the State (National Oceanic and Atmospheric Administration, 1993). The rain changed to snow in the western parts of the State.

Runoff from the rain caused widespread moderate flooding across much of the eastern half of Virginia. There were three distinct areas of flooding where the recurrence intervals of the maximum discharges were greater than 10 years. One area was for streams tributary to the Potomac River in the extreme northern corner of the State, a second was for streams in the headwaters of the Rappahannock River Basin, and a third for streams in the middle Roanoke and upper Appomattox River Basins (fig. 62). The most severe flooding occurred in the upper Rappahannock River Basin where the maximum discharges had a recurrence interval of about 20 years. The magnitudes of the maximum stages and discharges at selected streamflow-gaging stations are summarized in table 39.

Elsewhere in the State, the recurrence intervals of maximum discharges in the James, Shenandoah, Anna, Meherrin, Nottoway, and parts of the Roanoke River Basins were between 2 and 10 years. In southwestern Virginia, maximum discharges had recurrence intervals of less than 2 years. Major damage from the flooding was road closures and power outages. There were no deaths reported.

MARCH 23–30, 1993

A maximum discharge of 19,000 cubic feet per second was recorded at the streamflow-gaging station on Walker Creek at Bane, Virginia (site 6, fig. 63), on March 24, 1993. This was the second highest discharge recorded at this site since systematic data collection began in 1938. The calculated recurrence interval was in excess of 100 years. Maximum stage and discharge data for March 23–24 at selected stations in the New River and adjacent basins are summarized in table 40.

Precipitation for March 23–25 was only 2 to 3 inches over the central part of the New River Basin including the Walker Creek Basin. Precipitation for the surrounding area was less than 2 inches. However, the ground was well saturated from five episodes of precipitation earlier in the month, and local rainfall amounts could have been greater. The runoff from the March 23–25 precipitation produced the highest maximum discharges for the 1993 water year at 15 streamflow-gaging stations in the New River Basin, at 20 stations in the Tennessee River Basin, at 5 stations in the Big Sandy and Roanoke River Basins, and at 1 station in the upper James River Basin. The greatest flooding occurred on tributaries in the lower New River Basin where maximum discharges with recurrence intervals of 8 to more than 100 years were recorded. Elsewhere the flooding was less severe. In the Tennessee and upper New River Basins, maximum discharges had recurrence intervals of less than 2 years to about 10 years. In the Big Sandy, upper James, and upper Roanoke River Basins, maximum discharges had recurrence intervals of less than 2 to about 5 years. The approximate extent of the 10-year recurrence-interval flooding and location of these stations are shown in figure 63.

The March 23–25 rainfall was not the final precipitation and subsequent runoff for the month, as another storm occurred on March 27–30. The cumulative effect of the frequent precipitation resulted in greater-than-normal flows for the month of March. Numerous streamflow-gaging stations across the State recorded new maximum monthly mean flows for the month. At long-term stations on the James River, the monthly mean flows were the highest for March since at least 1899. In the Tennessee and Big Sandy River Basins, the monthly mean flows were the highest since 1975, and in the Roanoke and New River Basins flows were the highest since 1930.

REFERENCE

National Oceanic and Atmospheric Administration, 1993, Climatological data, Virginia: Asheville, N.C., National Climatic Data Center, v. 103, no. 3, 23 p.

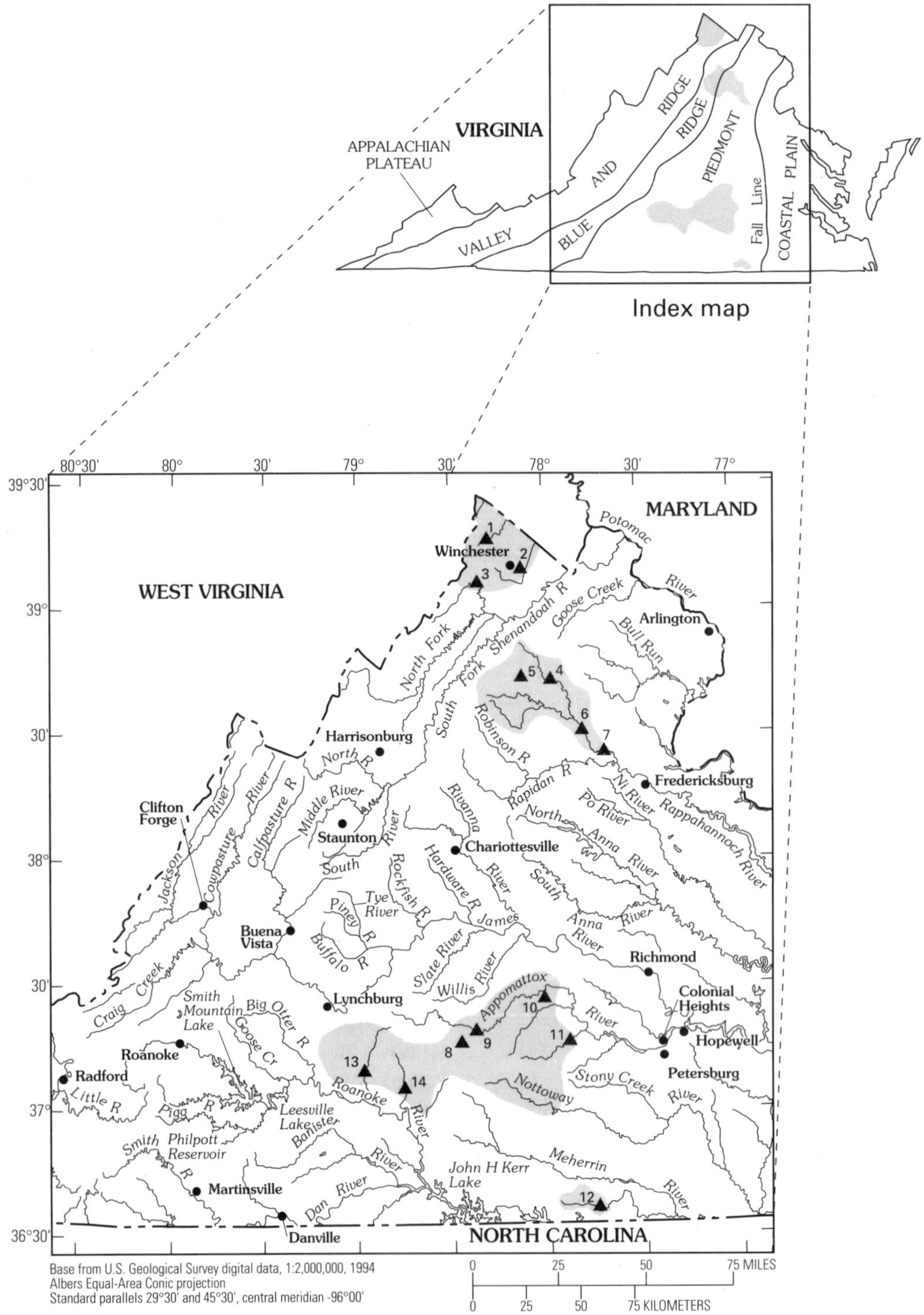

EXPLANATION

Approximate extent of 10-year or greater recurrence-interval flooding

▲3 **Flood-determination site**—Number corresponds to that in table 39

Figure 62. Approximate extent of 10-year or greater recurrence-interval flooding during March 4–8, 1993, location of selected streamflow-gaging stations, and physiographic provinces in eastern Virginia.

Table 39. Maximum stages and discharges prior to and during March 4–8, 1993, in eastern Virginia

[mi^2, square miles; ft, feet above an arbitrary datum; ft^3/s, cubic feet per second; --, not determined or not applicable. Source: Recurrence intervals calculated from U.S. Geological Survey data. Other data from U.S. Geological Survey reports or data bases]

				Maximum prior to March 4–8, 1993				Maximum during March 4–8, 1993			
Site no. (fig. 62)	Station no.	Stream and place of determination	Drainage area (mi^2)	Period	Year	Stage (ft)	Discharge (ft^3/s)	Day	Stage (ft)	Discharge (ft^3/s)	Discharge recurrence interval (years)
1	01613900	Hogue Creek near Hayfield, VA	15.0	1961–93	1972	8.85	2,760	4	7.27	2,320	20
2	01615000	Opequon Creek near Berryville, VA	57.4	1942,	1942	18.40	--	4	11.07	5,990	15
				1944–93	1988	13.47	12,600				
3	01634500	Cedar Creek near Winchester, VA	103	1936,	1936	25.00	18,000	4	16.15	9,090	15
				1938–93	1942	27.00	22,000				
4	01662000	Rappahannock River near Warrenton, VA	195	1942–93	1942	23.50	32,000	4	20.03	13,200	20
5	01662800	Battle Run near Laurel Mills, VA	27.6	1959–93	1976	13.90	9,120	4	11.18	3,270	10
6	01664000	Rappahannock River at Remington, VA	620	1942–93	1942	30.00	90,000	5	21.96	36,600	20
7	01668000	Rappahannock River near Fredericksburg, VA	1,596	1908–93	1942	25.90	140,000	5	15.54	57,800	10
8	02039000	Buffalo Creek near Hampden Sydney, VA	69.7	1940,	1940	15.00	--	4	10.22	5,310	15
				1947–93	1972	12.38	9,160				
9	020395003	Appomattox River at Farmville, VA	303	1926–93	1972	29.70	33,100	5	20.78	11,700	10
10	02040500	Flat Creek near Amelia, VA	73.0	1947,	1947	6.90	1,090	4	11.24	3,930	20
				1954–70,	1961	9.21	3,300				
				1972–93	1972	15.81	10,600				
11	02041000	Deep Creek near Mannboro, VA	158	1940,	1940	14.80	10,000	5	14.32	8,980	10
				1947–93	1972	24.04	15,000				
12	02052500	Fountains Creek near Brink, VA	65.2	1940,	1940	18.50	2,720	4	17.97	3,980	10
				1954–93	1972	24.14	16,000				
13	02064000	Falling River near Naruna, VA	173	1930–93	1972	29.21	32,600	4	21.01	11,400	15
14	02065500	Cub Creek at Phenix, VA	98.0	1940,	1940	17.50	4,000	5	14.38	5,550	15
				1947–93	1972	[1]20.37	7,380				
					1987	19.31	10,600				

[1] Backwater.

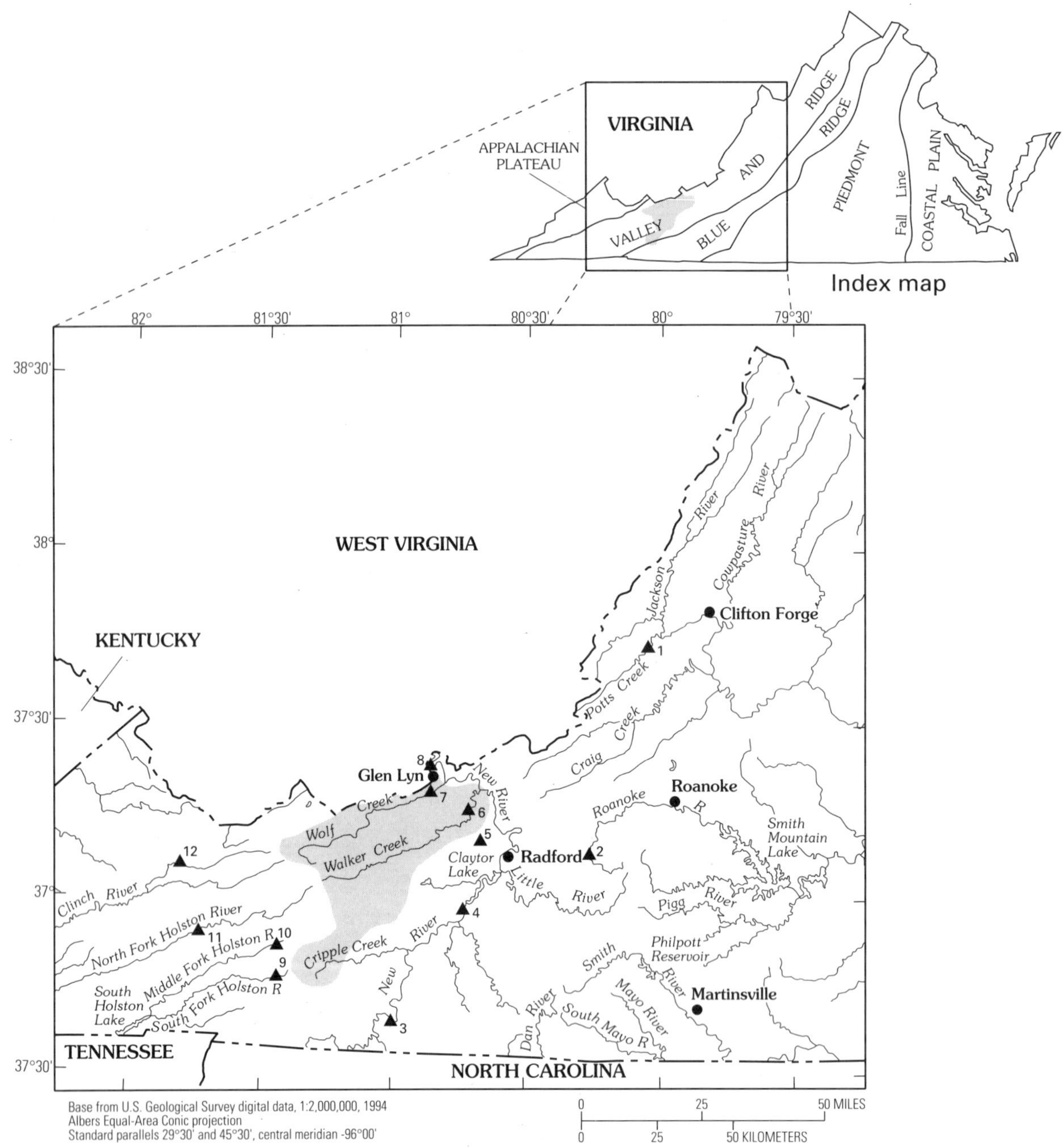

EXPLANATION

Approximate extent of 10-year or greater recurrence-interval flooding

Flood-determination site—Number corresponds to that in table 40

Figure 63. Approximate extent of 10-year or greater recurrence-interval flooding during March 23–24, 1993, and location of selected streamflow-gaging stations in western Virginia.

Table 40. Maximum stages and discharges prior to and during March 23–24, 1993, in western Virginia

[mi^2, square miles; ft, feet above an arbitrary datum; ft^3/s, cubic feet per second; --, not determined or not applicable; >, greater than. Source: Recurrence intervals calculated from U.S. Geological Survey data. Other data from U.S. Geological Survey reports or data bases]

				Maximum prior to March 23–24, 1993				Maximum during March 23–24, 1993			
Site no. (fig. 63)	Station no.	Stream and place of determination	Drainage area (mi^2)	Period	Year	Stage (ft)	Discharge (ft^3/s)	Day	Stage (ft)	Discharge (ft^3/s)	Discharge recurrence interval (years)
1	02014000	Potts Creek near Covington, VA	153	1878,	1878	12.0	--	24	9.06	5,080	5
				1929–56,	1935	10.10	7,510				
				1966–93	1985	13.46	15,400				
2	02053800	South Fork Roanoke River near Shawsville, VA	110	1961–93	1972	11.12	14,200	24	5.93	4,570	5
3	03164000	New River near Galax, VA	1,131	1930–93	1940	25.7	141,000	24	8.66	31,300	5
4	03168000	New River at Allisonia, VA	2,202	1930–93	1940	23.42	185,000	24	11.69	61,600	8
5	03168750	Thorne Springs Branch near Dublin, VA	4.77	1957–93	1973	8.01	2,200	23	2.66	246	5
6	03173000	Walker Creek at Bane, VA	305	1878,	1878	23.5	40,000	24	17.43	19,000	>100
				1938–93	1992	19.28	25,000				
7	03175500	Wolf Creek near Narrows, VA	223	1909–16,	1916	13.0	11,000	24	11.88	11,100	35
				1938–93	1957	13.8	12,900				
8	03176500	New River at Glen Lyn, VA	3,768	1878,	1878	33.1	240,000	24	17.47	100,000	8
				1915–93	1940	[1]27.50	[1]226,000				
9	03471200	South Fork Holston River at Teas, VA	31.1	1967–93	1977	--	5,410	23	14.01	2,420	10
10	03473500	Middle Fork Holston River at Groseclose, VA	7.39	1948–93	1953	7.42	813	23	4.30	220	3
11	03488000	North Fork Holston River near Saltville, VA	222	1862,	1862	15	22,000	24	9.74	8,770	5
				1921–93	1957	13.20	16,500				
12	03521500	Clinch River at Richlands, VA	137	1901,	1901	21.3	11,500	24	10.78	4,080	3
				1946–93	1957	19.3	9,640				

[1]Flow in the New River downstream from Clayton Lake has been regulated since 1939.

March 8–12, 1993, in East-Central Nebraska

By J.A. Boohar

On March 9, 1993, severe flooding occurred because of ice jamming on the Platte River in Sarpy County (fig. 64). Ice on the Elkhorn River had loosened during warm weather in early March, sending water and ice into the still-frozen Platte River near Gretna. A 2-mile ice jam forced waters out of the banks of the Elkhorn River along various points from West Point down to its confluence with the Platte River.

The Elkhorn River at West Point (site 8, table 41) had a discharge of 28,000 cubic feet per second at a stage of 18.60 feet, which was determined to have a 10-year recurrence interval. Although the March maximum for this streamflow-gaging station was not the maximum for the water year, the stage was the maximum for the period of record. The Elkhorn River at Waterloo (site 12) had a discharge of 33,500 cubic feet per second at a stage of 15.76 feet and was calculated to have an approximate 10-year recurrence interval. Flow from tributary streams and the ice jam on the Platte River near Gretna caused backwater conditions that contributed to the flooding along the Elkhorn River.

A western Sarpy County levee at the confluence of the Elkhorn and Platte Rivers upstream from the ice jam near Gretna ruptured and allowed water to escape the channel and flood cabins and about 8,000 acres of farmland downstream from the ice jam. Along the west bank of the Platte River farther downstream, water supplies from the city of Lincoln well field near Ashland were threatened when two of three water mains washed out. The remaining main held after the line was shored up by workers. The West Sarpy County Drainage District created a channel in the jam using dynamite. This released pressure against the levee and allowed repairs to take place. The continuous-record streamflow-gaging station, Platte River near Ashland (site 13), recorded a maximum stage of 19.23 feet on March 10. Discharge for this peak, including bypass flow, was estimated to be 130,000 cubic feet per second and was determined to be greater than the 100-year recurrence interval.

At Columbus, near the confluence of the Loup and the Platte Rivers, a 0.5-mile ice jam caused flooding and evacuation of some residents from the Columbus area. Above-freezing temperatures caused melting on the Cedar River and neighboring streams that drain into the Loup River and eventually into the Platte. Highway 81, south of Columbus, sustained extensive damage and was closed to traffic.

Upstream along the Platte River, floodwater from the Wood River rose to a depth of 2 feet in the town of Wood River and forced about 20 people from their homes. The Wood River near Alda (site 1) had a discharge of 1,300 cubic feet per second at a stage of 11.81 feet on March 12. The volume of water draining into the Platte River, and eventually entering the Missouri River, created backwater at the Missouri River. The flow at site 14, Platte River at Louisville, 16.5 miles upstream from the confluence of the Missouri River, was determined to be 143,000 cubic feet per second, just 1,000 cubic feet per second less than the maximum for the period of record, at a stage of 11.49 feet. This discharge was determined to have greater than a 50-year recurrence interval.

Site 18, West Fork Big Blue River near Dorchester, recorded a stage of 21.71 feet at 0530 on March 11. Discharge estimated for this stage was 12,400 cubic feet per second, which was a new maximum discharge for the period of record although the stage was not a new maximum. Flooding on the West Fork Big Blue River near Beaver Crossing resulted in two fatalities.

Of the 40 counties that reported flooding, the following 12 had major flooding: Buffalo, Butler, Colfax, Cuming, Dodge, western Douglas, Hall, Nance, Platte, Sarpy, Saunders, and Seward. Damages estimated by the Civil Defense and the Federal Emergency Management Agency in five categories—agriculture, highway system, public assistance, individual assistance, and soil erosion—totaled more than $16.6 million in the 40 counties (Federal Emergency Management Agency, 1993).

REFERENCE

Federal Emergency Management Agency, 1993, Interagency hazard mitigation team report for Nebraska: FEMA–983DR–NE.

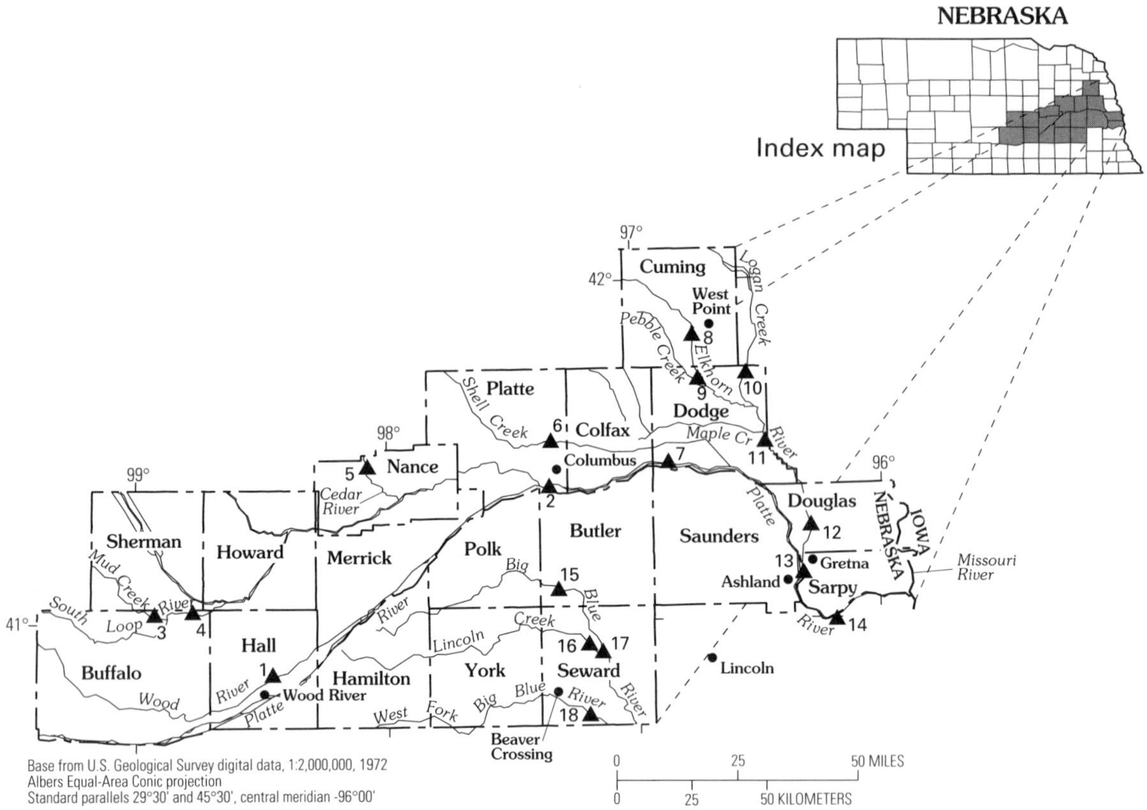

EXPLANATION

▲17 **Flood-determination site**—Number corresponds to that in table 41

Figure 64. Location of flood-determination sites for flood of March 8–12, 1993, in east-central Nebraska.

Table 41. Maximum stages and discharges prior to and during March 8–12, 1993, in east-central Nebraska

[mi^2, square miles; ft, feet above an arbitrary datum; ft^3/s, cubic feet per second; >, greater than. Source: Recurrence intervals calculated from U.S. Geological Survey data. Other data from U.S. Geological Survey reports or data bases]

				Maximum prior to March 8, 1993				Maximum during March 8–12, 1993			
Site no. (fig. 64)	Station no.	Stream and place of determination	Drainage area (mi^2)	Period	Year	Stage (ft)	Discharge (ft^3/s)	Day	Stage (ft)	Discharge (ft^3/s)	Discharge recurrence interval (years)
1	06772000	Wood River near Alda, NE	628	1953–93	1967	12.22	1,630	12	11.81	1,300	20
2	06774000	Platte River near Duncan, NE	60,900	1895–09,	1905	[1]6.50	44,100	11	7.86	18,000	10
				1928–93	1987	6.78					
3	06783500	Mud Creek near Sweetwater, NE	707	1946–93	1947	23.20	27,000	9	18.95	2,070	5
								7	19.50		
4	06784000	South Loup River at St. Michael, NE	2,350	1943–93	1968	11.00	27,500	9	10.32	9,970	7
					1947	12.00					
5	06792000	Cedar River near Fullerton, NE	1,220	1931–32, 1940–93	1966	16.90	64,700	8	9.40	8,000	9
6	06795500	Shell Creek near Columbus, NE	270	1947–75, 1977–93	1990	22.76	8,000	9	21.53	3,600	8
7	06796000	Platte River at North Bend, NE	77,100	1949–93	1960	10.04	112,000	10	9.71	97,800	50
					1978	15.55		8	10.97		
8	06799350	Elkhorn River at West Point, NE	5,100	1972–93	1969	13.21	33,000	9	18.60	28,000	10
					1978	16.09					
9	06799385	Pebble Creek near Scribner, NE	204	1978–93	1991	24.15	27,900	8	18.52	4,070	2
10	06799500	Logan Creek near Uehling, NE	1,030	1941–93	1971	20.15	25,200	9	17.96	10,300	4
11	06800000	Maple Creek near Nickerson, NE	450	1951–93	1990	[1]16.30	11,600	9	14.26	6,270	6
					1984	[1]17.65					
12	06800500	Elkhorn River at Waterloo, NE	6,900	1928–93	1944	[1]16.60	100,000	11	15.76	33,500	12
13	06801000	Platte River near Ashland, NE	84,200	1928–53, 1988–93	1944	[1]8.10	107,000	10	19.23	[2]130,000	>100
14	06805500	Platte River at Louisville, NE	85,800	1953–93	1984	11.34	144,000	10	11.49	143,000	>50
					1960	12.45					
15	06879900	Big Blue River at Surprise, NE	345	1964–93	1965	11.52	10,700	10	9.57	2,360	3

Table 41. Maximum stages and discharges prior to and during March 8–12, 1993, in east-central Nebraska—Continued

Site no. (fig. 64)	Station no.	Stream and place of determination	Drainage area (mi^2)	Maximum prior to March 8, 1993				Maximum during March 8–12, 1993			
				Period	Year	Stage (ft)	Discharge (ft^3/s)	Day	Stage (ft)	Discharge (ft^3/s)	Discharge recurrence interval (years)
16	06880000	Lincoln Creek near Seward, NE	446	1953–93	1957	20.53	10,100	10	18.35	3,280	6
17	06880500	Big Blue River at Seward, NE	1,099	1953–93	1957	22.34	15,300	11	20.86	7,640	7
					1967	[1]22.83					
18	06880800	West Fork Big Blue River near Dorchester, NE	1,206	1958–93	1986	22.62	11,800	11	21.71	[2]12,400	20

[1]Site and datum then in use.
[2]Estimated.

April 1993, in New York

By Carolyn O. Szabo

A severe winter storm deposited large amounts of snow over New York State on March 13–14. Many areas received three to five times greater-than-normal snowfall for March. Syracuse reported approximately 43 inches for the 2-day period. Of that amount, more than 35 inches fell within 24 hours (Cornell University, 1993). More than 40 inches of snow fell in extreme northern and northeastern New York. By March 17, the snowpack in the Oswego River Basin (fig. 65) ranged from 30 to 40 inches, with a water equivalent of more than 10 inches in some areas.

April 1993 was New York's third wettest April on record. Statewide precipitation for April averaged 5.42 inches, 165 percent of normal for the month. Several rainstorms, combined with the deep snowpack and warm temperatures, caused rivers and lakes to rise slowly but continuously during the month. Flood watches and warnings were issued throughout the State as rivers continued to rise. Flooding was most severe on the Black and Oswego Rivers, east and southeast of Lake Ontario, respectively (fig. 66).

Most flooding in the Black River Basin occurred April 11–12. Record river stages and discharges were recorded throughout the downstream reaches of the basin from Watson to Dexter (fig. 66) where the river stage was more than 1 foot higher than the record-high level reached during the flood of December 1984–January 1985. The streamflow-gaging station on Black River at Watertown (site 9) recorded the maximum stage and discharge for the period of record (table 42).

The Oswego River experienced flooding throughout April. Peak stages at various lakes and rivers in the basin occurred April 24–27 and were similar to those recorded during the floods from Hurricane Agnes in 1972 (table 42). The Seneca River, most of which is included in the New York State Barge Canal system, begins at the outlet of Seneca Lake and meanders northeastward through the lowlands south of Lake Ontario until it joins with the Oneida River, near Phoenix, to become the Oswego River (fig. 66). Because the channel slope is relatively flat, the Seneca River frequently overflows its banks as a result of excessive runoff. During the spring of 1993, river levels remained high throughout April and into the early part of May, damaging many homes and businesses along the river.

The level of destruction in the Oswego and Black River Basins, although not considered a national disaster area, resulted in local states of emergencies. The Governor of New York obtained a disaster declaration under the Small Business Administration (SBA) program. Seventeen counties (4 of which were declared SBA disaster areas and 13 contiguous) (fig. 67) were judged eligible to receive support under various SBA programs (New York State Disaster Preparedness Commission, 1993).

REFERENCES

Cornell University, 1993, New York climate: Northeast Regional Climate Center Bulletin, v. 93, no. 3, p. 1.

New York State Disaster Preparedness Commission, 1993, After action report, Nor'easter '92, World Trade Center Explosion, Blizzard '93, Flooding '93: 25 p.

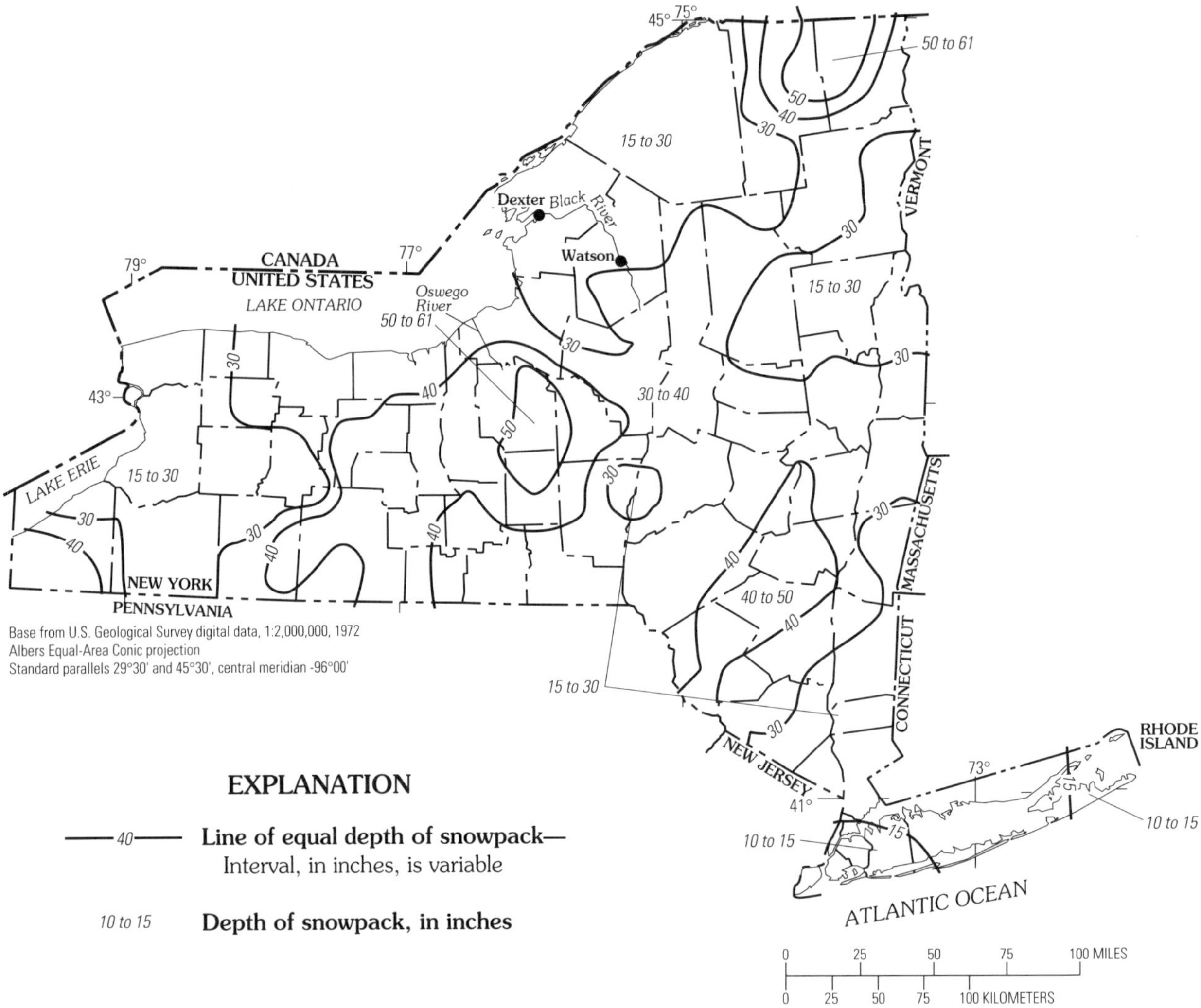

Figure 65. Lines of equal depth of snowpack, March 1993, in New York (data from Cornell University, 1993).

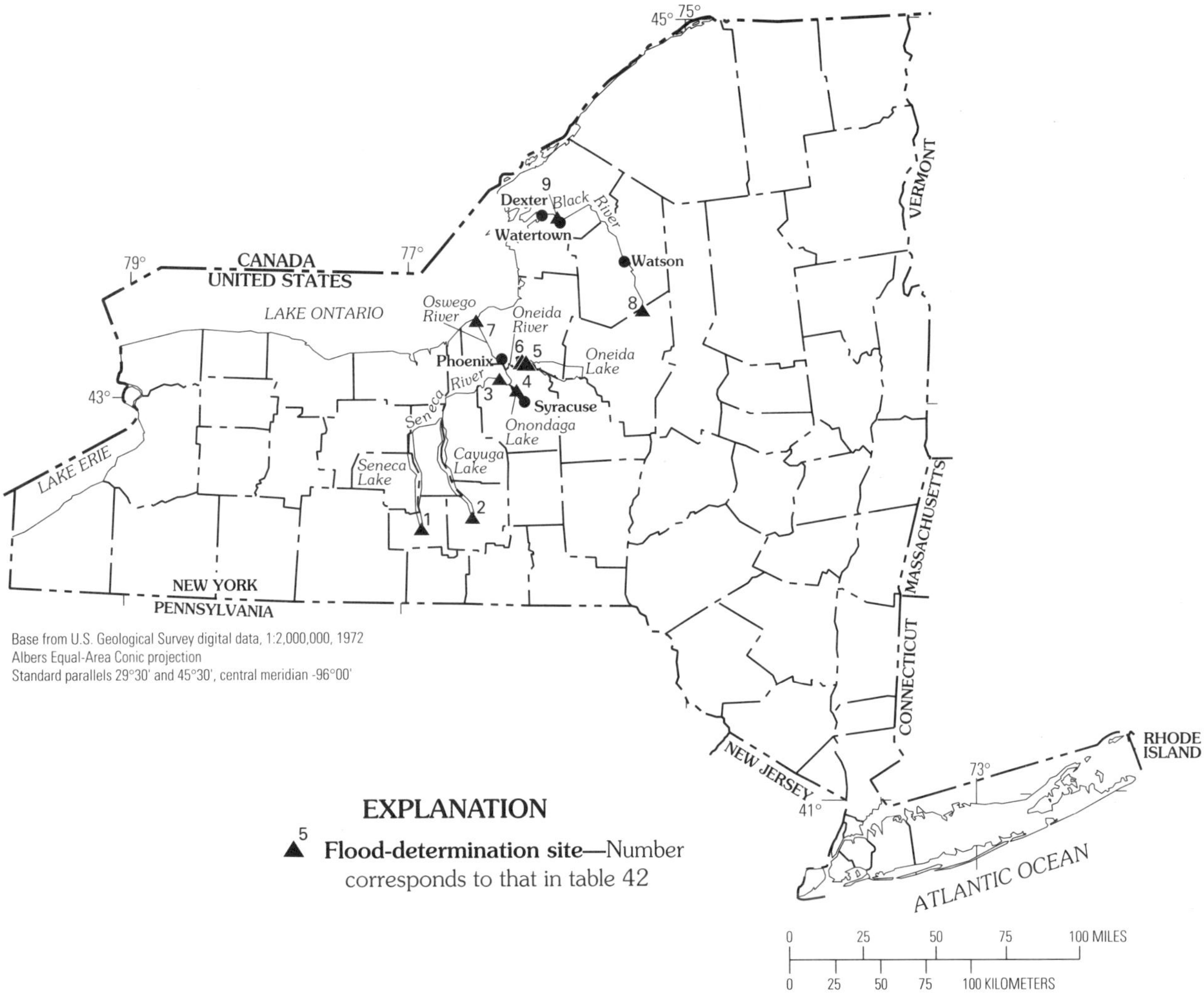

Figure 66. Location of selected U.S. Geological Survey streamflow-gaging stations in New York used to determine maximum stages and discharges for April 1993.

Table 42. Maximum stages and discharges prior to and during April 1993, in New York

[mi^2, square miles; ft, for rivers—feet above an arbitrary datum, for lakes—feet above sea level; ft^3/s, cubic feet per second; --, not determined or not applicable; >, greater than. Source: Recurrence intervals calculated from U.S. Geological Survey data. Other data from U.S. Geological Survey reports or data bases]

				Maximum prior to April 1993				Maximum during April 1993			
Site no. (fig. 66)	Station no.	Stream and place of determination	Drainage area (mi^2)	Period	Year	Stage (ft)	Discharge (ft^3/s)	Day	Stage (ft)	Discharge (ft^3/s)	Discharge recurrence interval (years)
1	04232400	Seneca Lake at Watkins Glen, NY	704	1956–93	1972	448.88	--	26, 27	448.95	--	--
2	04233500	Cayuga Lake at Ithaca, NY	1,564	1905–25, 1956–93	1972	386.33	--	26	386.46	--	--
3	04237500	Seneca River at Baldwinsville, NY	3,138	1949–93	1960, 1972	9.21 --	[1]17,200	27	--	[1]18,100	50–100
4	04240495	Onondaga Lake at Liverpool, NY	285	1970–93	1972	369.21	--	26, 27	369.78	--	--
5	04246000	Oneida Lake at Brewerton, NY	1,382	1936, 1951–93	1936	373.50	--	24	373.14	--	--
6	04246500	Oneida River at Caughdenoy, NY	1,382	1903–12, 1948–93	1903	--	[1]13,800	24	--	[1]11,300	25–50
7	04249000	Oswego River at Oswego, NY	5,100	1900–06, 1933–93	1936	13.10	37,500	27	13.13	36,900	50–100
8	04252500	Black River near Boonville, NY	304	1911–93	1982, 1984	11.31 11.40	12,800 12,800	11	10.79	10,300	7
9	04260500	Black River at Watertown, NY	1,864	1921–93	1984	13.01	36,500	12	14.20	42,600	>100

[1]Mean daily discharge.

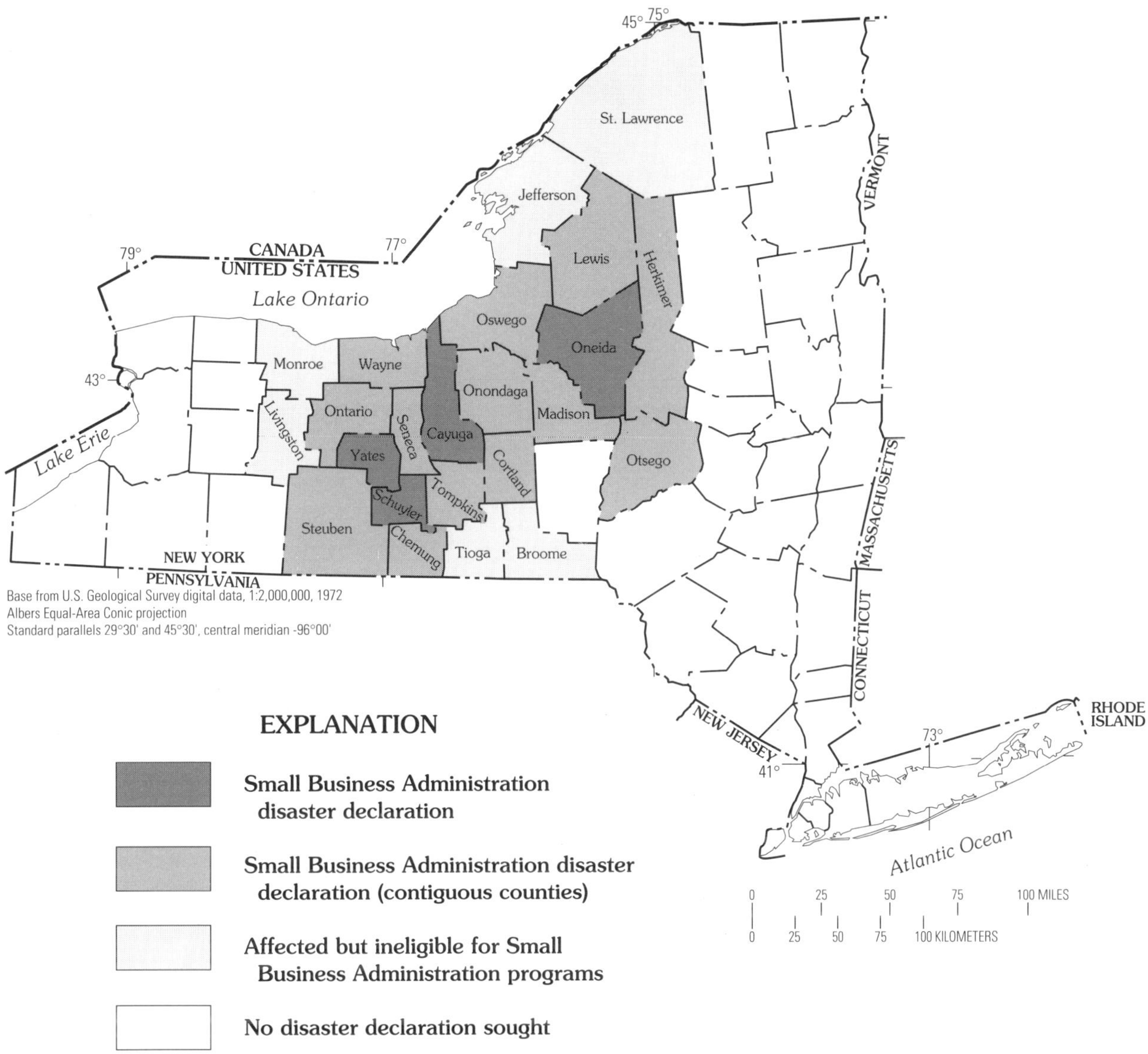

Figure 67. Counties for which Small Business Administration disaster declarations were sought following flood of April 1993, in New York.

May–September 1993, in Southeastern Kansas

By Seth E. Studley

Excessive precipitation fell across Kansas from May–September 1993 with more than the annual average falling during the 5-month period. Maximum precipitation amounts occurred in Marion and Cherokee Counties (fig. 68). In 1993, mean annual discharge at 81 of the State's streamflow-gaging stations exceeded the previous maximum mean annual discharge.

Kansas is divided into two major drainage basins—the Missouri River Basin to the north and the Arkansas River Basin to the south. This article will cover only the Arkansas River Basin because a separate article titled "Floods of the Upper Mississippi River Basin, Spring and Summer 1993" provides information on the Missouri River Basin.

May thunderstorms in south-central and southeastern Kansas produced substantial precipitation that caused flooding in the lower Arkansas River Basin and its tributaries. The Verdigris River near Virgil (site 14, fig. 68), North Cottonwood River below Marion Lake (site 16), and the Cottonwood River at Marion (site 17) all had maximum peak discharges for the period of record (table 43). In addition to these sites, the Ninnescah River near Peck (site 11) and the Arkansas River at Arkansas City (site 13) had notable maximum peak discharges.

During the latter part of July, maximum peak discharges for the period of record were recorded at 10 streamflow-gaging stations (sites 1, 2, 3, 6, 10, and 14–18). Rattlesnake Creek near Zenith (site 3) had a maximum discharge in excess of the 100-year recurrence interval. Lightning Creek near McCune (site 18) had a maximum discharge of record on September 25, with a discharge of 67,500 cubic feet per second at a stage of 19.79 feet. The flood was a result of extremely excessive rainfall on September 24 and 25, during which the precipitation gage at Girard (fig. 68) recorded 15.84 inches on September 25 (National Oceanic and Atmospheric Administration, 1993).

Damage in the area was estimated at $6.5 million with two lives lost. Two hundred fifty-four houses were damaged, several thousand head of livestock were lost, more than 120,000 acres of farmland crops were damaged, and nearly $1 million worth of farm machinery was destroyed (Federal Emergency Management Agency, 1993).

REFERENCES

Federal Emergency Management Agency, 1993, Interagency hazard mitigation team report for Kansas: FEMA–100DR–KS.

National Oceanic and Atmospheric Administration, 1993, Climatological data, Kansas, September 1993: Asheville, N.C., National Climatic Data Center, v. 107, no. 9, 31 p.

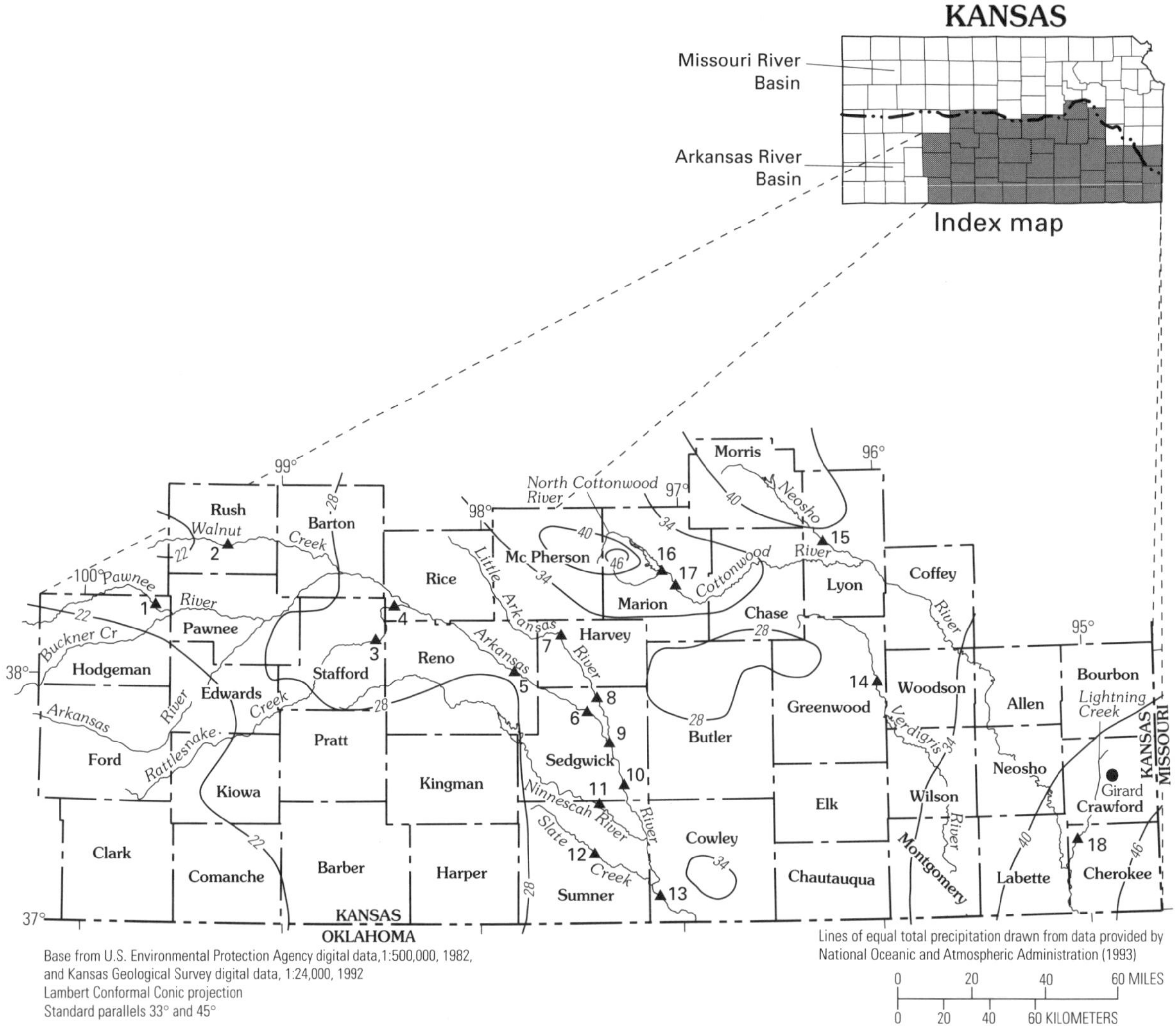

EXPLANATION

—34— **Line of equal total precipitation—** Interval 6 inches

▲13 **Flood-determination site—**Number corresponds to that in table 43

Figure 68. Lines of equal total precipitation and location of flood-determination sites for floods of May–September 1993, in southeastern Kansas.

Table 43. Maximum stages and discharges prior to and during May–September 1993, in southeastern Kansas

[mi[2], square miles; ft, feet above an arbitrary datum; ft[3]/s, cubic feet per second; >, greater than; --, not determined or not applicable. Source: Recurrence intervals calculated from U.S. Geological Survey data. Other data from U.S. Geological Survey reports or data bases]

				Maximum prior to May–September 1993				Maximum during May–September 1993			
Site no. (fig. 68)	Station no.	Stream and place of determination	Drainage area (mi²)	Period	Year	Stage (ft)	Discharge (ft³/s)	Date (month/day)	Stage (ft)	Discharge (ft³/s)	Discharge recurrence interval (years)
1	07140850	Pawnee River near Burdett, KS	1,091	1982–93	1987	14.50	1,550	7/21	27.38	4,290	20
2	07141780	Walnut Creek near Rush Center, KS	1,256	1970–93	1970	24.89	5,020	7/21	34.00	5,790	25
3	07142575	Rattlesnake Creek near Zenith, KS	1,052	1973–93	1973	9.95	18,200	7/18	9.90	29,300	>100
4	07142620	Rattlesnake Creek near Raymond, KS	1,167	1960–93	1973	8.74	2,140	7/20	8.63	1,520	--
5	07143330	Arkansas River near Hutchinson, KS	30,910	1960–93	1973	12.95	24,700	7/23	12.33	15,700	--
6	07143375	Arkansas River near Maize, KS	39,110	1987–93	1987	15.01	16,700	7/15	16.84	44,900	--
7	07143665	Little Arkansas River at Alta Mills, KS	736	1973–93	1973	27.42	30,100	7/14	27.30	21,000	10–25
8	07144200	Little Arkansas River at Valley Center, KS	1,327	1922–93	1945	22.05	32,000	7/15	--	29,300	25–50
9	07144300	Arkansas River at Wichita, KS	40,490	1934–93	1979	--	48,400	7/15	--	43,800	50–100
10	07144550	Arkansas River at Derby, KS	40,830	1968–93	1979	15.87	51,700	7/15	16.19	54,200	25–50
11	07145500	Ninnescah River near Peck, KS	2,129	1937–93	1957	21.85	38,200	5/10	18.85	21,600	5–10
12	07145700	Slate Creek near Wellington, KS	154	1969–93	1975	25.82	28,500	5/10	22.56	7,530	10–25
13	07146500	Arkansas River at Arkansas City, KS	43,713	1902–06, 1921–93	1923	28.43	103,000	5/11	27.62	78,600	10–25
14	07165750	Verdigris River near Virgil, KS	312	1989–93	1992	17.90	8,370	5/9	18.14	8,520	--
15	07179730	Neosho River near Americus, KS	622	1963–93	1985	27.43	17,000	7/22	27.84	17,400	--
16	07179795	North Cottonwood River below Marion Lake, KS	200	1968–93	1971	--	3,390	5/26	18.19	4,530	--
17	07180200	Cottonwood River at Marion, KS	502	1984–93	1985	30.23	16,100	5/8	31.40	24,800	--
18	07184000	Lightning Creek near McCune, KS	197	1939–46, 1959–93	1986	18.48	31,000	9/25	19.79	67,500	>100

May 8–14 and September 25–27, 1993, in Oklahoma

By Darrell M. Walters *and* Robert L. Tortorelli

Floods of May 8–14 affected most of Oklahoma except for the Panhandle and the southeast. From September 25–27, floods affected the extreme northeastern part of the State. These areas and selected flood-determination and reservoir-measurement sites are shown in figure 69.

FLOODS OF MAY 8–14, 1993

From May 5–10, 1993, a slow-moving storm system generated widespread strong thunderstorms over most of Oklahoma. The storm system produced intense local rainfall on May 8 over an area that was already saturated from previous rainfall. More than 5 inches of rain fell in 3 hours in southwestern Oklahoma City (National Oceanic and Atmospheric Administration, 1993), which caused flash flooding along Twin, Brock, and Lightning Creeks (sites 27–29, fig. 69) and claimed four lives. Flooding was reported in Guthrie, Kingfisher, Skiatook, and Sperry (near sites 7, 5, 13, and 14, fig. 69). More than 2,700 homes were damaged statewide, at least 1,000 of them in Oklahoma County (Oklahoma Climatological Survey, 1993a). A total of 43 of the 77 counties in the State were declared eligible for emergency assistance (Oklahoma Climatological Survey, 1994).

Several locations throughout the flooded area (fig. 69) reported 24-hour rainfall totals greater than 6 inches from May 8–10 (Oklahoma Climatological Survey, 1993a). Will Rogers Airport, in southwestern Oklahoma City, received 7.06 inches of rain during the 24-hour period from 1900 May 7 to 1900 May 8, of which 5.28 inches fell during the 3-hour period from 1500 to 1800 May 8 (National Oceanic and Atmospheric Administration, 1993) (fig. 70). According to statistics in U.S. Weather Bureau Technical Paper No. 40 (Hershfield, 1961), this 3-hour total was very near the 100-year frequency of 5.38 inches (fig. 71).

Excessive rainfall from the May storm system produced high stages and discharges of record at several streamflow-gaging stations (table 44). New maximum discharges for the period of record were recorded at 13 streamflow-gaging stations (sites 1, 6, 9, 12, 15, 16, 18, 19, 27–30, and 41). Many other streams had discharges with recurrence intervals of 10 years or more.

The small urban creeks in southwest Oklahoma City (sites 27–29) had extremely high stages and discharges. Twin Creek discharge (site 27) was more than three times the 100-year discharge; Brock Creek discharge (site 28) was more than two times the 100-year discharge, and Lightning Creek discharge (site 29, table 44) was nearly two times the 100-year discharge (Federal Emergency Management Agency, 1988).

Many of the rivers and their tributaries in the flooded area are regulated by reservoirs. Large amounts of runoff produced new maximum elevations and contents for the period of record at five lakes at sites 46, 48, 52, 55, and 59 (table 45). The flood-pool storage exceeded design quantities for Kaw, Keystone, and Eufaula Lakes and Lake Altus (sites 46, 48, 56, and 57).

FLOOD OF SEPTEMBER 25–27, 1993

Excessive rainfall during September 24 and 25 in extreme northeastern Oklahoma, southeastern Kansas, and southwestern Missouri produced flooding on the Neosho and Spring Rivers (sites 17–19, fig. 69) in Ottawa County (Oklahoma Climatological Survey, 1993b). About 500 structures were evacuated in the county, including 300 in Miami, Oklahoma (Oklahoma Climatological Survey, 1994). Flooding also was reported in Craig and Nowata Counties.

The 24-hour rainfall totals in extreme northeastern Oklahoma reached 5.5 to 6.5 inches in several towns. Larger amounts were reported in some locations in southeastern Kansas, which received as much as 18 inches in 24 hours. Rainfall totals in southwestern Missouri were as much as 15 inches (Oklahoma Climatological Survey, 1993b).

On September 26, streamflow at the streamflow-gaging station at Spring River near Quapaw (site 19) reached a record stage and discharge, with a recurrence interval of more than 100 years. Streamflow at the streamflow-gaging station at Tar Creek at 22nd Street at Miami (site 18) also reached a record for stage and

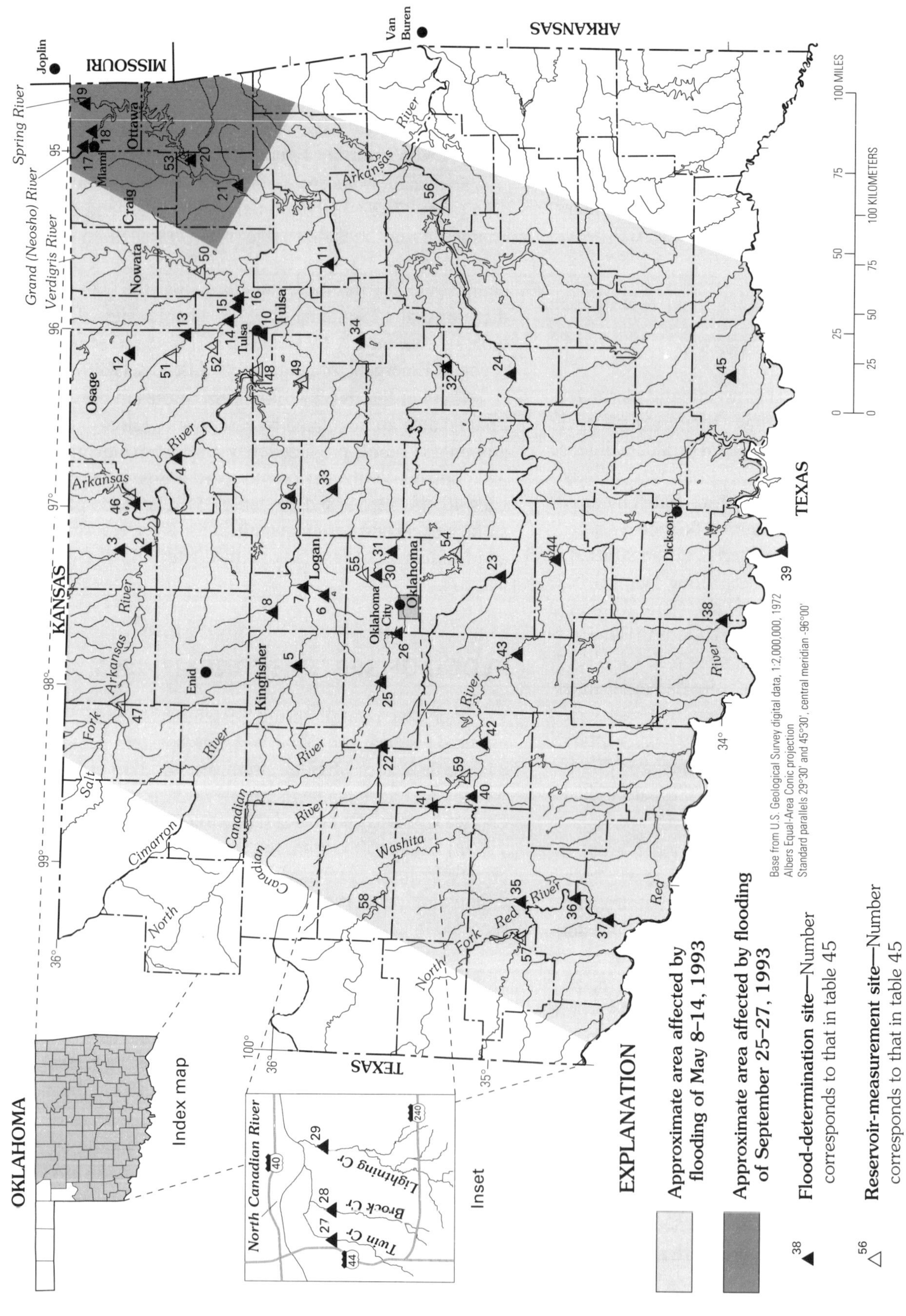

Figure 69. Location of flood-determination and reservoir-measurement sites in Oklahoma for floods of May and September 1993.

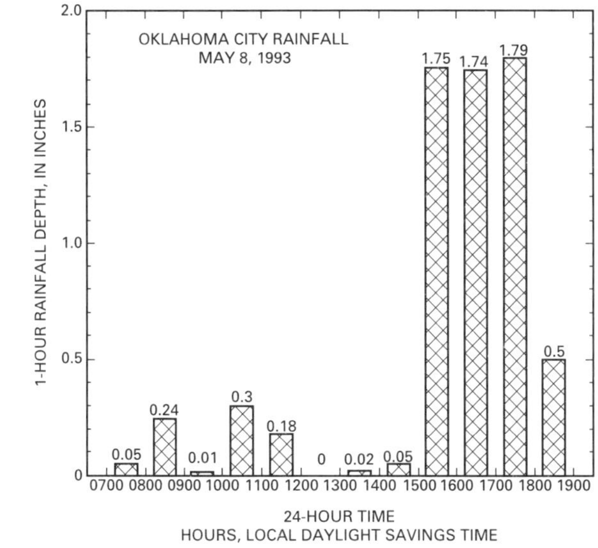

Figure 70. 1-hour rainfall depths at National Weather Service gage at Will Rogers Airport, Oklahoma City, Oklahoma, May 8, 1993 (data from National Oceanic and Atmospheric Administration, 1993).

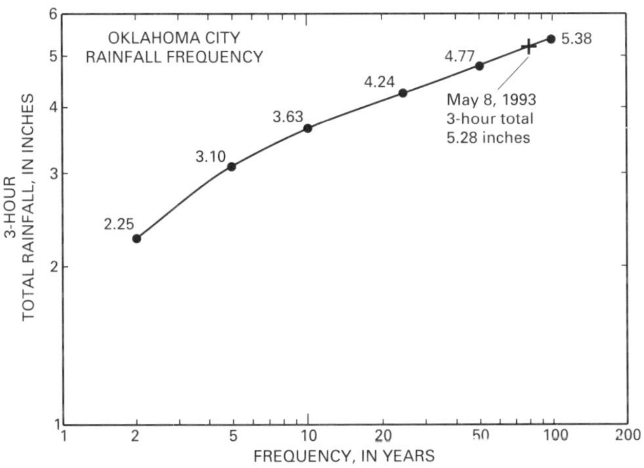

Figure 71. 3-hour storm-rainfall frequency at Oklahoma City, Oklahoma (Hershfield, 1961; National Oceanic and Atmospheric Administration, 1993).

discharge on September 26, with a 25-year recurrence interval. The flood-pool storage exceeded design quantities for the Lake O' the Cherokees (site 53).

REFERENCES

Federal Emergency Management Agency, 1988, City of Oklahoma City, Oklahoma: Washington, D.C., Federal Emergency Management Agency Flood Insurance Study, community number 400027, 196 p.

Hershfield, D.M., 1961, Rainfall frequency atlas of the United States for durations from 30 minutes to 24 hours and return periods from 1 to 100 years: U.S. Weather Bureau Technical Paper No. 40, p. 53–63.

National Oceanic and Atmospheric Administration, 1993, Hourly precipitation data, Oklahoma, May 1993: Asheville, N.C., National Climatic Data Center, v. 43, no. 5, p. 11.

Oklahoma Climatological Survey, 1993a, Oklahoma monthly summary May 1993: Norman, Oklahoma Climatological Survey, p. 2.

———1993b, Oklahoma monthly summary September 1993: Norman, Oklahoma Climatological Survey, p. 2.

———1994, Oklahoma annual summary 1993: Norman, Oklahoma Climatological Survey, p. 3.

Table 44. Maximum stages and discharges prior to and during May and September 1993, in Oklahoma

[mi^2, square miles; ft, feet above an arbitrary datum; ft^3/s, cubic feet per second; --, not determined or not applicable; >, greater than. Source: Recurrence intervals from U.S. Geological Survey data; other data from U.S. Geological Survey reports or data bases]

					Maximum prior to May 1993			Maximum during May and September 1993			
Site no. (fig. 69)	Station no.	Stream and place of determination	Drainage area (mi^2)	Period	Year	Stage (ft)	Discharge (ft^3/s)	Date (month/day)	Stage (ft)	Discharge (ft^3/s)	Discharg recurrence interval (years)
1	07148140	Arkansas River near Ponca City, OK	46,530	1976–93	1986	[1]13.97	39,300	5/14	20.11	62,900	[2]100
2	07151000	Salt Fork Arkansas River at Tonkawa, OK	4,528	1936–93	1973	28.98	97,300	5/10	25.51	45,500	25
3	07152000	Chikaskia River near Blackwell, OK	1,859	1936–93	1942	[1]34.28	85,000	5/10	34.31	53,300	10
4	07152500	Arkansas River at Ralston, OK	54,465	1923–93	1973	22.98	211,000	5/12	19.76	139,000	[2]25
5	07159100	Cimarron River near Dover, OK	15,713	1973-93	1986	26.10	123,000	5/10	22.49	63,500	10
6	07159750	Cottonwood Creek near Seward, OK	320	1973–82, 1990–93	1976	[1]23.99	29,900	5/09	32.74	46,100	25
7	07160000	Cimarron River at Guthrie, OK	6,892	1938–76, 1983–93	1957	[1]18.58	158,000	5/10	17.73	93,100	25
8	07160500	Skeleton Creek near Lovell, OK	410	1950–93	1957	[1]34.58	75,200	5/09	35.47	30,300	10–25
9	07161450	Cimarron River near Ripley, OK	17,979	1988–93	1990	[1]21.43	55,200	5/10	28.36	141,000	--
10	07164500	Arkansas River at Tulsa, OK	74,615	1905–93	1986	25.21	307,000	5/14	17.48	149,000	[2]25
11	07165570	Arkansas River near Haskell, OK	75,473	1973–93	1986	22.82	259,000	5/14	19.28	150,000	[2]10–25
12	07174600	Sand Creek at Okesa, OK	139	1959–93	1974	[1]28.60	19,500	5/09	24.29	20,200	25–50
13	07176500	Bird Creek near Avant, OK	364	1945–93	1959	31.40	32,400	5/09	28.03	27,500	10–25
					1974	32.03	--				
14	07177500	Bird Creek near Sperry, OK	905	1939–93	1959	32.60	90,000	5/10	29.88	30,600	10
15	07178000	Bird Creek near Owasso, OK	1,022	1929–31, 1935–39, 1987–93	1938	[1]26.20	19,700	5/11	26.94	31,700	10–25

Table 44. Maximum stages and discharges prior to and during May and September 1993, in Oklahoma—Continued

Site no. (fig. 69)	Station no.	Stream and place of determination	Drainage area (mi^2)	Maximum prior to May 1993: Period	Year	Stage (ft)	Discharge (ft^3/s)	Maximum during May and September 1993: Date (month/day)	Stage (ft)	Discharge (ft^3/s)	Discharg recurrence interval (years)
16	07178200	Bird Creek at State Highway 266 near Catoosa, OK	1,103	1988–93	1990	28.39	18,500	5/11	33.22	27,400	--
17	07185000	Neosho River near Commerce, OK	5,876	1939–93	1951	34.03	267,000	9/27	24.09	75,600	[2]25
18	07185095	Tar Creek at 22nd Street bridge at Miami, OK	44.7	1984–93	1985	14.13	9,040	9/25	14.70	12,400	50–100
19	07188000	Spring River near Quapaw, OK	2,510	1939–93	1943	[1]43.40	190,000	9/26	46.60	230,000	>100
20	07190500	Neosho River near Langley, OK	10,335	1940–93	1943	45.50	300,000	9/27	34.26	170,000	[2]10–25
21	07191500	Neosho River near Chouteau, OK	11,534	1938–93	1946	45.00	400,000	9/27	--	136,000	[2]25
22	07228500	Canadian River at Bridgeport, OK	25,276	1945–64, 1970–93	1948	38.95	150,000	5/09	17.32	51,200	[2]10–25
23	07229200	Canadian River at Purcell, OK	25,939	1960–61, 1980–83, 1986–93	1987	14.75	102,000	5/09	14.42	84,100	10–25
24	07231500	Canadian River at Calvin, OK	27,952	1906, 1935–93	1950	17.35	174,000	5/10	15.32	93,400	[2]10
25	07239500	North Canadian River near El Reno, OK	13,042	1903–07, 1938–93	1941 1990	15.98 18.53	15,000 7,630	5/10	21.41	14,600	[2]50–100
26	07241000	North Canadian River below Lake Overholser near Oklahoma City, OK	13,222	1953–93	1987	29.85	18,700	5/10	25.39	13,900	[2]10–25
27	07241210	Twin Creek at SW 29 Street at Oklahoma City, OK	3.23	--	--	--	[4]3,040	5/08	[3]1,223.30	10,000	>100
28	07241225	Brock Creek at SW 29 Street at Oklahoma City, OK	4.74	--	--	--	[4]5,210	5/08	[3]1,206.1	12,000	>100
29	07241255	Lightning Creek at SW 25 Street at Oklahoma City, OK	13.9	--	--	--	[4]8,020	5/08	[3]1,194.8	15,000	>100
30	07241520	North Canadian River at Britton Road at Oklahoma City, OK	13,413	1989–93	1989	23.48	31,300	5/09	24.80	38,100	--

Table 44. Maximum stages and discharges prior to and during May and September 1993, in Oklahoma—Continued

Site no. (fig. 69)	Station no.	Stream and place of determination	Drainage area (mi²)	Maximum prior to May 1993: Period	Year	Stage (ft)	Discharge (ft³/s)	Maximum during May and September 1993: Date (month/day)	Stage (ft)	Discharge (ft³/s)	Discharge recurrence interval (years)
31	07241550	North Canadian River near Harrah, OK	13,501	1969–93	1987	[1]19.60	27,200	5/09	21.90	25,500	[2]25–50
32	07242000	North Canadian River near Wetumka, OK	14,290	1938–93	1945	26.40	66,000	5/14	19.29	29,000	[2]10–25
33	07243000	Dry Creek near Kendrick, OK	69	1956–93	1974	24.20	18,000	5/08	22.37	11,100	10–25
34	07243500	Deep Fork near Beggs, OK	2,018	1938–93	1943	34.55	66,800	5/14	30.03	30,700	10
35	07304500	Elk Creek near Hobart, OK	549	1949–93	1986	31.65	28,000	5/09	31.65	23,600	25–50
36	07305000	North Fork Red River near Headrick, OK	4,244	1905–07, 1938–93	1986	19.07	59,000	5/10	18.83	56,100	[2]100
37	07307028	North Fork Red River near Tipton, OK	4,691	1985–93	1986	19.15	57,200	5/10	19.18	51,200	[2]100
38	07315700	Mud Creek near Courtney, OK	572	1961–93	1987 1990	33.14 [2]32.41	38,800 49,600	5/09	32.30	47,700	25
39	07316000	Red River near Gainesville, TX, OK	30,782	1936–93	1987	40.08	265,000	5/11	30.99	117,000	[2]10–25
40	07325500	Washita River at Carnegie, OK	3,129	1912, 1921, 1923, 1934–36, 1938–93	1949 1983	[2]26.21 29.67	50,000 25,200	5/10	30.68	25,500	[2]25
41	07325800	Cobb Creek near Eakly, OK	132	1969–93	1986	24.38	10,000	5/08	21.93	10,900	[2]25
42	07326500	Washita River at Anadarko, OK	3,656	1903-07, 1936–37, 1964–93	1983	25.20	44,700	5/11	25.03	28,300	[2]25–50
43	07328100	Washita River at Alex, OK	4,787	1940–52, 1955, 1957, 1965–93	1993	28.70	23,400	5/13	18.71	17,800	[2]25
44	07328500	Washita River near Pauls Valley, OK	5,330	1938–93	1950 1987	29.88 [2]28.72	30,000 43,600	 5/10	 21.49	 31,200	 [2]25
45	07332500	Blue River Near Blue, OK	476	1936–93	1981	44.20	65,200	5/09	35.15	34,100	10–25

[1]Site and datum then in use. [2]Recurrence interval based on period of regulated discharge.
[3]Elevation in feet above sea level. [4]Flood Insurance Study 100-year recurrence-interval flood discharge (Federal Emergency Management Agency, 1988).

Table 45. Maximum elevations and contents of selected reservoirs prior to and during May and September 1993, in Oklahoma

[mi^2, square miles; ft, feet above sea level; acre-ft, acre-feet]

Site no. (fig. 69)	Station no.	Station name	Drainage area (mi^2)	Maximum prior to May 1993				Maximum during May and September 1993		
				Period	Year	Elevation (ft)	Contents (acre-ft)	Date (month/day)	Elevation (ft)	Contents (acre-ft)
46	07148130	Kaw Lake near Ponca City, OK	46,530	1976–93	1986	1,045.52	1,387,000	5/13	1,047.12	1,434,000
47	07150000	Great Salt Plains Lake near Jet, OK	3,200	1941–93	1951	1,134.38	189,400	5/12	1,133.90	160,100
48	07164200	Keystone Lake near Sand Springs, OK	74,506	1964–93	1974	754.86	1,886,000	5/14	756.49	1,876,000
49	07165000	Heyburn Lake near Heyburn, OK	123	1951–93	1974	776.85	32,210	5/09	773.23	24,000
50	07171300	Oologah Lake Near Oologah, OK	4,339	1963–93	1986	664.90	1,751,000	5/20	658.55	1,417,000
51	07176460	Birch Lake near Barnsdall, OK	66	1977–93	1986	769.04	47,400	5/19	765.02	40,200
52	07177400	Skiatook Lake near Skiatook, OK	354	1985–93	1990	723.31	427,100	5/24	724.20	438,200
53	07190000	Lake O' The Cherokees at Langley, OK	10,298	1940–93	1957	755.27	2,213,000	9/28	754.52	2,168,000
54	07229900	Lake Thunderbird near Norman, OK	256	1965–93	1990	1,048.38	187,400	5/13	1,034.90	152,200
55	07242340	Arcadia Lake near Arcadia, OK	105	1987–93	1990	1,014.63	45,980	5/19	1,021.16	63,680
56	07244800	Eufaula Lake near Brooken, OK	47,522	1964–93	1990	599.72	4,237,000	5/17	597.44	3,891,000
57	07302500	Lake Altus at Lugert, OK	2,515	1944–93	1951	1,562.10	170,600	5/10	1,561.55	151,000
58	07324300	Foss Reservoir near Foss, OK	1,496	1961–93	1989	1,647.35	217,300	5/19	1,643.60	189,100
59	07325900	Fort Cobb Reservoir near Fort Cobb, OK	304	1959–93	1989	1,349.89	116,500	5/17	1,351.07	122,700

June 7–9, 1993, in Northeastern Illinois

By G.O. Balding

Severe weather that was experienced in Illinois during June 1993 was more severe than all the 1993 spring months combined. Wayne Wendland, State Climatologist, stated that June was the seventh wettest June since 1895 (Illinois State Water Survey, 1993, p. 2). Thunderstorms producing large amounts of precipitation hit the Chicago area on June 7–8. The city of Burr Ridge recorded an unofficial 4.63 inches of rain in about 2 hours, and an unofficial 6.03 inches (50-year recurrence interval) fell during the 24-hour period of June 7–8 (Illinois State Water Survey, 1993, p. 2). Thunderstorms on June 7, in northeastern Illinois, produced 50- to 70-mile-per-hour winds and 3/4- to 1-inch hailstones; on June 8, there were sightings of tornadoes and funnel clouds (Wendland and Dennison, 1993, p. 3). Total precipitation at selected precipitation gages in northeastern Illinois ranged from 1.81 to 5.11 inches (table 46).

The excessive rainfall on June 7 and 8 caused Sawmill Creek near Lemont (site 10, table 47, fig. 72) to reach a maximum discharge of 1,600 cubic feet per second, which was estimated to correspond to a 100-year recurrence interval. Other streams in the southern Chicago suburbs had maximum discharges ranging from less than a 2-year to less than a 10-year recurrence interval.

Table 46. Precipitation totals for June 7–9, 1993, at selected precipitation gages in northeastern Illinois

[Data from National Oceanic and Atmospheric Administration, 1993]

Site no. (fig. 72)	Latitude, longitude	Precipitation gage	Precipitation (inches)
1	41°45', 88°21'	Aurora	3.12
2	41°24', 88°17'	Channahon Dresden Island	2.30
3	42°00', 87°53'	Chicago O'Hare WSO AP	1.81
4	41°44', 87°46'	Chicago Midway AP 3 SW	5.11
5	41°30', 88°06'	Joliet Brandon Road Dam	2.22
6	41°30', 87°41'	Park Forest	4.01
7	41°20', 87°48'	Peotone	2.88
8	41°49', 88°04'	Wheaton 3 SE	2.53

The Chicago Tribune newspaper reported that the downpour flooded parts of the southern and southwestern Chicago suburbs leaving streets flooded to a depth of 4 feet and closing parts of expressways in Chicago for up to 4 hours (Thornton and Kass, 1993). Sewers on the southside of Chicago quickly reached capacity. Raw sewage was reported coming up through the storm drains and into the basements of homes (in many parts of Chicago, storm sewers also are sanitary sewers). The Des Plaines River became so swollen at Lockport that the Chicago Sanitary and Ship Canal could not effectively drain into it. Less than 1 inch of rain fell at O'Hare International Airport (site 3) on June 7, yet about 250 inbound and outbound flights were canceled. In Blue Island, the roof of a 50,000-square-foot warehouse collapsed, apparently under the weight of accumulated rainwater. Even though discharge at the streamflow-gaging station on Sawmill Creek near Lemont (site 10) was nearly equivalent to a 100-year recurrence interval, the Lemont Reporter (1993) stated that there was no major flooding in Lemont.

Total property damage as a result of the flooding is unknown. On June 8, 1993, the Governor of Illinois declared Cook and Du Page Counties State disaster areas (Swanson, 1993).

REFERENCES

Illinois State Water Survey, 1993, Illinois water and climate summary—June 1993: Champaign, Ill., 12 p.

Lemont Reporter, 1993, Heavy rains take toll in suburbs: Lemont, Ill., June 9, 1993, p. 1.

National Oceanic and Atmospheric Administration, 1993, Climatological data, Illinois, August 1993: U.S. Department of Commerce, v. 98, no. 8, 35 p.

Swanson, Stevenson, 1993, Deep Tunnel gets its fill from rain that won't quit: Chicago, Ill., Chicago Tribune, June 9, 1993, p. 1.

Thornton, Jerry, and Kass, John, 1993, June river—rain wider than a mile: Chicago, Ill., Chicago Tribune, June 8, 1993, p. 1.

Wendland, W.M., and Dennison, Jean, 1993, Weather and climate impacts in the Midwest: Champaign, Ill., Midwestern Climate Center, v. III, no. 7, 8 p.

Table 47. Maximum stages and discharges prior to and during June 7–9, 1993, in northeastern Illinois

[mi^2, square miles; ft, feet above an arbitrary datum; ft^3/s, cubic feet per second; <, less than; --, not determined or not applicable. Source: Recurrence intervals calculated from U.S. Geological Survey data. Other data from U.S. Geological Survey reports or data bases]

				Maximum prior to June 1993				Maximum during June 7–9, 1993			
Site no. (fig. 72)	Station no.	Stream and place of determination	Drainage area (mi^2)	Period	Year	Stage (ft)	Discharge (ft^3/s)	Day	Stage (ft)	Discharge (ft^3/s)	Discharge recurrence interval (years)
9	05533000	Flag Creek near Willow Springs, IL	16.5	1951–93	1961	13.71	2,680	7	8.13	791	2–5
10	05533400	Sawmill Creek near Lemont, IL	13.0	1960–79, 1985–93	1990	15.46	1,730	7	15.60	1,600	100
11	05536235	Deer Creek near Chicago Heights, IL	23.1	1948–93	1957	11.75	1,380	8	11.29	830	5–10
12	05536275	Thorn Creek at Thornton, IL	104	1948–93	1957	16.00	4,700	9	13.81	2,610	2–5
13	05536290	Little Calumet River at South Holland, IL	208	1947–93	1957	20.11	4,440	9	17.58	3,020	2–5
14	05537500	Long Run near Lemont, IL	20.9	1951–93	1954	9.91	3,160	8	5.89	490	<2
15	05540091	Spring Brook at Forest Preserve near Warrenville, IL	6.83	1991–93	1993	10.63	194	8	10.42	176	--
16	05540095	West Branch Du Page River near Warrenville, IL	90.4	1968–93	1987	5.85	3,050	8	3.41	920	<2
17	05540130	West Branch Du Page River near Naperville, IL	123	1988–93	1991	9.58	3,420	8	8.20	1,530	--
18	05540195	St. Joseph Creek at Route 34 at Lisle, IL	11.1	1988–93	1990	11.30	938	7	7.51	301	--
19	05540250	East Branch Du Page River at Bollingbrook, IL	75.8	1961–70, 1988–93	1990	22.67	1,990	9	20.71	1,100	--
20	05540275	Spring Brook at 87th Street near Naperville, IL	9.90	1987–93	1990	7.68	694	8	5.78	153	--

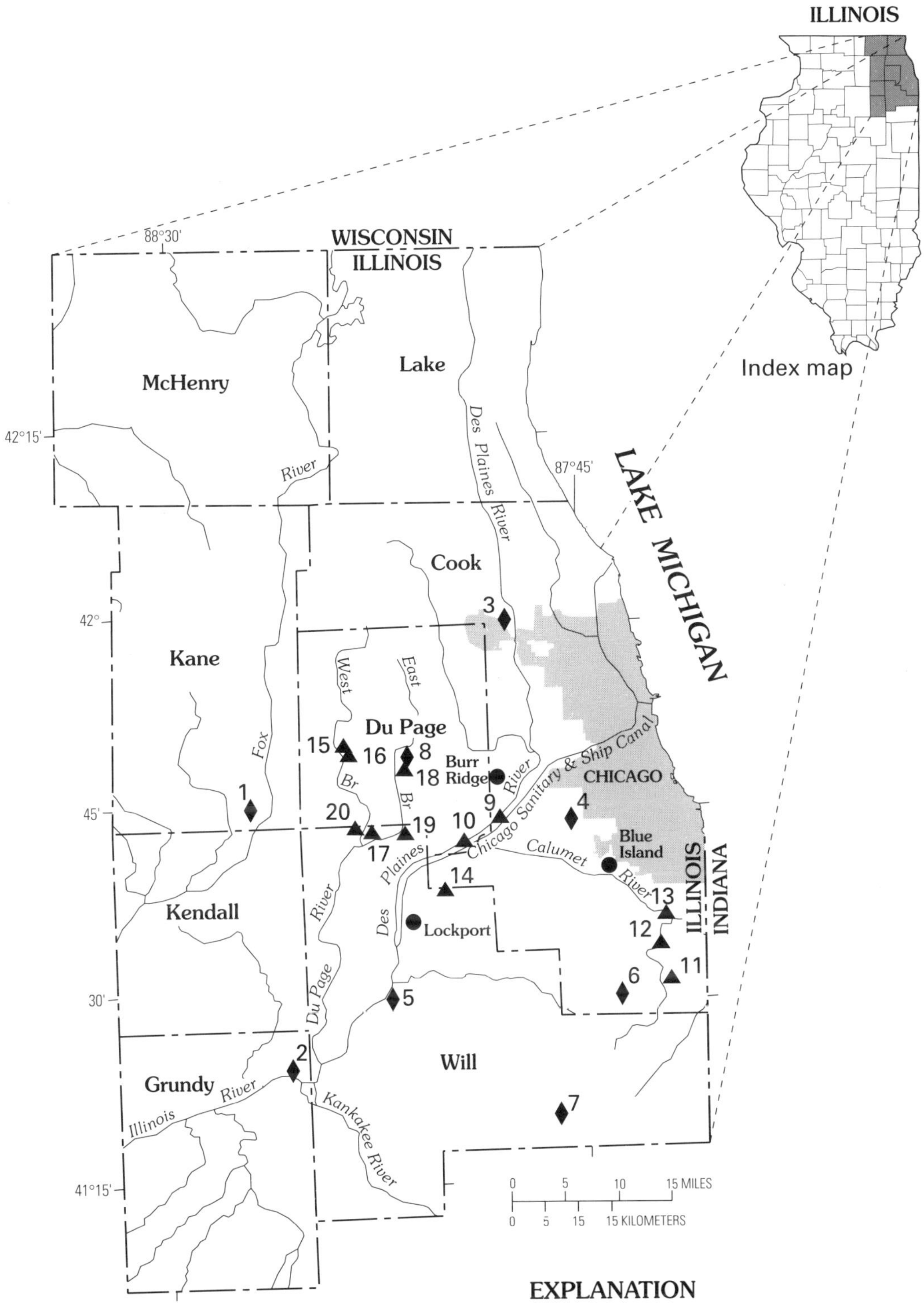

Base from U.S. Geological Survey 1:2,000,000
Albers Equal Area Conic projection
Standard parallels 33° and 45°, central meridian -89°

EXPLANATION

▲4 **Flood-determination site**—Number corresponds to that in table 47

♦2 **Precipitation gage**—Number corresponds to that in table 46

Figure 72. Location of flood-determination and precipitation gage sites for flood of June 7–9, 1993, in northeastern Illinois.

June 8–9, 1993, in Northern Indiana, and August 17, 1993, in West-Central Indiana

By Donald V. Arvin

JUNE 8-9, 1993

Rainy conditions during the last 3 days of May and the first 5 days of June 1993 produced 3 to 4 inches of precipitation, saturating the soils in northern Indiana. Then on June 7 and 8, rainfall totalling 5 to 8 inches fell on these water-soaked soils and caused flooding in many of the region's river basins (National Oceanic and Atmospheric Administration, 1993a). Affected areas included Lake, Porter, LaPorte, St. Joseph, Elkhart, Lagrange, and Jasper Counties (fig. 73). Flooding caused many local roads and city streets to be closed. Levees along the Kankakee River in southern LaPorte County were overtopped.

On June 9, discharges exceeded the 100-year recurrence interval at the streamflow-gaging station on Trail Creek at Michigan City (site 2, fig. 73), which monitors streamflow from an area of 54.1 square miles. The instantaneous maximum discharge of 4,240 cubic feet per second was the largest discharge recorded at the station since data collection began in June 1969.

Major flooding also occurred in other basins across northern Indiana (table 48). The Little Elkhart River at Middlebury (site 3) had a discharge with a recurrence interval between 70 and 80 years. The Iroquois River at Rosebud (site 5) discharge had a 30-year recurrence interval, and Hart Ditch at Munster (site 6) had a discharge with a 10-year recurrence interval.

The wet conditions that prevailed prior to the rains of June 7 and 8 markedly increased the effect of the flooding. On June 2 and 3, the period-of-record high water level of 20.98 feet occurred at the lake-level recording gage on Pine Lake at LaPorte (site 7).

AUGUST 17, 1993

Flash floods along small streams in west-central Indiana on August 17, 1993, caused substantial damage to private property, railroads, and public highways. The areas most affected were Montgomery, Vermillion, Parke, Putnam, Vigo, Clay, Owen, Morgan, Monroe, Brown, and Bartholomew Counties, although other nearby areas also sustained damage. Evacuations occurred in the cities of Rosedale, Cloverdale, Quincy, Paragon, Trevlac, and Nashville, Indiana (fig. 73).

The flooding of August 17 was preceded by excessive rainfall of 6 to 8 inches on August 12 (National Oceanic and Atmospheric Administration, 1993b). Then late during the evening of August 16 and early August 17, the area received another 2 to 11 inches of rain. The area affected was a 10–15 mile wide area extending from Clinton to just west of Columbus.

Floodwaters from Doe Creek swept away two homes and severely damaged a city bridge. U.S. Highway 231 was overtopped by Doe Creek as one lane and the supporting embankment were ripped away. Croys Creek submerged parts of Interstate 70 in Putnam County, forcing closure of that main roadway for several hours. Surrounding areas experienced similar damage. Railroad tracks were washed away near Paragon, Quincy, Fontanet, and Cloverdale. Many roads were closed and received extensive damage. In some areas, fields of 8-foot corn stalks were entirely submerged.

After the intense rainstorm ended on August 17, the flash-flood water levels fell quickly. As the water drained to the receiving streams, the maximum discharges tended to dissipate, causing lesser effects on the larger streams. The streamflow-gaging station near Reelsville (site 8) recorded a gage height of 22.11 feet. This corresponded to a discharge of 10,200 cubic feet per second, which had a recurrence interval of 5–10 years.

REFERENCES

National Oceanic and Atmospheric Administration, 1993a, Monthly report of river and flood conditions for June 1993: Indianapolis, Indiana, 3 p.

———1993b, Monthly report of river and flood conditions for August 1993: Indianapolis, Indiana, 4 p.

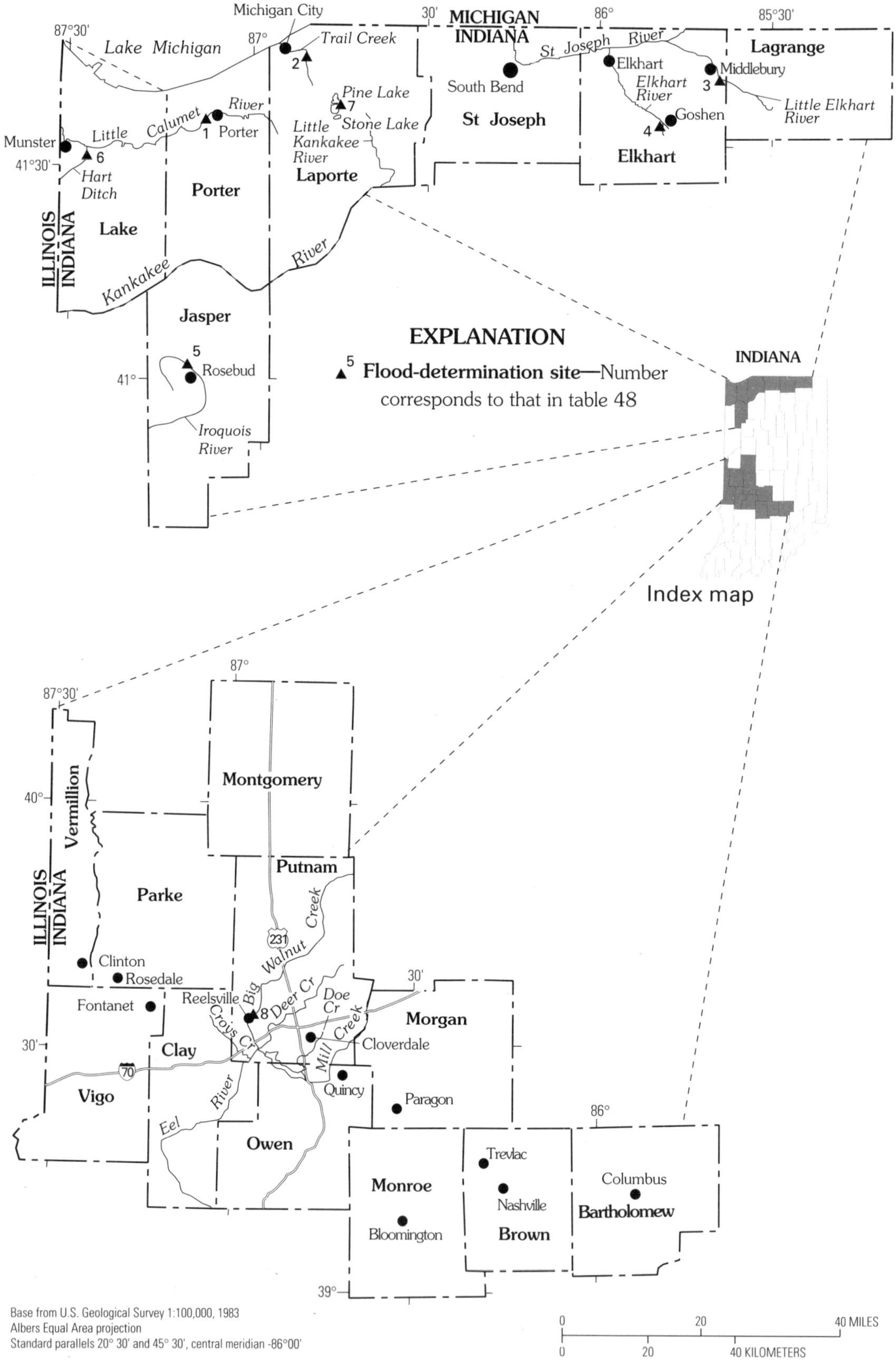

Figure 73. Location of flood-determination sites for flood of June 8–9, 1993, in northern Indiana, and August 17, 1993, in west-central Indiana.

Table 48. Maximum stages and discharges prior to and during June 8–9 and August 17, 1993, in northern and west-central Indiana

[mi^2, square miles; ft, feet above an arbitrary datum; ft^3/s, cubic feet per second; >, greater than; --, not determined or not applicable. Source: Recurrence intervals calculated from U.S. Geological Survey data. Other data from U.S. Geological Survey reports or data bases]

				Maximum prior to June 1993				Maximum during June and August 1993			
Site no. (fig. 73)	Station no.	Stream and place of determination	Drainage area (mi^2)	Period	Year	Stage (ft)	Discharge (ft^3/s)	Date (month/day)	Stage (ft)	Discharge (ft^3/s)	Discharge recurrence interval (years)
1	04094000	Little Calumet River at Porter, IN	66.2	1945–93	1990	10.93	3,880	6/9	9.73	2,220	10
					1954	11.66	3,110				
2	04095300	Trail Creek at Michigan City, IN	54.1	1969–93	1990	12.79	3,880	6/9	12.97	4,240	>100
3	04099808	Little Elkhart River at Middlebury, IN	97.6	1979–93	1985	10.52	2,470	6/9	10.20	2,180	70–80
4	04100500	Elkhart River at Goshen, IN	594	1931–93	1985	11.87	6,360	6/9	10.09	4,870	15–20
					1982	11.94	6,180				
5	05521000	Iroquois River at Rosebud, IN	35.6	1948–93	1990	7.93	656	6/9	5.83	396	30
					1959	8.86	383				
6	05536190	Hart Ditch at Munster, IN	70.7	1942–93	1990	8.72	3,010	6/8	7.23	2,360	10
7	05515220	Pine Lake at La Porte, IN	10.7	1946–75, 1980–93	1983	20.81	--	6/2, 3	20.98	--	--
8	03357500	Big Walnut Creek near Reelsville, IN	588	1950–93	1990	--	12,200	8/7	22.11	10,200	5–10

July and August 1993, in Eastern and South-Central North Dakota

By Gregg J. Wiche, R.E. Harkness, *and* Tara Williams-Sether

The same weather patterns that caused flooding in the nine-State area of the upper Mississippi River Basin during the spring and summer of 1993 also produced widespread flooding of unprecedented magnitude in the drainage of the Hudson Bay Basin in North Dakota (fig. 74). The most severe flooding occurred in the Devils Lake, Red River of the North, and upper James River Basins and on several tributaries to the Missouri River. Urban flooding affected Bismarck, Jamestown, Valley City, Fargo, and other smaller communities. As a result of the flooding, 39 counties in North Dakota were declared disaster areas, about 10,000 people received $22.4 million in Federal assistance, and damage to public infrastructure was about $22.5 million (Russell Staiger, North Dakota Division of Emergency Management, oral commun., 1994). Agricultural losses caused by the flooding have been estimated at $500 million (Federal Emergency Management Agency, 1993).

Severe flooding in the Devils Lake, Sheyenne River, and James River Basins (fig. 74) was preceded by near-normal precipitation during April and May and much-greater-than-normal precipitation during June; drainage areas of the tributaries to the Missouri River received near-normal precipitation from May until the last week in June (fig. 75). The rate of accumulation of precipitation at Devils Lake indicates that many small-to-moderate rainstorms occurred during July; however, at Cooperstown, Jamestown, and Bismarck (fig. 75), individual large storms produced from 30 to 50 percent of the July precipitation total. The large amounts of precipitation (6.30 inches at Cooperstown on July 25, 6.35 inches at Jamestown on July 15, and 5.55 inches at Bismarck on July 15 to 16) from the individual storms caused urban flooding. In other parts of the flooded area, precipitation from the individual storms totaled more than 3.0 inches, and 5-day precipitation totaled more than 5.0 inches (National Oceanic and Atmospheric Administration, 1993). Thus, the distribution of precipitation in July, as well as the antecedent conditions, affected flooding.

The most severe flooding in the Devils Lake Basin (a closed basin) occurred in the Edmore Coulee and Edmore Coulee tributary subbasins. Maximum stages and discharges for the period of record occurred at Edmore Coulee near Edmore (site 2) and Edmore Coulee tributary near Webster (site 3) (table 49). Runoff into Devils Lake from Big Coulee and Channel A began in late June and continued throughout the winter of 1993–94 as the chain of lakes upstream from Devils Lake slowly drained. Runoff during 1993 produced the second largest inflow to Devils Lake in about 110 years. As a result of the inflow, the lake rose 5.2 feet from March–December 1993.

Although maximum discharges during July were less than the maximum discharges for the period of record, maximum or near-maximum stages occurred at streamflow-gaging stations on the Sheyenne River (sites 7–9) (table 49). During the 1993 flood, the presence of summer vegetation increased the resistance to flow, so that higher stages were required for a given discharge. Also, widespread floods are unusual during the growing season, and the extent and magnitude of the flooding throughout North Dakota was greater than might be expected on the basis of the computed recurrence intervals for maximum discharges (table 49).

The most severe flooding in the James River Basin occurred in and around Jamestown and upstream from Jamestown Reservoir. Maximum discharge on the James River near Manfred (site 14) was greater than the 100-year recurrence interval and also exceeded the maximums that occurred during the spring snowmelt floods of 1969 and 1979. Maximum discharge on the James River at Jamestown (site 16) had a 50-year recurrence interval for the regulated period. The maximum discharge at Jamestown is unusual in that most of the runoff contributing to the maximum came from areas in and adjacent to Jamestown. Little flooding was attributed to releases from Pipestem or Jamestown Reservoirs.

REFERENCES

Federal Emergency Management Agency, 1993, Interagency hazard mitigation team report for North Dakota: FEMA–1001DR–ND.

National Oceanic and Atmospheric Administration, 1993, Climatological data, North Dakota, July and August 1993: Asheville, N.C., National Climatic Data Center, v. 102, no. 7–8.

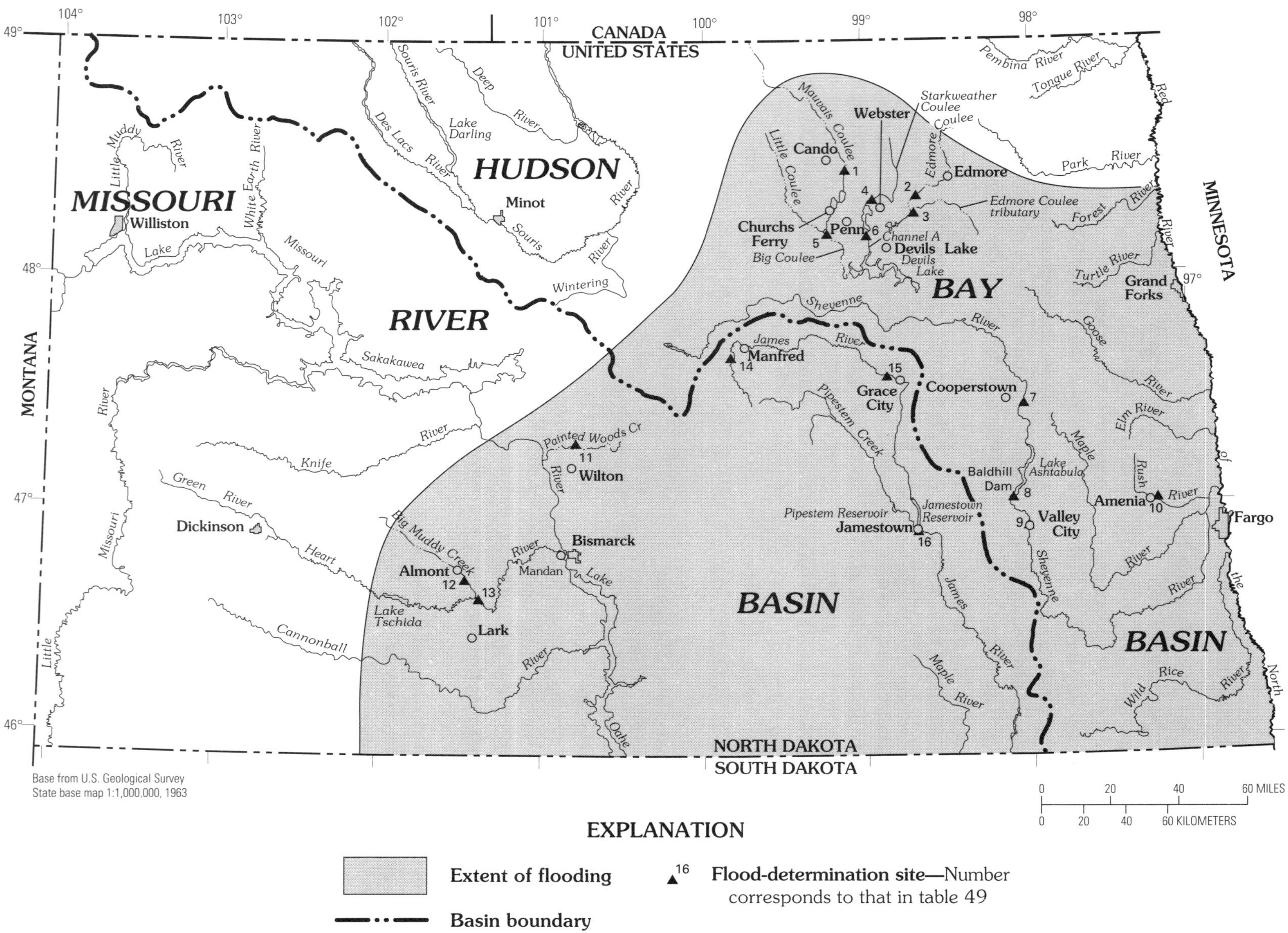

Figure 74. Extent of flooding and location of flood-determination sites for July–August 1993, in eastern and south-central North Dakota.

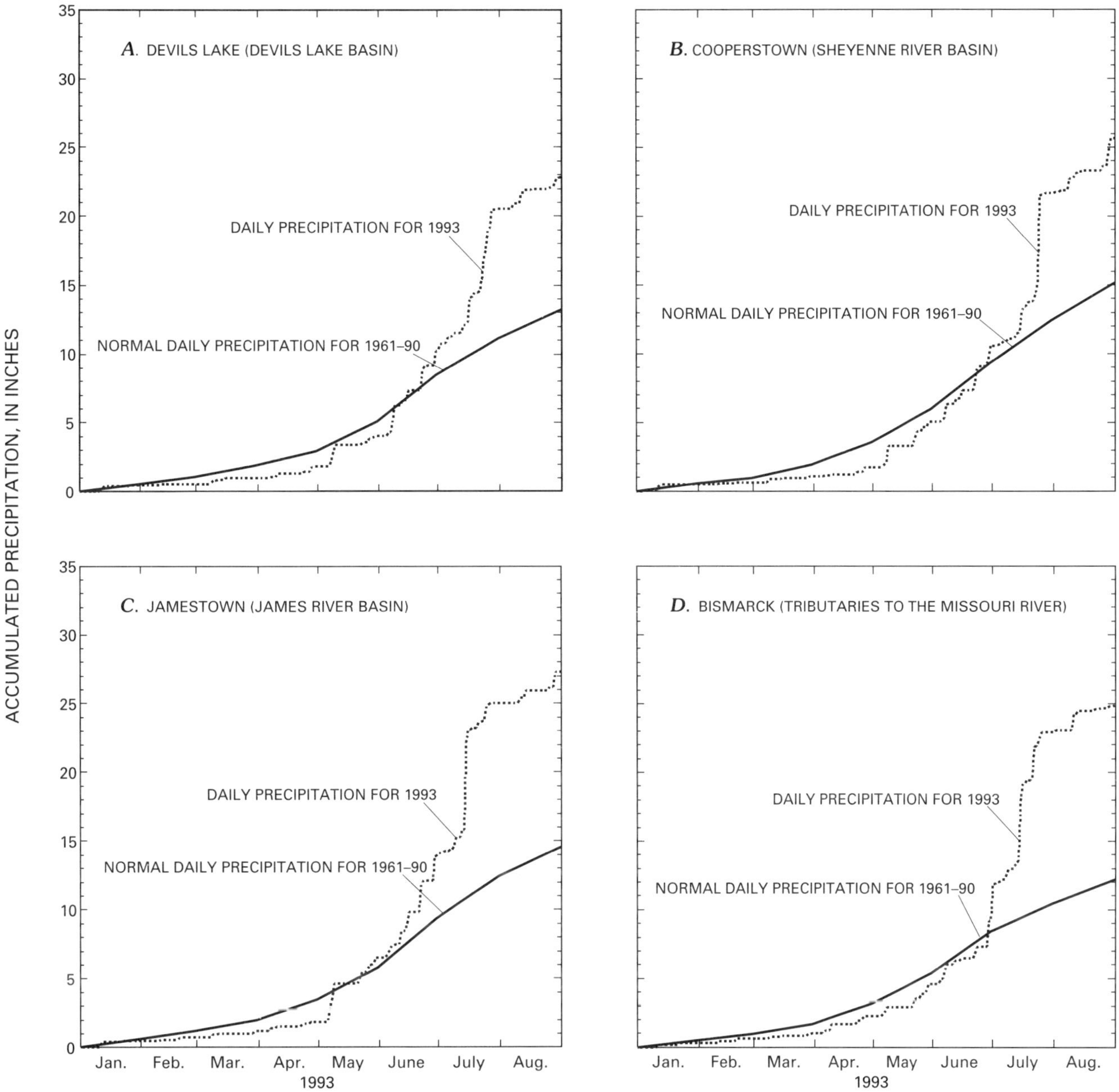

Figure 75. Accumulated daily precipitation for January–August 1993 and accumulated normal daily precipitation for January–August 1961–90 at *(A)* Devils Lake, *(B)* Cooperstown, *(C)* Jamestown, and *(D)* Bismarck, North Dakota (data from National Oceanic and Atmospheric Administration, 1993).

Table 49. Maximum stages and discharges prior to and during July and August 1993, in eastern and south-central North Dakota

[mi^2, square miles; ft, feet above an arbitrary datum; ft^3/s, cubic feet per second; <, less than; --, not determined or not applicable; >, greater than. Source: Recurrence intervals calculated from U.S. Geological Survey data. Other data from U.S. Geological Survey reports or data bases]

				Maximum prior to July and August 1993				Maximum during July and August 1993			
Site no. (fig. 74)	Station no.	Stream and place of determination	Drainage area (mi^2)	Period	Year	Stage (ft)	Discharge (ft^3/s)	Date (month/ day)	Stage (ft)	Discharge (ft^3/s)	Discharge recurrence interval (years)
1	05056100	Mauvais Coulee near Cando, ND	387	1956–93	1979	11.18	2,660	8/2	8.83	711	<10
2	05056200	Edmore Coulee near Edmore, ND	382	1956–93	1979	87.10	1,110	7/30	87.76	1,180	10
3	05056215	Edmore Coulee tributary near Webster, ND	148	1986–93	1987	72.48	739	8/2	75.06	1,330	--
4	05056239	Starkweather Coulee near Webster, ND	310	1979–93	1987	--	570	7/30	8.43	335	<10
					1989	10.05	--				
5	05056400	Big Coulee near Churchs Ferry, ND	[1]1,620	1950–93	1979	7.59	1,420	8/30	6.94	348	<10
6	05056410	Channel A near Penn, ND	930	1983–93	1987	42.87	[2]1,090	8/15	43.67	1,560	10
7	05057000	Sheyenne River near Cooperstown, ND	6,470	1944–93	1950	18.69	7,830	7/25	18.33	2,780	<10
8	05058000	Sheyenne River below Baldhill Dam, ND	7,470	1949–93	1979	36.26	4,740	7/28	34.98	3,720	[3]10
9	05058500	Sheyenne River at Valley City, ND	7,810	1938–75,	1948	--	4,580	7/16	[1]18.05	--	[3]25
				1979–93	1969	17.62	--	7/28	--	3,830	--
10	05060500	Rush River at Amenia, ND	116	1946–93	1966	12.15	--	7/17	10.23	2,970	50
					1979		3,490				
11	06341800	Painted Woods Creek near Wilton, ND	427	1958–81,	1979	9.64	4,050	7/23	8.13	1,580	10
				1982–93							
12	06347500	Big Muddy Creek near Almont, ND	456	1946–73,	1950	30.70	20,200	7/23	30.99	8,390	25
				1991–93							
13	06348000	Heart River near Lark, ND	2,750	1946–93	1950	20.70	29,200	7/23	16.85	12,100	[3]10
14	06467600	James River near Manfred, ND	253	1955–93	1979	9.20	2,000	7/23	9.40	2,700	>100
15	06468170	James River near Grace City, ND	1,060	1968–93	1969	12.00	3,100	7/28	13.49	3,520	10
16	06470000	James River at Jamestown, ND	2,820	1928–33,	1950	[5]15.82	[5]6,390	7/16	13.58	1,300	[3]50
				1935,	1983	[6]8.76	[6]996				
				1937–39,							
				1943–93							

[1]Drainage area was reduced from about 2,510 to 1,620 square miles by the completion of Channel A in 1979.
[2]A maximum discharge of about 1,600 cubic feet per second occurred in May 1979 based on streamflow measurements made by the North Dakota State Water Commission.
[3]Discharge recurrence interval for regulated period.
[4]Maximum stage occurred on July 16.
[5]Maximum before 1973 regulation.
[6]Maximum after 1973 regulation.

July 13–17, 1993, in Central Mississippi

By W. Trent Baldwin

Excessive rainfall from scattered thunderstorms during July 1993 produced flooding in central Mississippi. The city of Vicksburg in west-central Mississippi was the hardest hit as it received 6.6 inches of rainfall in a 3-hour period beginning at 1900 on July 11 (National Oceanic and Atmospheric Administration, 1993). The 100-year, 3-hour rainfall for Vicksburg is 5.4 inches according to Hershfield (1961).

Maximum stages and discharges were documented at eight flood-determination sites (fig. 76) that experienced at least a 2-year recurrence interval and are summarized in table 50. At Red Cane Creek tributary near Pisgah (site 6), the maximum peak discharge was the largest for the period of record. The flood was more severe at the flood-determination sites with the smaller drainage areas (sites 3 and 6), which indicates the intense but isolated nature of the thunderstorms.

REFERENCES

Hershfield, D.M., 1961, Rainfall frequency atlas of the United States: U.S. Department of Commerce, Weather Bureau Technical Paper No. 40, 115 p.

National Oceanic and Atmospheric Administration, 1993, Climatological data, Mississippi: Asheville, N.C., National Climatic Data Center, v. 98, no. 7, 27 p.

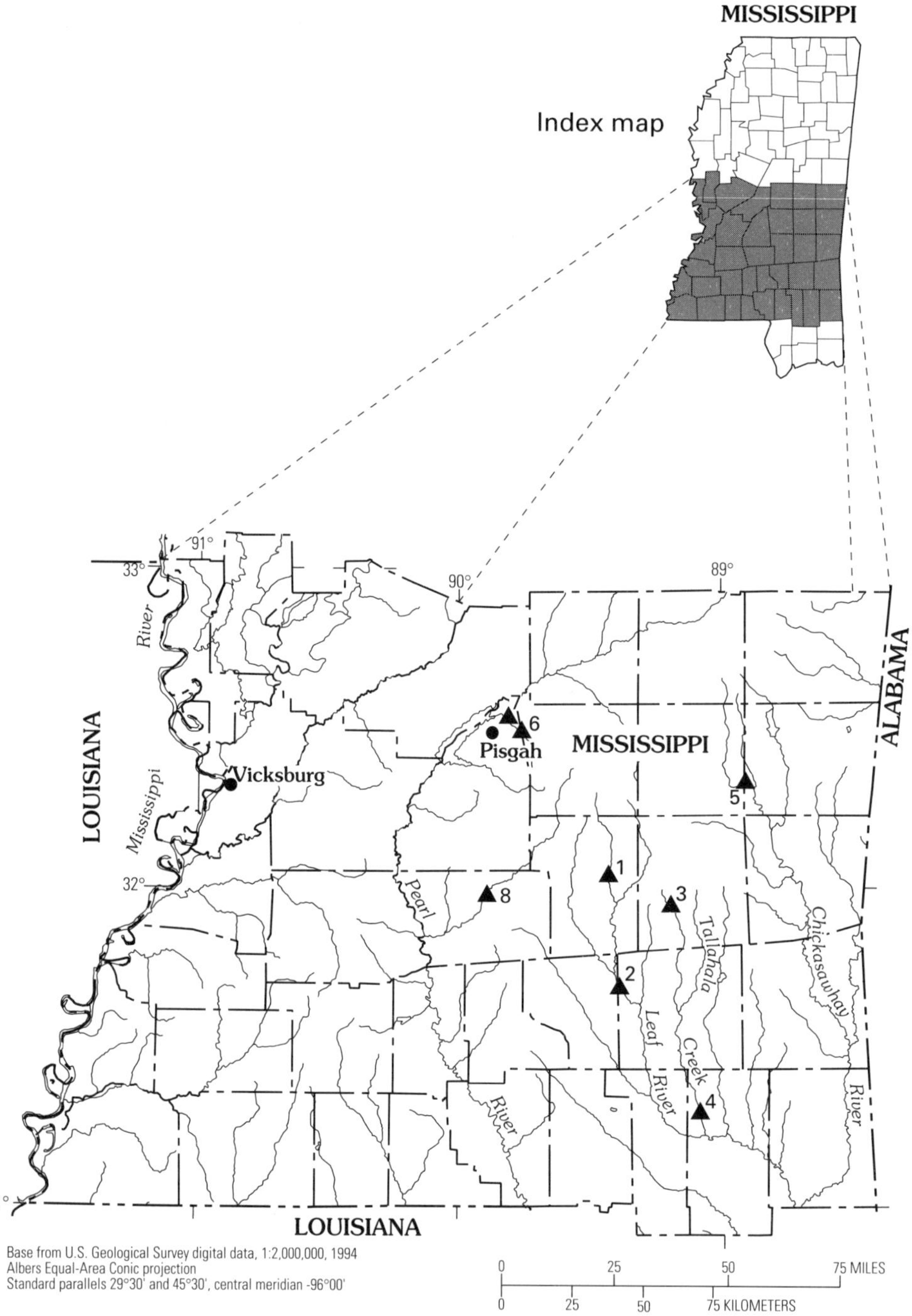

EXPLANATION

▲1 **Flood-determination site**—Number corresponds to that in table 50

Figure 76. Location of flood-determination sites for flood of July 13–17, 1993, in central Mississippi.

Table 50. Maximum stages and discharges prior to and during July 13–17, 1993, in central Mississippi

[mi^2, square miles; ft, feet above an arbitrary datum; ft^3/s, cubic feet per second; --, not determined or not applicable. Source: Recurrence intervals calculated from U.S. Geological Survey data. Other data from U.S. Geological Survey reports or data bases]

Site no. (fig. 76)	Station no.	Stream and place of determination	Drainage area (mi^2)	Maximum prior to July 1993: Period	Year	Stage (ft)	Discharge (ft^3/s)	Maximum during July 13–17, 1993: Day	Stage (ft)	Discharge (ft^3/s)	Discharge recurrence interval (years)
		Pascagoula River Basin									
1	02471100	Leaf River near Raleigh, MS	143	1856, 1900, 1940–93	1856 1974	-- 28.17	[1]17,000 17,000	14	23.53	8,150	5
2	02472000	Leaf River near Collins, MS	743	1856, 1900, 1939–93	1856 1974	33.00 32.60	56,000 54,200	15	24.89	21,500	4
3	02473850	Tallahala Creek tributary at Lake Como, MS	3.21	1964–93	1964	12.27	3,120	13	11.49	2,380	60
4	02474500	Tallahala Creek near Runnelstown, MS	612	1900, 1919, 1940–93	1900	[2]30.50	38,000	17	19.75	8,970	2
5	02475500	Chunky River near Chunky, MS	369	1939–93	1979	26.64	40,900	14	17.93	9,570	2
		Pearl River Basin									
6	02484750	Red Cane Creek tributary near Pisgah, MS	.10	1965–93	1983	6.98	134	14	7.38	143	25
7	02484760	Fannegusha Creek near Sand Hill, MS	52.3	1971–93	1980	13.35	9,000	14	11.96	4,910	4
8	02487500	Strong River at D'Lo, MS	425	1900, 1929–93	1983	33.48	26,400	14	24.89	8,270	2

[1]Estimated.

[2]Site and datum then in use.

August 12, 1993, in Eastern Illinois

By G.O. Balding

Excessive rainfall during the early morning hours of August 12, 1993, occurred over the adjacent cities of Champaign and Urbana, Illinois (fig. 77). Rain showers on August 9 produced about 0.5 inch at the Urbana precipitation gage. Rain continued each day with 0.26 inch falling on August 10 and 1.56 inches falling on August 11. Just before midnight on August 11, a very localized and slow-moving thunderstorm system produced 4.77 inches of rain in less than 2 hours and 6.30 inches of rain in 3 hours. The storm had a recurrence interval exceeding 100 years (the magnitude of a 100-year, 3-hour storm at that location is 4.23 inches) (Wayne M. Wendland, State Climatologist, written commun., 1994). Precipitation totals during August 9–12 at selected precipitation gages are listed in table 51.

The streamflow-gaging station, Boneyard Creek at Urbana (site 9), on the University of Illinois campus (fig. 77), recorded a maximum discharge for the period of record of 905 cubic feet per second from the 4.46-square-mile urban drainage area (table 52). Other continuous-record streamflow-gaging stations nearest to the Boneyard Creek at Urbana station recorded streamflow discharges having flood recurrence intervals of less than 2 years, which demonstrates the localized nature of the flooding.

Flooding and sewer backup occurred throughout the Champaign-Urbana area and other nearby communities. The News-Gazette reported that the excessive rainfall forced sandbagging in southwestern Champaign, closed the State highway south of Savoy, flooded streets in Tolono and around Campustown (University of Illinois), and flooded basements and street-level businesses in Champaign-Urbana that had never been flooded before (Cook and Kline, 1993). Many residences in the Champaign-Urbana area received substantial flood damage either by sewer backup, overland flooding, or from seepage into basements and crawl spaces that exceeded the capacity of residential sump pumps. On the University of Illinois campus, the basement of the new Grainger Engineering Library was filled with 6 feet of water and was 1 of 36 buildings on campus that had flooded basements (Wurth, 1993).

The University of Illinois experienced about $4 million of damage to engineering equipment and buildings (Wurth, 1994). Residential and commercial property damage from the flooding is not known.

REFERENCES

Cook, Anne, and Kline, Greg, 1993, Area gets taste of flooding: Champaign, Ill., The News-Gazette, August 12, 1993, p. A–1.

National Oceanic and Atmospheric Administration, 1993, Climatological data, Illinois, August 1993: U.S. Department of Commerce, v. 98, no. 8, 27 p.

Wurth, Julie, 1993, UI cleaning up, assessing damage: Champaign, Ill., The News-Gazette, August 13, 1993, p. A–1.

———1994, Plans take new angle on Boneyard: Champaign, Ill., The News-Gazette, October 2, 1994, p. A–1.

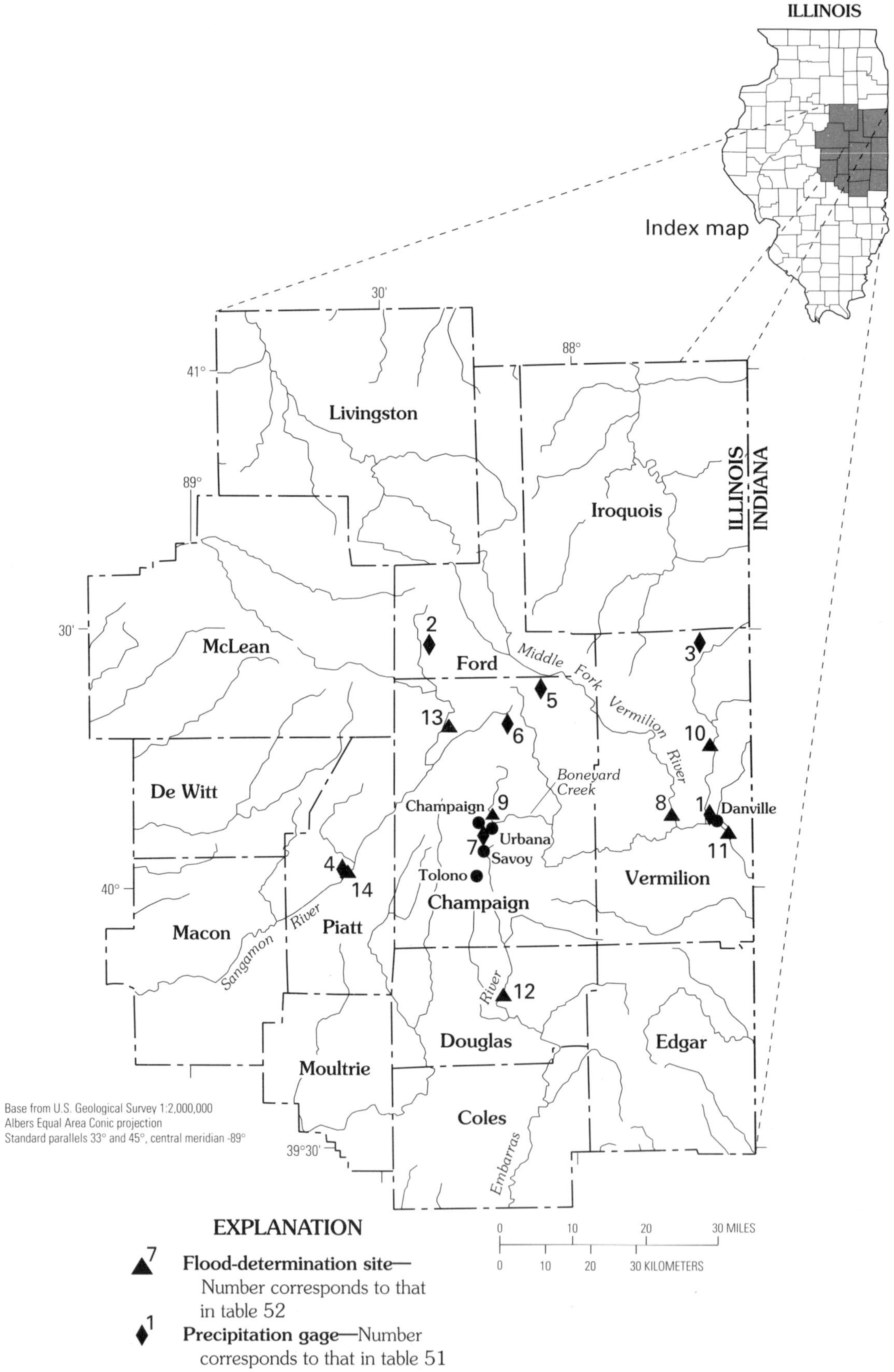

Figure 77. Location of flood-determination and precipitation gage sites for flood of August 12, 1993, in eastern Illinois.

Table 51. Precipitation totals for August 9–12, 1993, at selected precipitation gages in eastern Illinois

[Data from National Oceanic and Atmospheric Administration, 1993]

Site no. (fig. 77)	Latitude, longitude	Precipitation gage	Precipitation (inches)
1	40°08', 87°39'	Danville	0.85
2	40°28', 88°22'	Gibson City 1 E	1.28
3	40°28', 87°40'	Hoopeston 1 NE	.29
4	40°02', 88°36'	Monticello 2	1.17
5	40°23', 88°05'	Paxton	1.82
6	40°19', 88°10'	Rantoul	2.74
7	40°06', 88°14'	Urbana	7.58

Table 52. Maximum stages and discharges prior to and during August 12, 1993, in eastern Illinois

[mi^2, square miles; ft, feet above an arbitrary datum; ft^3/s, cubic feet per second; <, less than; >, greater than; --, not determined or not applicable. Source: Recurrence intervals calculated from U.S. Geological Survey data. Other data from U.S. Geological Survey reports or data bases]

				Maximum prior to August 12, 1993				Maximum on August 12, 1993		
Site no. (fig. 77)	Station no.	Stream and place of determination	Drainage area (mi^2)	Period	Year	Stage (ft)	Discharge (ft^3/s)	Stage (ft)	Discharge (ft^3/s)	Discharge recurrence interval (years)
8	03336646	Middle Fork Vermilion River above Oakwood, IL	432	1978–93	1990	15.96	12,000	2.53	329	<2
9	03337000	Boneyard Creek at Urbana, IL	4.46	1948–93	1979	8.94	982	9.93	905	>100
10	03338780	North Fork Vermilion River near Bismarck, IL	262	1970–73, 1988–93	1990	21.97	20,100	5.17	157	--
11	03339000	Vermilion River near Danville, IL	1,290	1914–21, 1928–93	1939	28.59	48,700	4.62	1,160	<2
12	03343400	Embarras River near Camargo, IL	186	1960–93	1979	16.38	6,240	8.12	694	<2
13	05570910	Sangamon River at Fisher, IL	240	1978–93	1985	18.30	8,550	7.80	291	<2
14	05572000	Sangamon River at Monticello, IL	550	1908–12, 1914–93	1926	18.50	19,000	8.96	553	<2

August 31, 1993, Storm Surge and Flood of Hurricane Emily on Hatteras Island, North Carolina

By J. Curtis Weaver *and* Thomas J. Zembrzuski, Jr.

THE STORM

Hurricane Emily began as a tropical depression in the Atlantic Ocean on August 22, 1993 (National Oceanic and Atmospheric Administration, 1993). It reached tropical storm strength 400 miles south of Bermuda on August 25, and by the next day was upgraded to a hurricane (fig. 78). Emily's strength dropped briefly to a tropical storm early August 27 but by the end of the day had returned to hurricane force. Its westerly track veered northwesterly as it steadily intensified and slowly moved toward the south-central coastal States. The storm achieved category-three strength (maximum sustained winds greater than 110 miles per hour) at 1500 on August 31. The storm's center passed within 25 miles of Hatteras Island, North Carolina (fig. 78), at approximately 1800 on August 31 before turning north and then northeastward away from land. Maximum wind gusts measured by the National Weather Service (NWS) during the storm ranged from 98 miles per hour at Buxton, North Carolina, to 132 miles per hour at Diamond Shoals, North Carolina, approximately 14 miles southeast of Cape Hatteras. An unofficial measurement of a maximum wind gust of 107 miles per hour was made at a commercial establishment in Buxton on Hatteras Island (W. DeMaurice, National Weather Service, oral commun., October 1993).

Although the eye of the hurricane did not make landfall, the western edge of the eye wall moved over part of Hatteras Island. As the eye approached the island, the storm's counterclockwise winds blew first

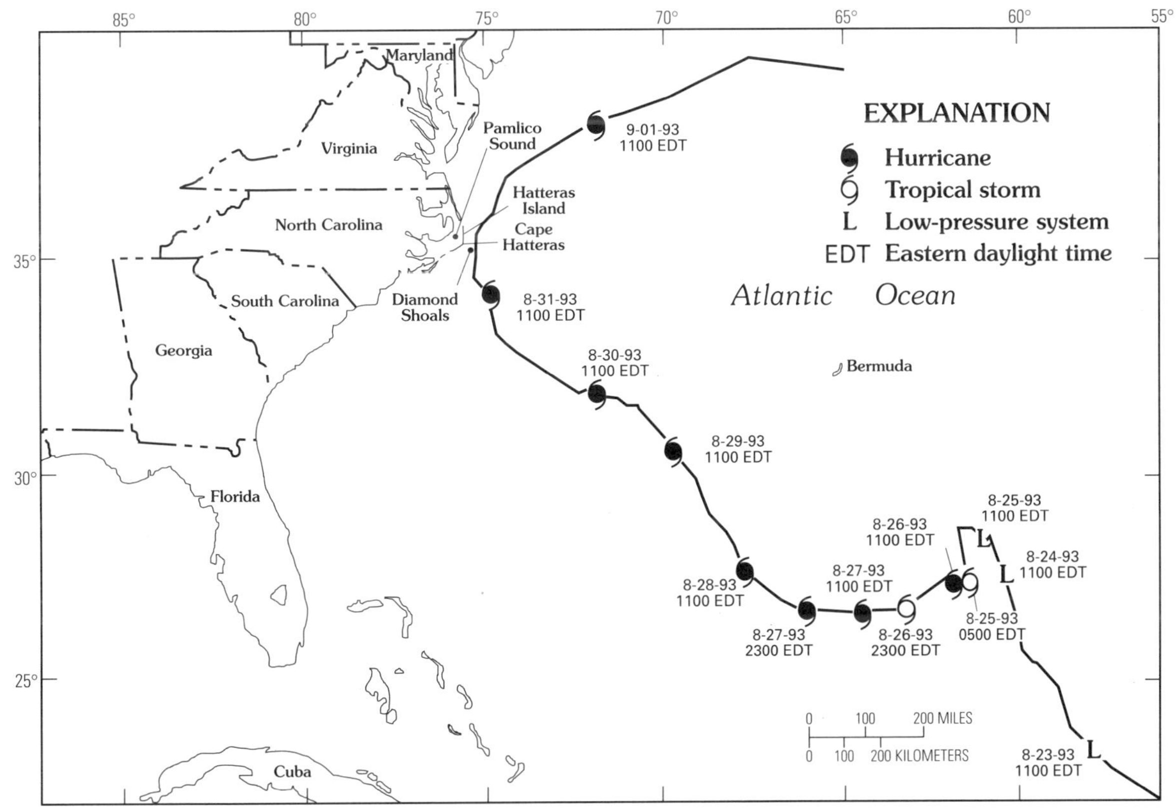

Figure 78. Path of Hurricane Emily, August 23–September 1, 1993, south-central coastal States (modified from National Oceanic and Atmospheric Administration, 1993).

from the north and then north-northwest across Pamlico Sound, pushing water onto the sound side of the island. The area most affected by the resulting storm surge was between the southern tip of the island at Hatteras and Little Kinnakeet, North Carolina (fig. 79). Although the NWS station at Cape Hatteras reported 7.51 inches of rain from the storm, rainfall was not a factor in the flooding. No more than 1 inch of precipitation was reported at any of the NWS precipitation gages in the counties adjacent to Pamlico Sound.

THE FLOOD

Documentation of notable coastal floods provides scientists and flood-plain managers with useful data and technical information for an improved understanding of the hazard and for improved management of flood-prone areas. Of particular importance is documentation of the duration of flooding (and rates of rise and recession of floodwaters), maximum flood elevations and depths, and delineation of the extent of storm-surge flooding.

Duration of Flooding

The most severe flooding on Hatteras Island occurred between 1600 and 2200 on August 31, but the duration of flooding on the island varied, depending on location and ground elevation. Although there are no active tide gages on southern Hatteras Island to provide official documentation, a discontinued U.S. Army Corps of Engineers (COE) recording tide gage located in Pamlico Sound near Buxton (fig. 79) was reactivated by a volunteer observer prior to the arrival of the storm. It appears that the stilling-well gage had produced satisfactory graphic record up to a point when interference between the float and the weight prevented it from recording higher stages. A reproduction of the graphic record is shown in figure 80; annotation of the axes and estimated record were added to the original strip chart by the NWS (W. DeMaurice, National Weather Service, written commun., October 1993). The tide-gage datum is not known. A survey of the outside staff gage located next to the discontinued recording gage produced a datum correction of -1.23 feet. This correction can be used to convert chart readings to sea-level datum. The actual maximum stage at the gage was later verified by using a nearby high-water mark (mark 42, table 53) surveyed to sea-level datum.

The tide gage indicated that the level of Pamlico Sound began to rise rapidly by 1500 on August 31 in response to a change in the wind direction and an increase in wind speed (fig. 80). The maximum rate of rise was 2.7 feet per hour between 1630 and 1730, and the maximum level was estimated to have occurred about 1830. The total rise was more than 7 feet in 4 hours. The gage record indicated that by 2200, the water level had declined more than 4 feet to an elevation at which most of the flooded area was again above water.

It is reasonable to assume that the timing of the flood peak was similar throughout the flooded area. In Buxton, the total rise in water level was in the range of 9 to 10 feet. A simple extrapolation of the maximum rate of rise at the reactivated tide gage against the total rise in Buxton suggests that the maximum rate of rise in Buxton was in the range of 3.5 to 4.0 feet per hour.

Flood Elevations

High-water marks were flagged during the 3 days following the hurricane. Numerous "still-water" marks, usually fine debris or seed lines, were found deposited on the interior walls of poorly sealed buildings and sheds. Long, nearly unbroken 1- to 4-foot thick rows of mixed waterborne debris and seagrass were deposited along the sound side of the dunes, especially in the undeveloped expanses of the Cape Hatteras National Seashore between villages on the southern end of the island.

Of a total of 108 marks flagged, 62 were chosen to provide a wide representation of elevations in the flooded area. Level circuits were run to each high-water mark from benchmarks tied to established U.S. Coast and Geodetic Survey and North Carolina Geodetic Survey benchmarks in the area.

Data for each of the 62 surveyed high-water marks are tabulated to show latitude, longitude, type and quality of mark, as well as the water-level and ground-surface elevations (table 53). The location and elevation of selected marks are in figure 81.

In the Avon vicinity, still-water surface elevations ranged from 5.92 feet (mark 1, table 53) at the Little Kinnakeet Coast Guard Station (north of Avon) to 7.62 feet (mark 5) within the village boundary. The maximum water depth (determined from table 53 as the difference between water-surface and ground-surface elevation) surveyed was 5.54 feet (mark 7). Flood levels along the sound side of the dunes averaged about 7.5 feet.

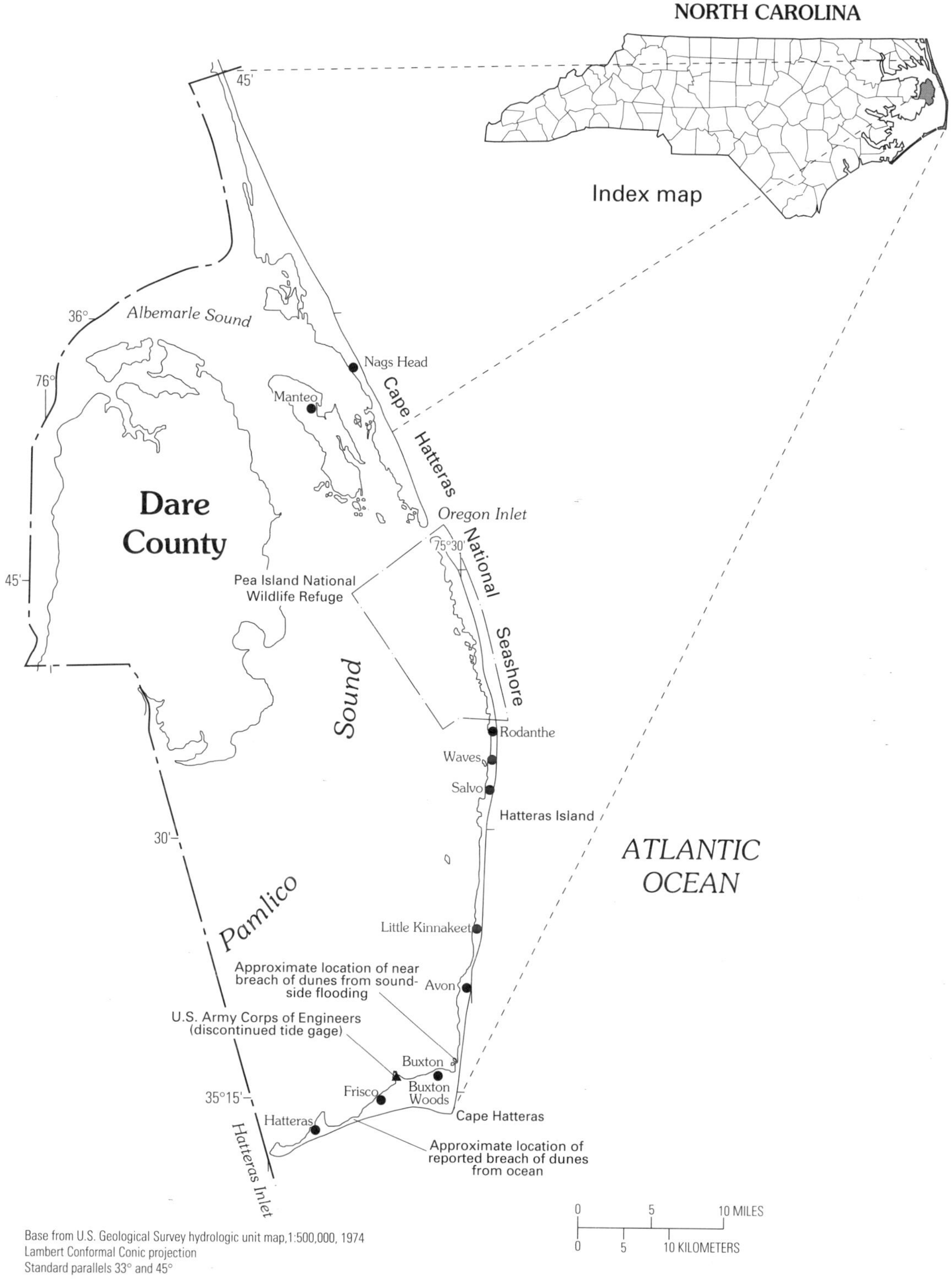

Figure 79. Selected features of Hatteras Island, North Carolina.

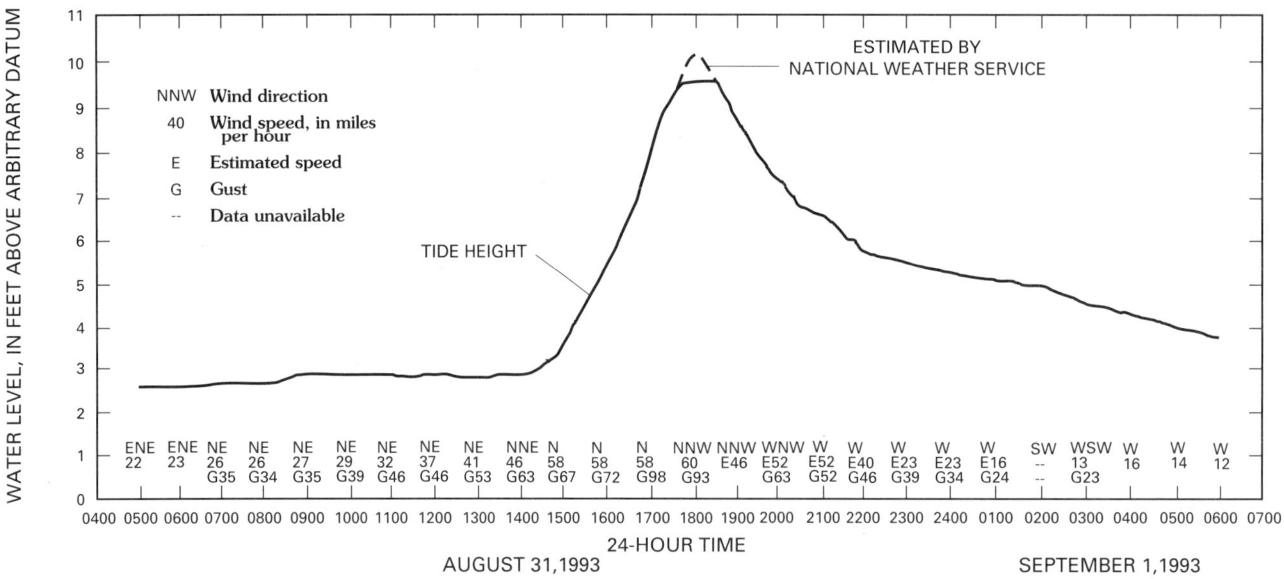

Figure 80. Reproduction of original strip chart from discontinued U.S. Army Corps of Engineers tide gage at Buxton, North Carolina, August 31–September 1, 1993. (Source: Original strip chart from Joe Ortner, Buxton, N.C.; wind direction and velocity, and estimated record from W. Demaurice, National Weather Service, Buxton, N.C.)

Along the Cape Hatteras National Seashore between Avon and Buxton, only the dunes remained above water during the storm surge. Flood levels increased from north to south, ranging from 7.6 feet (mark 13) near Avon to 11.5 feet (mark 23) near the Buxton village limits. The deposition of waterborne debris by flooding from the sounds indicated that a near breach of the dunes occurred about 0.5 mile north of Buxton.

The highest still-water elevations from Hurricane Emily occurred in the village of Buxton in structures closest to the Pamlico Sound shoreline where a maximum of 10.54 feet was surveyed (mark 25). The Cape Hatteras School was inundated to a level of 9.30 feet (mark 36). The maximum water depth surveyed in the area was 6.1 feet (mark 29). Flood levels declined considerably inland (south) of the sound, where a line of secondary dunes restricted access of the storm surge into Buxton Woods. Further movement of floodwaters into Buxton Woods was impeded by dense underbrush and undulating topography.

Flood levels continued to decline west of Buxton to the village of Frisco, where the previously mentioned reactivated COE tide gage recorded most of the rising and falling limbs of the storm-surge hydrograph (fig. 80). The maximum stage of the water-surface peak shown in figure 80 was estimated by the NWS to be slightly more than 10 feet. Applying the datum correction (-1.23 feet) suggests that the actual water-surface peak at the discontinued U.S. Army Corps of Engineers tide gage was in the range of 8.7 to 9.0 feet above sea level. This is in reasonable agreement with the elevation (8.31 feet) of a high-water mark (mark 42) surveyed inside a house located next to the reactivated tide gage. The maximum elevation surveyed in the vicinity of Frisco was 8.7 feet (mark 47). The deepest floodwaters noted in this area were about 4 feet (mark 44).

Two small breaches of the dunes from the ocean side (fig. 79) occurred in a stretch of the Cape Hatteras National Seashore between the villages of Frisco and Hatteras (Federal Emergency Management Agency, 1994), even though flood levels continued to decrease to the west of Frisco. A maximum elevation of 8.3 feet (mark 57) was surveyed along the dunes, but the highest still-water elevation surveyed was 7.02 feet (mark 54) above sea level. In Hatteras village, surveyed still-water marks of better than poor quality ranged from 6.4 (mark 59) to 6.98 feet (mark 57) above sea level. The maximum water depth surveyed in the village was nearly 3.6 feet (mark 57).

Delineation of the Extent of Storm-Surge Flooding

The general area of inundation extended from north of Avon village to the southern tip of Hatteras Island (fig. 81). Onsite reconnaissance and interviews

Table 53. Location and descriptions of high-water marks and water- and ground-surface elevations for the Hurricane Emily flood, Hatteras Island, North Carolina, August 31, 1993

[Locations of selected high-water marks (HWM) are shown in figure 81]

HWM no. (fig. 81)	Nearest town	Latitude	Longitude	Type and quality[1] of HWM	Water-surface elevation (feet above sea level)	Ground-surface elevation (feet above sea level)
1	Avon	35° 24' 22"	76° 29' 29"	Seed line, excellent	5.92	[2]4.0
2	Avon	35° 22' 38"	75° 29' 40"	Drift line on dunes, poor	7.2	[3]7.2
3	Avon	35° 22' 35"	75° 29' 40"	Drift line on dunes, poor	7.6	[3]7.6
4	Avon	35° 22' 33"	75° 29' 40"	Drift line on dunes, poor	7.5	[3]7.5
5	Avon	35° 22' 17"	75° 29' 55"	Seed line, excellent	7.62	4.4
6	Avon	35° 21' 38"	75° 30' 28"	Seed line, good	7.3	[2]2.3
7	Avon	35° 21' 37"	75° 30' 28"	Seed line, excellent	7.54	[2]2.0
8	Avon	35° 21' 30"	75° 30' 09"	Seed line, excellent	7.53	3.8
9	Avon	35° 21' 30"	75° 30' 00"	Wash line, poor	7.5	[3]7.5
10	Avon	35° 21' 29"	75° 30' 03"	Seed line, good	7.6	6.0
11	Avon	35° 21' 07"	75° 30' 41"	Seed line, excellent	7.62	[2]5.2
12	Avon	35° 20' 35"	75° 30' 19"	Drift line, poor	7.5	[3]7.5
13	Avon	35° 20' 05"	75° 30' 20"	Drift line, poor	7.6	[3]7.6
14	Avon	35° 20' 04"	75° 30' 27"	Seed line, fair	7.5	[2]5.0
15	Avon	35° 19' 21"	75° 30' 34"	Seed line, good	8.5	5.3
16	Avon	35° 19' 20"	75° 30' 32"	Debris line, poor	8.5	[3]8.5
17	Avon	35° 19' 13"	75° 30' 33"	Drift line on dunes, poor	8.8	[3]8.8
18	Buxton	35° 17' 10"	75° 30' 54"	Drift line on dunes, poor	10.8	[3]10.8
19	Buxton	35° 16' 59"	75° 30' 56"	Drift line on dunes, poor	11.1	[3]11.1
20	Buxton	35° 16' 40"	75° 30' 59"	Drift line on dunes, poor	11.1	[3]11.1
21	Buxton (at near-breach)	35° 16' 37"	75° 31' 00"	Drift line on dunes, poor	[4]11.2	[3]11.2
22	Buxton	35° 16' 28"	75° 31' 01"	Drift line on dunes, poor	10.3	[3]10.3
23	Buxton	35° 16' 09"	75° 31' 06"	Drift line on dunes, poor	11.5	[3]11.5
24	Buxton	35° 16' 05"	75° 31' 08"	Drift line on dunes, poor	[4]10.3	[3]10.3
25	Buxton	35° 15' 59"	75° 31' 10"	Seed line, excellent	10.54	5.9
26	Buxton	35° 15' 53"	75° 31' 19"	Seed line, excellent	9.71	[2]6.4
27	Buxton	35° 15' 36"	75° 31' 40"	Seed line, good	8.4	7.5
28	Buxton	35° 15' 52"	75° 31' 42"	Seed line, excellent	8.68	5.8
29	Buxton	35° 16' 07"	75° 31' 43"	Water stain, good	10.1	4.0
30	Buxton	35° 16' 07"	75° 31' 59"	Seed line, good	9.5	7.0
31	Buxton	35° 15' 48"	75° 32' 27"	Seed line, fair	7.1	4.9
32	Buxton	35° 15' 49"	75° 32' 27"	Seed line, good	7.1	3.9
33	Buxton	35° 15' 53"	75° 32' 34"	Seed line, good	9.0	7.8
34	Buxton	35° 15' 39"	75° 32' 35"	Drift line, poor	6.3	[3]6.3
35	Buxton	35° 16' 09"	75° 32' 37"	Seed line, good	10.0	5.4
36	Buxton	35° 15' 54"	75° 33' 03"	Seed line, excellent	9.30	4.9
37	Buxton	35° 15' 52"	75° 33' 29"	Seed line, excellent	9.42	4.6
38	Buxton	35° 15' 49"	75° 33' 44"	Drift line near road, poor	[4]9.6	[3]9.6
39	Frisco	35° 15' 27"	75° 34' 59"	Drift line, poor	6.3	[3]6.3
40	Frisco	35° 15' 22"	75° 35' 08"	Seed line, good	6.4	5.5

Table 53. Location and descriptions of high-water marks and water- and ground-surface elevations for the Hurricane Emily flood, Hatteras Island, North Carolina, August 31, 1993—Continued

HWM no. (fig. 81)	Nearest town	Latitude	Longitude	Type and quality[1] of HWM	Water-surface elevation (feet above sea level)	Ground-surface elevation (feet above sea level)
41	Frisco	35° 15' 50"	75° 35' 23"	Seed line, good	5.9	4.2
42	Frisco	35° 15' 50"	75° 35' 35"	Seed line, excellent	8.31	3.9
43	Frisco	35° 14' 57"	75° 35' 56"	Seed line, excellent	7.88	4.3
44	Frisco	35° 15' 04"	75° 36' 10"	Seed line, good	8.0	3.8
45	Frisco	35° 14' 25"	75° 37' 11"	Seed line, excellent	8.05	4.5
46	Frisco	35° 14' 01"	75° 37' 32"	Seed line, good	8.1	5.9
47	Frisco	35° 13' 35"	75° 38' 11"	Seed line, good	8.7	5.4
48	Frisco	35° 13' 33"	75° 38' 14"	Drift line on dunes, poor	[4]8.3	[3]8.3
49	Frisco	35° 13' 22"	75° 38' 59"	Drift line on dunes, poor	7.8	[3]7.8
50	Frisco	35° 13' 17"	75° 39' 13"	Drift line on dunes, poor	7.7	[3]7.7
51	Frisco	35° 13' 12"	75° 39' 26"	Drift line on dunes, poor	7.3	[3]7.3
52	Hatteras	35° 13' 01"	75° 40' 03"	Drift line on dunes, poor	7.9	[3]7.9
53	Hatteras	35° 13' 00"	75° 40' 05"	Drift line on dunes, poor	[4]7.8	[3]7.8
54	Hatteras	35° 12' 59"	75° 41' 12"	Seed line, excellent	7.02	3.9
55	Hatteras	35° 12' 56"	75° 41' 18"	Seed line, good	7.0	3.3
56	Hatteras	35° 13' 07"	75° 41' 25"	Seed line, poor	5.8	[2]3.5
57	Hatteras	35° 13' 22"	75° 41' 04"	Seed line, excellent	6.98	[2]3.4
58	Hatteras	35° 13' 14"	75° 41' 27"	Seed line, fair	6.7	[2]4.0
59	Hatteras	35° 13' 07"	75° 41' 40"	Seed line, fair	6.4	4.4
60	Hatteras	35° 12' 46"	75° 41' 46"	Seed line, excellent	6.93	[2]3.3
61	Hatteras	35° 12' 32"	75° 41' 55"	Seed line, excellent	6.68	4.2
62	Hatteras	35° 12' 28"	75° 42' 10"	Seed line, excellent	6.68	[2]5.6

[1]Quality: "excellent," ± 0.05 foot; "good," ± 0.1 foot; "fair," ± 0.25 foot; "poor," greater than 0.25 foot.
[2]Estimate.
[3]Ground-surface elevation at drift line is same as water-surface elevation.
[4]Water-surface elevation is average of two marks.

with residents and Federal, State, and local officials were the basis for mapping the flooded area. The location and elevations of the 62 surveyed high-water marks were plotted on U.S. Geological Survey (USGS) 7.5-minute, 5-foot contour-interval topographic quadrangle maps to assist with the delineation of the inundated areas.

In most places, the extent of the storm surge was fairly easy to delineate. An obvious and mostly continuous drift line was deposited on the sound side of ocean-front dunes that extended north of Buxton and southwest of Frisco. The northern extent of flooding during Hurricane Emily was about 3 miles north of Avon village near the Little Kinnakeet Coast Guard Station (fig. 79). From this point to the northern edge of Buxton village, the storm surge completely inundated State Highway 12. Except for several structurally elevated properties, the entire village of Avon was inundated. Likewise, from the western end of Frisco village to the end of Hatteras Island, the storm surge crossed over Highway 12 and reached the sand dunes at the ocean. The entire village of Hatteras was flooded.

In the wide part of Hatteras Island between Buxton and Frisco, several factors made the extent of inundation into Buxton Woods more difficult to delineate. The ground elevation in much of this area ranges from 5 to 10 feet (with many notable exceptions of higher elevations), the same range as the water-surface elevations during the flood. Although the overall relief is low in most places, there is considerable undulation of the

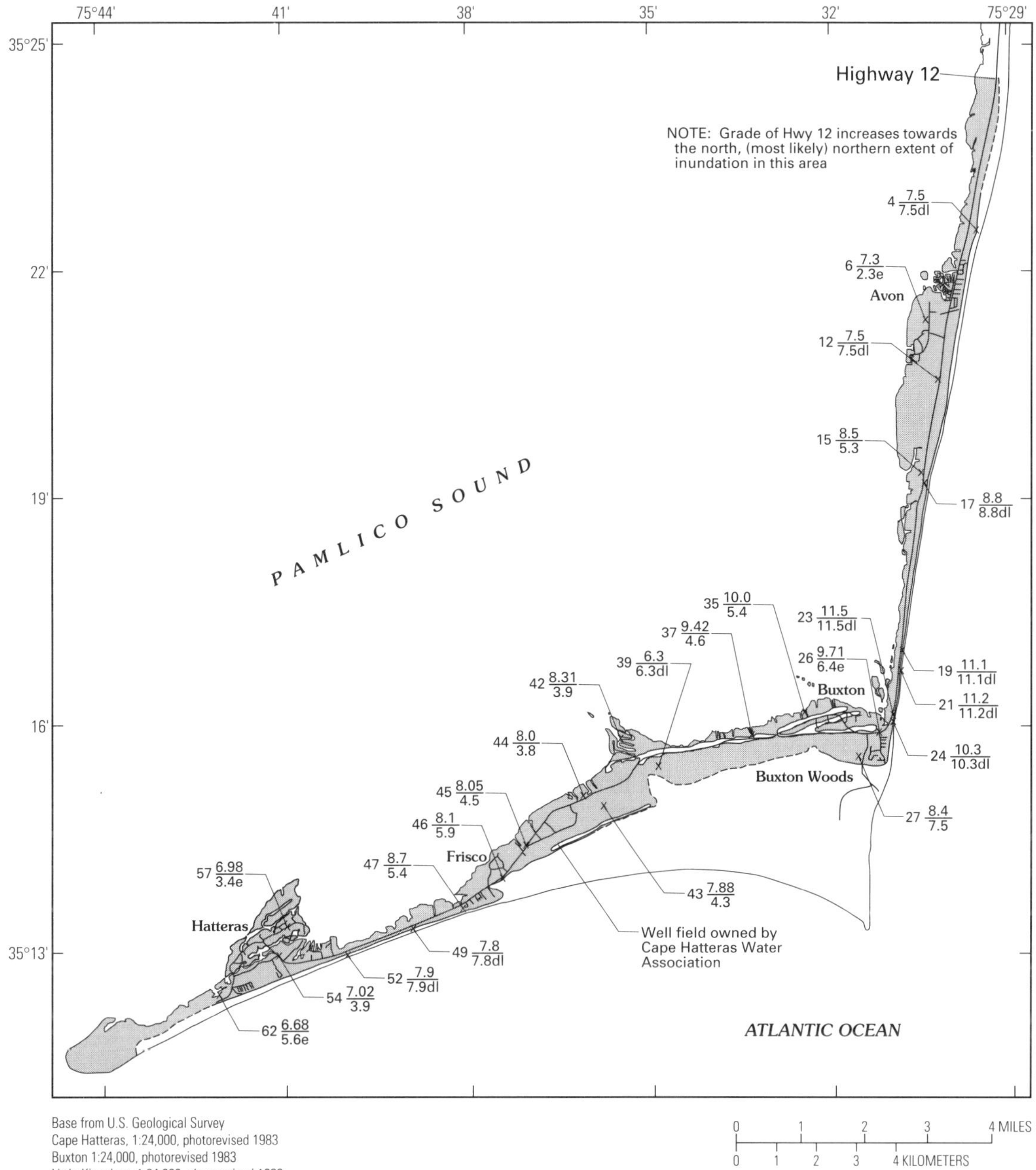

Base from U.S. Geological Survey
Cape Hatteras, 1:24,000, photorevised 1983
Buxton 1:24,000, photorevised 1983
Little Kinnakett, 1:24,000, photorevised 1983
Hatteras, 1:24,000, photoinspected 1987

EXPLANATION

Area inundated by storm surge

Area not inundated by storm surge

Boundary delineating extent of inundation from Pamlico Sound—Dashed where approximately located

x **Selected high-water mark (HWM)**

62 6.68/5.6e

HWM number—Number corresponds to that in table 53

Water-surface elevation, in feet above sea level—"Excellent" marks are listed at the precision of 0.01 foot; ground elevations and marks less than excellent are listed to the nearest 0.1 foot

Land-surface elevation, in feet above sea level— e is estimated elevation; dl is debris-line elevation

Figure 81. Location of and water-level elevations at selected high-water marks surveyed on Hatteras Island, North Carolina, for August 31, 1993, storm surge from Hurricane Emily.

topography. Natural wetlands cover most of the low-lying areas, and some artificial drainage ditches and canals connect wetland pockets to Pamlico Sound. Vegetation in Buxton Woods generally is quite dense, and the artificial drainage system is not extensive. These factors, as well as the fairly short duration of flooding, probably acted to attenuate the storm surge as it moved inland from Pamlico Sound into Buxton Woods. Definitive high-water marks could not be located in the interior of the island.

Although the high-water elevations are not known, other evidence indicates rather extensive intrusion of the storm surge into Buxton Woods, at least inland to the dunes that mark the edge of the Cape Hatteras Water Association (CHWA) well field (fig. 81). Specific-conductance measurements of well water were made during September 3 and 4 by the North Carolina Division of Environmental Management (T. Mew, Ground Water Section, North Carolina Department of Environment, Health, and Natural Resources, written commun., September 1993). These measurements suggest that saltwater intruded to the interior of the island.

The extent of storm-surge inundation is shown in figure 81. Areas where inundation was confirmed by reconnaissance or extrapolation from high-water mark elevations are delineated with a solid line. The areas where field surveys were inconclusive and where high-water elevations could not be extrapolated (notably, Buxton Woods) are delineated by dashed lines.

FLOOD DAMAGES

Although Hurricane Emily was not particularly notable in terms of wind damage, it will be remembered for extensive inundation from the Pamlico Sound. Many homes and buildings, while sustaining minor damage due to winds, were extensively damaged or destroyed by flooding. The Federal Emergency Management Agency (FEMA) reported 160 homes destroyed, 216 extensively damaged, and 144 that received minor damage (Federal Emergency Management Agency, 1994). Several commercial establishments and public buildings, including the Dare County Emergency Operations Center, also were flooded. The Cape Hatteras School sustained approximately $3.1 million in damages to structure and contents. In total, Hurricane Emily's damages were estimated to be $12.6 million (J. Self, Division of Emergency Management, North Carolina Department of Environment, Health, and Natural Resources, oral commun., March 1994). The Governor of North Carolina declared a state of disaster on September 3, 1993. A Presidential declaration of major disaster followed on September 10.

Approximately 160,000 tourists and residents were evacuated from the Cape Hatteras area prior to the storm. Although no deaths occurred during Hurricane Emily, three drownings occurred on the day after the storm when swimmers were lost in the heavy surf near Nags Head (fig. 79), more than 50 miles north of the flooded area, and in Virginia.

The saline floodwaters that covered Buxton Woods had a delayed effect on the quality of water pumped from the CHWA well field. Water produced and distributed by CHWA is a blend from several individual wells. Chloride content of water distributed to customers was 40 to 45 milligrams per liter prior to the storm. The chloride content rose slowly but steadily during the next 3 months to a maximum of 280 to 285 milligrams per liter, as the saltwater infiltrated the production zones of the wells, before declining during another 3 months to a level of about 60 milligrams per liter (J. Coleman, Cape Hatteras Water Association, oral commun., August 1994). Although individual wells were not tested for chloride, water from some probably had maximum chloride concentrations higher than 285 milligrams per liter.

COMPARISON WITH OTHER STORMS AND WITH 100-YEAR FLOOD ELEVATIONS

The flooding caused by Hurricane Emily was the most extensive in the recollection of long-time residents of the island. Flood levels surpassed those from other large storms, which occurred in September 1933, September 1944, and September 1985 (Hurricane Gloria). On the southern end of Hatteras Island, flood levels also were higher than levels generated during March 13-14, 1993, by the "storm of the century" (Federal Emergency Management Agency, 1994). Perhaps the last time that flooding was more severe was during the hurricane of August 1899 when the "entire island was covered with water to a depth of 4 to 10 feet; there were not more than four houses in which the tide did not rise to a depth of 1 to 4 feet" (Carney and Hardy, 1967).

The 100-year flood elevations on the sound side of the ocean-front dunes on Hatteras Island were determined by FEMA to range from 6.3 feet in the village of Hatteras to 8.6 feet in the area between the villages of

Buxton and Avon (Federal Emergency Management Agency, 1993). Hurricane Emily's still-water flood elevations exceeded the published (Federal Emergency Management Agency, 1993) 100-year flood levels in Hatteras, Frisco, and the parts of Buxton that were closest to Pamlico Sound by amounts ranging from a few tenths of a foot in Hatteras to about 2 feet in Buxton. In Avon, flood levels were either close to or slightly below the FEMA 100-year elevations. Flood levels were probably several feet below the FEMA 100-year elevations in the Buxton Woods area.

REFERENCES

Carney, C.B., and Hardy, A.V., 1967, North Carolina hurricanes: Raleigh, N.C., U.S. Department of Commerce Environmental Science Services Administration, Weather Bureau, 40 p.

Federal Emergency Management Agency, 1993, Flood insurance study, Dare County, North Carolina (unincorporated areas): Federal Emergency Management Agency, National Flood Insurance Program, Community Number 375348, 28 p.

———1994, Hurricane Emily: Federal Emergency Management Agency Interagency Hazard Mitigation Team Report FEMA–1003–DR–NC, Region IV, 50 p.

National Oceanic and Atmospheric Administration, 1993, Special climate summary—Hurricane Emily: Columbia, S.C., Southeast Regional Climate Center, 6 p.

September 22–26, 1993, in Northwestern and Central Missouri

By Rodney Southard

In Missouri, 1993 will be remembered for the devastation caused by the Missouri and Mississippi Rivers during the great flood of July–August 1993. However, record flooding also occurred during September throughout a large part of the State. After months of greater-than-normal rainfall, two series of thunderstorms swept across the State the third week of September. The first series of storms occurred September 22–23 and resulted in precipitation totals in excess of 5 inches in northwestern Missouri (National Oceanic and Atmospheric Administration, 1993). As these storms traveled southeast, they weakened and then redeveloped over east-central Missouri resulting in 3 to 6 inches of rainfall. The second series of storms on September 24 and 25 originated in southern Kansas and traveled to southwestern and east-central Missouri. These storms produced intense rainfall in a 75-mile-wide band across the State (fig. 82). Two-day rainfall totals in excess of 7 inches were recorded at several precipitation gages (table 54).

Because damage estimates for the September flooding were included in the totals of the Missouri and Mississippi main-stem floods, which occurred during July and August, no specific estimates exist for the September flooding. For flooding July–September 1993, 102 counties and 1 city were declared eligible for Federal disaster assistance. The estimated damage in Missouri totaled $4 billion, and at least 31 deaths have been attributed to the flooding (Jane Epperson, Missouri Department of Natural Resources, written commun., 1994). Many individuals were affected by the flooding, and about $42 million was disbursed for individual assistance to flood victims in Missouri by Federal Emergency Management Agency (FEMA), which is a large part of the total of $93 million for individual assistance in the upper Mississippi River Basin (Ron McCabe, Federal Emergency Management Agency, oral commun., 1994).

During September, numerous rivers were affected by the excess rainfall as streamflow-gaging stations or flood-determination sites measured record or near-record maximum discharges (fig. 83). The rainfall of September 22–23 resulted in devastating flooding in the Nodaway River Basin (site 6) in northwestern Missouri and in the Cuivre River Basin (site 5) in east-central Missouri. Flood-determination sites in both basins had discharge recurrence intervals in excess of 100 years (table 55). The September 24–25 storms resulted in more extensive flooding, and eight flood-determination sites had maximum discharges exceeding the 100-year recurrence interval, and nine sites established maximum discharges for the period of record. The most excessive precipitation occurred in south-central Missouri, and storm totals decreased eastward across the State. The Spring, Sac, James, and Pomme de Terre Rivers had maximum discharges substantially higher than the previous maximum discharges for the period of record. Two examples of the maximum discharges are the Spring River near Waco (site 33), which had a maximum discharge on September 26 of 1.47 times the 1943 maximum discharge of 103,000 cubic feet per second, and the Sac River near Dadeville (site 9), which had a maximum discharge 2.65 times greater than the 1986 maximum discharge of 13,600 cubic feet per second (table 55). Even though the storms dissipated as they moved eastward, the rain fell on ground saturated from the earlier storms of September 22 and 23, and the resulting flooding was severe in the Big River and Gasconade River Basins. Tributaries to the Gasconade River, such as the Big Piney River and Little Piney Creek, did not have extreme flooding, but the Gasconade River main stem had a major flood. The maximum discharge at the Jerome flood-determination site (site 20) had a 70-year recurrence interval. A number of other streams had substantial flooding because of the overlap of precipitation from both storms (table 55).

REFERENCES

Federal Emergency Management Agency, April 1993, Interagency hazard mitigation team report for Missouri: FEMA–995DR–MO, 28 p.

National Oceanic and Atmospheric Administration, 1993, Climatological data, Missouri, September 1993: U.S. Department of Commerce, v. 97, no. 9, 23 p.

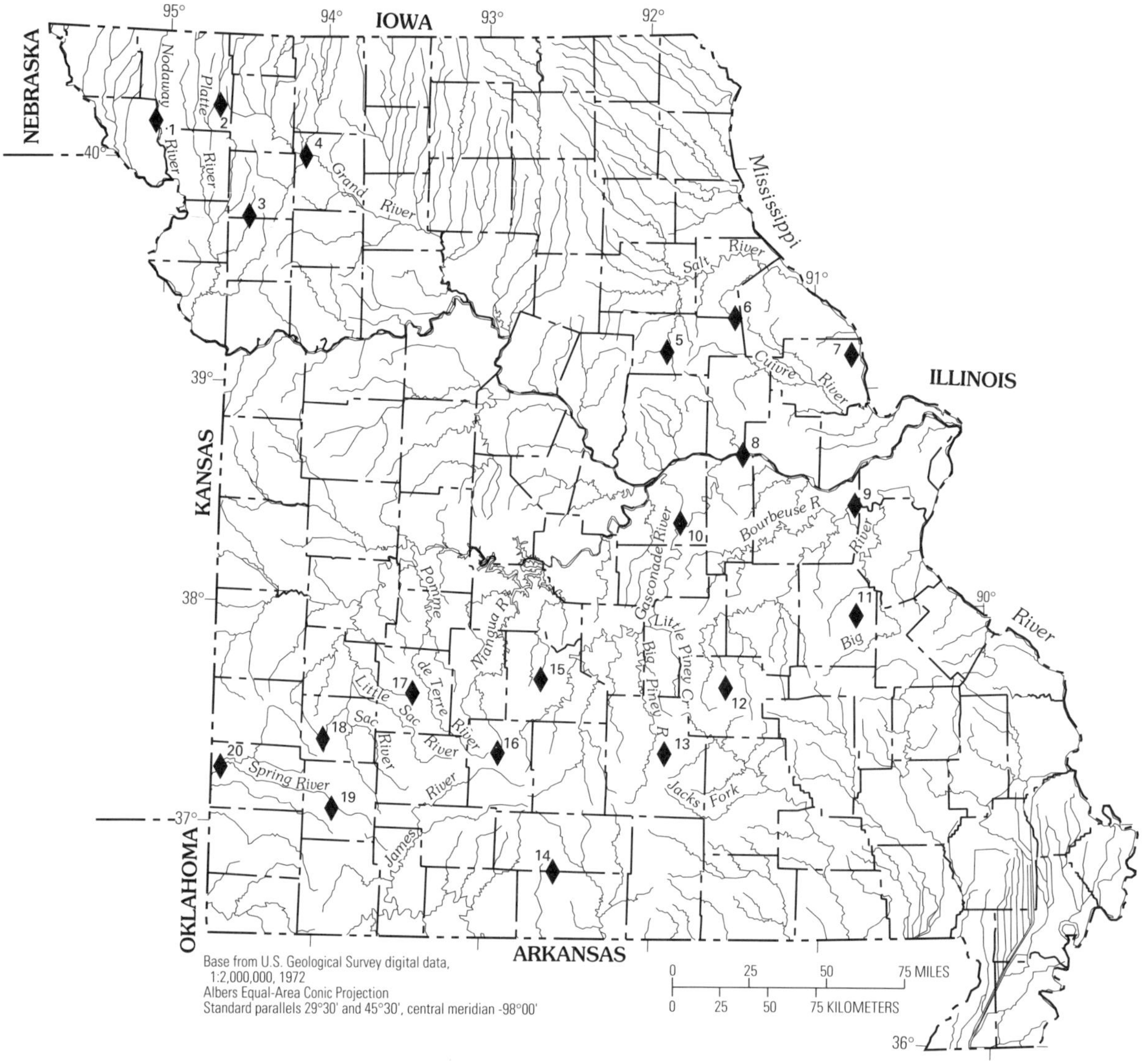

EXPLANATION

◆8 **Precipitation gage**—Number corresponds to that in table 54

Figure 82. Location of precipitation gages for flood of September 22–26, 1993, in northwestern and central Missouri.

Table 54. Precipitation totals for September 22–26, 1993, at selected precipitation gages in Missouri

[Data from National Oceanic and Atmospheric Administration, 1993]

Site no. (fig. 82)	Precipitation gage site	Precipitation, in inches					
		Sept. 22	Sept. 23	Sept. 24	Sept. 25	Sept. 26	Total
1	Graham 1 NW	5.30	0.30	0.31	0.49	0.06	6.46
2	Conception	5.10	.80	.38	.50	.05	6.83
3	Amity 7 WNW	6.70	.28	.30	.31	.05	7.64
4	Pattonsburg 2 S	2.59	1.14	.01	1.39	.11	5.24
5	Mexico	1.34	6.28	.22	.43	.35	8.62
6	Vandalia	2.39	3.10	.28	.59	0	6.36
7	Elsberry 1 S	.85	3.66	.19	0	.46	5.16
8	Hermann	0	4.45	.68	.77	.52	6.42
9	Pacific 1 S	0	3.04	1.74	.44	.65	5.87
10	Rich Fountain 3 E	0	4.83	.69	1.57	.07	7.16
11	Potosi 3 N	0	3.50	2.93	2.42	.29	9.14
12	Salem	0	.67	2.27	2.00	0	4.94
13	Houston 3 E	0	0	1.40	3.74	.50	5.64
14	Wasola	0	0	5.14	2.96	.19	8.29
15	Lebanon 2 W	0	0	3.15	3.58	.35	7.08
16	Marshfield	0	0	3.20	4.55	.55	8.30
17	Bolivar 1 NE	0	.01	4.47	4.80	.02	9.30
18	Lockwood	0	0	3.40	6.04	.03	9.47
19	Mt. Vernon M U SW CTR	.01	0	1.52	8.20	.02	9.75
20	Waco 2 E	0	0	.47	11.00	0	11.47

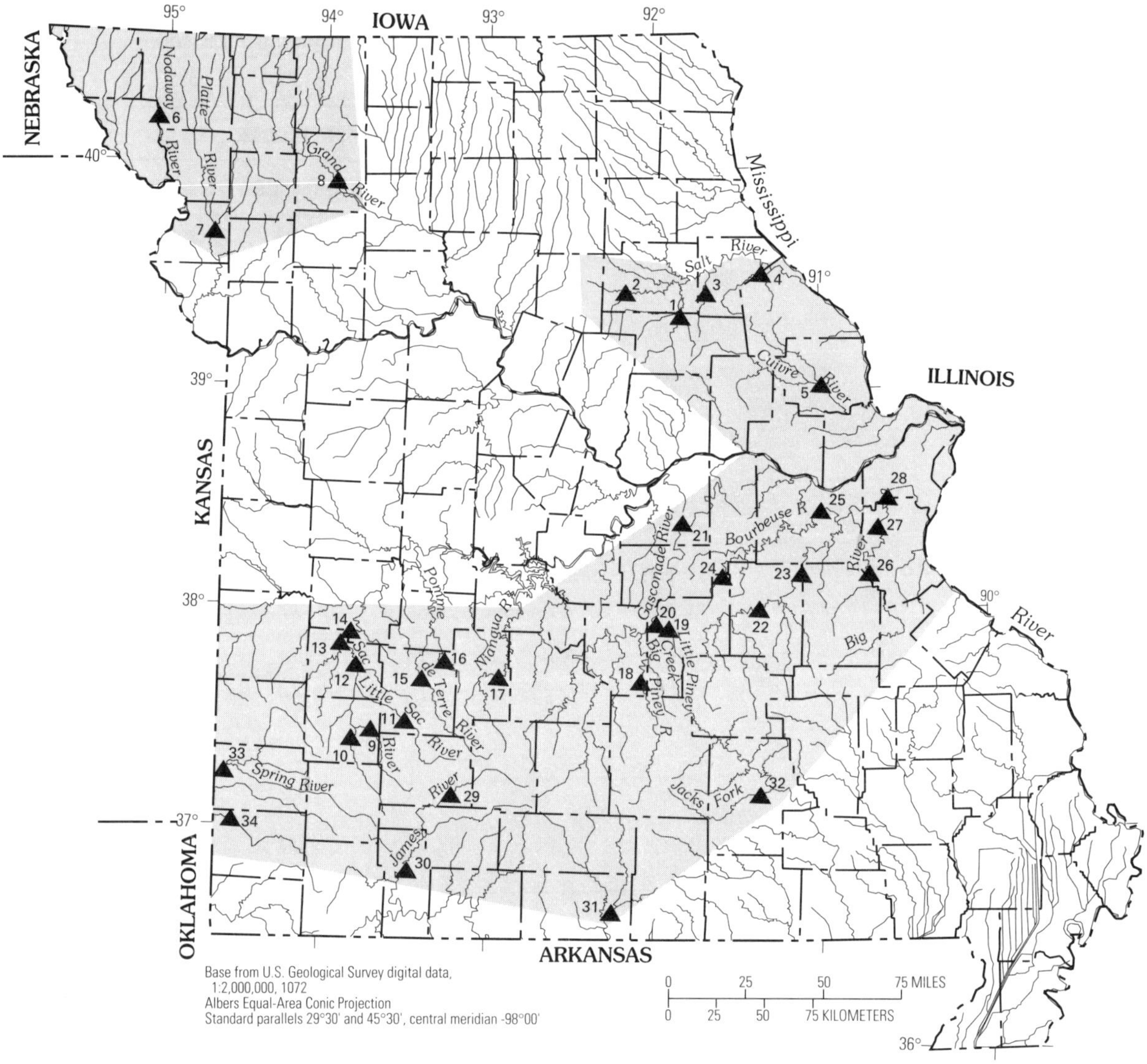

EXPLANATION

Flood area

▲34 Flood-determination site—Number corresponds to that in table 55

Figure 83. Location of flood-determination sites for flood of September 22–26, 1993, in Missouri.

Table 55. Maximum stages and discharges prior to and during September 1993, in northwestern and central Missouri

[mi^2, square miles; ft, feet above an arbitrary datum; ft^3/s, cubic feet per second; >, greater than; --, not determined or not applicable. Source: Recurrence intervals calculated from U.S. Geological Survey data; other data from U.S. Geological Survey reports or data bases]

				Maximum prior to September 1993				Maximum during September 1993			
Site no. (fig. 83)	Station no.	Stream and place of determination	Drainage area (mi^2)	Period	Year	Stage (ft)	Discharge (ft^3/s)	Day	Stage (ft)	Discharge (ft^3/s)	Discharge recurrence interval (years)
1	05504800	South Fork Salt River above Santa Fe, MO	233	1940–93	1969	28.24	28,800	23	28.66	31,800	>100
2	05506800	Elk Fork Salt River near Madison, MO	200	1969–93	1973	33.4	42,300	23	26.81	17,500	9
3	05507600	Lick Creek at Perry, MO	104	1980–93	1983	21.30	9,360	23	21.96	10,900	19
4	05508805	Spencer Creek below Plum Creek near Frankford, MO	206	1980–93	1981	16.86	16,200	22	18.54	20,300	>100
5	05514500	Cuivre River near Troy, MO	903	1922–72, 1979–93	1941	33.4	120,000	23	32.17	101,000	>100
6	06817700	Nodaway River near Graham, MO	1,380	1983–93	1989	23.34	26,600	22	26.89	90,700	>100
7	06820500	Platte River near Agency, MO	1,760	1924–30, 1932–93	1993	36.07	60,800	23	31.84	38,500	20
8	06897500	Grand River near Gallatin, MO	2,250	1921–93	1993	41.5	89,800	24	35.5	50,200	13
9	06918440	Sac River near Dadeville, MO	257	1966–93	1986	20.83	13,600	25	27.56	36,100	>100
10	06918460	Turnback Creek above Greenfield, MO	252	1965–93	1986	23.74	44,000	25	26.34	42,700	35
11	06918740	Little Sac River near Morrisville, MO	237	1969–93	1972	21.95	22,300	25	23.33	29,100	>100
12	06919020	Sac River at Highway J below Stockton, MO	1,292	1974–93	1985	24.91	–	25	23.71	13,300	--
					1986	–	14,800				
13	06919500	Cedar Creek near Pleasant View, MO	420	1923–26, 1949–93	1958	27.35	37,000	25	25.44	22,700	12
14	06919900	Sac River near Caplinger Mills, MO	1,810	1975–93	1986	30.00	60,000	25	29.93	51,200	--
15	06921070	Pomme de Terre River near Polk, MO	276	1969–93	1986	23.08	23,100	24	27.10	34,300	>100

Table 55. Maximum stages and discharges prior to and during September 1993, in northwestern and central Missouri—Continued

Site no. (fig. 83)	Station no.	Stream and place of determination	Drainage area (mi^2)	Maximum prior to September 1993				Maximum during September 1993			
				Period	Year	Stage (ft)	Discharge (ft^3/s)	Day	Stage (ft)	Discharge (ft^3/s)	Discharge recurrence interval (years)
16	06921200	Lindley Creek near Polk, MO	112	1957–93	1961	23.60	–	24	20.49	19,400	17
					1986	–	31,900				
17	06923250	Niangua River at Windyville, MO	377	1991–93	1992	10.44	4,700	24	24.36	44,700	>100
18	06930000	Big Piney River near Big Piney, MO	560	1922–82, 1988–93	1942	20.7	32,700	26	19.66	30,000	11
19	06932000	Little Piney Creek at Newburg, MO	200	1929–93	1985	16.6	–	24	11.78	9,700	3
					1946	–	32,500				
20	06933500	Gasconade River at Jerome, MO	2,840	1903–06, 1923–93	1982	31.34	136,000	27	29.60	110,000	70
21	06934000	Gasconade River near Rich Fountain, MO	3,180	1922–59, 1987–93	1982	33.27	134,000	28	31.28	106,000	50
22	07013000	Meramec River near Steelville, MO	781	1923–93	1985	26.15	51,200	26	17.51	22,300	3
23	07014500	Meramec River near Sullivan, MO	1,475	1921–33, 1944–93	1945	32.0	77,300	26	22.14	34,200	4
24	07015720	Bourbeuse River near High Gate, MO	135	1965–93	1982	23.65	49,300	24	17.56	13,400	2
25	07016500	Bourbeuse River at Union, MO	808	1921–93	1982	33.80	73,300	25	23.84	29,100	13
26	07018100	Big River near Richwoods, MO	735	1943–93	1957	27.15	55,800	23	30.33	59,800	50
27	07018500	Big River at Byrnesville, MO	917	1922–93	1985	26.47	43,000	25	29.37	63,600	>100
28	07019000	Meramec River near Eureka, MO	3,788	1903–06, 1922–93	1982	42.89	145,000	26	36.72	95,200	19
29	07050700	James River near Springfield, MO	246	1956–93	1957	18.20	24,800	25	19.45	41,100	>100
30	07052500	James River at Galena, MO	987	1922–93	1943	29.82	52,700	25	33.46	73,200	>100
31	07057500	North Fork River near Tecumseh, MO	561	1945–93	1985	28.10	133,000	25	22.60	52,500	19
32	07066000	Jacks Fork at Eminence, MO	398	1922–93	1985	17.58	55,800	25	16.32	44,100	20
33	07186000	Spring River near Waco, MO	1,164	1924–93	1943	30.94	103,000	26	34.06	151,000	>100
34	07187000	Shoal Creek above Joplin, MO	427	1942–93	1943	[1]16.8	[1]62,100	25	16.84	19,900	8

[1]Site and datum then in use.

INDEX

C

D

E

F

M

N

O

P

Q

R

S

T

U

V

W

Y

Z

☆U.S. GOVERNMENT PRINTING OFFICE: 1999–746-694